Timber Design

for the Civil and Structural PE Exams

Seventh Edition

Robert H. Kim, MSCE, PE
and Jai B. Kim, PhD, PE
with Parker E. Terril, BSCE, MSRE

Professional Publications, Inc. • Belmont, California

Benefit by Registering this Book with PPI

- Get book updates and corrections
- Hear the latest exam news
- Obtain exclusive exam tips and strategies
- Receive special discounts

Register your book at **www.ppi2pass.com/register**.

Report Errors and View Corrections for this Book

PPI is grateful to every reader who notifies us of a possible error. Your feedback allows us to improve the quality and accuracy of our products. You can report errata and view corrections at **www.ppi2pass.com/errata**.

Timber Design for the Civil and Structural PE Exams
Seventh Edition

Current printing of this edition: 2

Printing History

edition number	printing number	update
6	1	New edition. Copyright update.
7	1	New edition. Copyright update.
7	2	Minor corrections.

Printed in the United States of America

PPI
1250 Fifth Avenue, Belmont, CA 94002
(650) 593-9119
www.ppi2pass.com

ISBN: 978-1-59126-176-6

Library of Congress Control Number: 2009930227

Table of Contents

Chapter 3
Lumber Size Categories and Allowable Design Stress

Chapter 4
Beam Design: Sawn Lumber

Chapter 5
Beam Design: Glued Laminated Timber

Chapter 6
Axial Members, and Combined Bending and Axial Loading

Chapter 7
Mechanical Connections

Chapter 8
Nails and Spikes

Chapter 9
Bolts

Chapter 10
Lag Screws and Wood Screws

Chapter 11
Split Rings and Shear Plates

Chapter 12
Plywood and Nonplywood Structural Panels

Chapter 13
Practice Problems

List of Tables

List of Figures

Preface and Acknowledgments

We have developed this seventh edition of *Timber Design for the Civil and Structural PE Exams* to align with the 2005 edition of *National Design Specification for Wood Construction* (NDS) as well as the 1997 edition of *Plywood Design Specification* (PDS). This book will primarily serve engineers studying to take the National Council of Examiners for Engineering and Surveying (NCEES) Principles and Practice of Engineering (PE) civil or structural exams. It is also suitable as a classroom text for civil engineering seniors and graduate students.

This seventh edition owes much to its predecessors and to the guidance of PPI. Thanks go to the team at PPI that included Sarah Hubbard, director of new product development; Cathy Schrott, director of production; Tom Bergstrom, technical illustrator; Kate Hayes, typesetter; Amy Schwertman, cover designer; Courtnee Crystal, project editor; Megan Synnestvedt, proofreader and calculation checker; and Jenny Lindeburg King, editorial project supervisor.

We also wish to acknowledge Bucknell University in Lewisburg, Pennsylvania, for providing a conducive environment for the writing of this book, and the students in our CENG 403 Wood Engineering course at Bucknell for their comments, corrections, and suggestions. Thanks in particular go to Matthew Dawson and Nicole Hervol who helped work on extra problems, and to Parker Terril, BSCE, MSRE, who made significant revisions to the new edition.

Finally, we express gratitude to Mrs. Yung J. Kim, mother of Robert and wife of Jai, who provided the necessary support and motivation to write this book.

Should you find an error in this book, we hope two things happen. First, that you will let us know about it by using the errata section on the PPI website at **www.ppi2pass.com/errata**. Second, that you will learn something from the error—we know we will! We would appreciate constructive comments, suggestions for improvement, and recommendations for expansion.

Good luck on the exam!

Robert H. Kim, MSCE, PE
Jai B. Kim, PhD, PE

Introduction

The primary function of *Timber Design for the Civil and Structural PE Exams* is to serve as a study reference for practicing engineers and students preparing to take the National Council of Examiners for Engineering and Surveying (NCEES) Principles and Practice of Engineering (PE) civil and structural exams. As such, this book will guide you through the application of the 2005 edition of the *National Design Specification for Wood Construction* (NDS), which you must have at your side as you work this book's problems. Be aware, however, that while the NDS is incorporated into many major building codes and structural timber specifications, there may be codes and specifications that differ from, and take priority over, the specifications in the NDS. In practice, you should check with the governing jurisdiction to confirm which codes and specifications must be followed. In addition to the NDS, you may need to consult other references, many of which are identified in this book's References section, for more comprehensive explanations of timber theory.

The 2005 NDS requires that designs follow the provisions of either the allowable stress design (ASD) method or the load and resistance factor design (LRFD) method. The inclusion of the LRFD provisions is new in the 2005 edition of the NDS, and as such, the ASD continues to be the design method of choice for many designers. To encourage acceptance and use of LRFD methods, and to allow you to compare the two methods, some of this book's examples use both ASD and LRFD methods. For ASD, adjustment factors were used. For LRFD, the applicable factors were used in accordance with the NDS Appendix N.

Chapters 1 through 12 describe and summarize the key elements of the NDS (and the *Plywood Design Specification* (PDS) in Ch. 12) and Ch. 13 consists of practice problems to further reinforce these key elements. Each chapter in this book covers a separate NDS subject, so you may use this book to brush up on a few specific subjects and only study a few particular chapters. Or, you may study the book in its entirety. Do note, however, that the chapters frequently build on concepts and information that have been set out in earlier chapters. Therefore, you can use this book most effectively by studying the chapters in order. The book's chapters are meant to explain and clarify the NDS, but frequently assume that you can refer directly to the NDS itself when necessary. Among the book's examples are references to NDS tables ("NDS Tbl. . . "), sections ("NDS Sec. . . "), figures ("NDS Fig. . . "), equations ("NDS Eq. . . . "), and adjustment factors ("NDS Adj Fac. . . "). References to the *NDS Supplement* begin with "NDS Supp."

Throughout the book, example problems demonstrate the standard design principles, methods, and formulas needed to tackle problems commonly encountered on the exam. Take your time with these and make sure you understand each example before moving ahead. Keep in mind, though, that in actual design situations there are often several correct solutions to the same problem.

If you are a practicing engineer, engineering student, or instructor

Although this book is primarily intended to aid in exam preparation, it is also a valuable aid to engineers and can serve as a classroom text for civil engineering undergraduate and graduate students. For anyone using this book, the design examples serve as a step-by-step, comprehensive guide to timber design using the NDS.

If you are an examinee

If you are preparing to take either the NCEES civil or structural PE exam, work all of the examples and practice problems in this book to learn how to apply the principles presented. By solving the problems in this book, you will gain a better understanding of the timber design elements that may appear on the exams. By reviewing the solutions, you will learn efficient problem-solving methods that may benefit you in a timed situation.

About the NCEES exams

The civil PE exam consists of two sessions, each lasting four hours and consisting of 40 multiple choice questions. The morning (breadth) session of the exam may contain general timber design-related problems. The structural afternoon (depth) session of the exam may include more in-depth timber design-related problems.

The structural I PE exam consists of two sessions, each lasting four hours and consisting of 40 multiple choice questions of about equal difficulty, including several timber design-related problems. For each session of both the civil and the structural I PE exams, the problems typically require an average of six minutes to work.

The structural II PE exam also consists of two four hour sessions, one in the morning and one in the afternoon, but its questions are quite different from those of the civil and structural I PE exams. In each structural II PE exam session, you are given four scenario-type problems (two bridge and two building), for a total of eight problems. You must work either all four of the bridge problems or all four of the building problems (two in the morning and two in the afternoon). The two afternoon problems also contain seismic provisions. Of the four problems that must be worked, one will emphasize wood structures. The first three practice problems in Ch. 13 of this book are scenario-based problems related to timber design intended to illustrate to some degree the type of problems likely to appear on the structural II exam.

Note that NCEES has announced that it will replace the structural I and structural II exams with a new 16-hour structural PE exam in April 2011. For more information, go to www.ncees.org.

When it comes to taking the exams, it's important to bring the editions of the design standards that the exam is based on. Check PPI's website at **www.ppi2pass.com/civil** or **www.ppi2pass.com/structural** for current information and answers to frequently asked questions about the civil or structural PE exams. Furthermore, since the NCEES exams are open-book tests, you may want to mark this book, the NDS, and any other references you intend to use, and tab pages that include critical information such as tables and commonly used equations. Check with your engineering state board for restrictions regarding removable tabs and impermissible references. For a link to your state board, visit **www.ppi2pass.com/stateboards**.

References

Primary References

The following references were used to create this book and can serve as primary references for working timber design-related problems for the NCEES civil and structural PE exams. (If the reference is referred to in this book by a name other than its title, the alternative name is indicated in brackets following the reference.)

American Forest & Paper Association (AF&PA). *ASD/LRFD National Design Specification for Wood Construction with Commentary and Supplement: Design Values for Wood Construction.* Washington, DC: AF&PA American Wood Council, 2005. [NDS]

American Institute of Timber Construction (AITC). *Timber Construction Manual.* 5th ed. New York: John Wiley & Sons, 2005.

American Plywood Association (APA). *Design/Construction Guide: Diaphragms and Shear Walls.* Tacoma, WA: APA—The Engineered Wood Association, 2007.

American Plywood Association (APA), *Plywood Design Specification.* Tacoma, WA: APA—The Engineered Wood Association, 1998. [PDS]

American Plywood Association (APA), *Plywood Design Specification, Supplement 2: Design and Fabrication of Plywood-Lumber Beams.* Tacoma, WA: APA—The Engineered Wood Association, 1992. [PDS Supp 2]

American Society of Civil Engineers (ASCE), *Minimum Design Loads for Buildings and Other Structures.* New York: ASCE, 2006. [ASCE/SEI7-05]

U.S. Department of Agriculture, Forest Products Laboratory. *Wood Handbook: Wood as an Engineering Material.* Honolulu, HI: University Press of the Pacific, 2006.

Supplementary References

The following additional references were used to create this book.

Publications

American Association of State Highway and Transportation Officials (AASHTO), *Standard Specifications for Highway Bridges.* 17th ed. Washington, DC: AASHTO, 2002.

American Plywood Association (APA), *Panel Design Specification.* Tacoma, WA: APA—The Engineered Wood Association, 2008.

Breyer, Donald E., Kelly E. Cobeen, Kenneth J. Fridley, and David G. Pollock, *Design of Wood Structures—ASD.* 5th ed. New York: McGraw-Hill, 2007.

Gurfinkel, German. *Wood Engineering.* 2nd ed. Dubuque, IA: Kendall/Hunt Publishing, 1981.

Kim, Jai B., and Robert H. Kim, "Oak A-Frame Timber Bridges Meeting the Modern Deflection Requirement," Transportation Research Record No. 1319, Washington DC, 1991.

Proceedings

Barron, Kevin, and Jai B. Kim. "Non-Linear Modeling of the Heel Joint of Metal Plate Connected Roof Trusses," in *6th World Conference on Timber Engineering* (Proceedings). British Columbia, Canada, 2000.

Brill-Edwards, Chris, and Jai B. Kim. "Analog Model of Metal Plated Wood Truss Heel Joint and Force Comparison on Analog Truss Model Implementation Heel Joint Model" and "Comparison of Full-Scale Truss Test Results with Those by Analytical Methods" in *1st RILEM Symposium on Timber Engineering* (Proceedings). Stockholm, Sweden: RILEM Publications, 1999.

Cabler, S., J. A. Guinther, and Jai B. Kim. "Stiffness Evaluation of Metal Plates Connected on Wood Members" in *5th World Conference on Timber Engineering* (Proceedings). Montreux, Switzerland, 1998.

Kim, Jai B., and Robert H. Kim, "Design and Full-Scale Tests of 30 Ft. A-Frame Timber Oak Bridge" in *International Association for Bridge and Structural Engineering Conference on Innovative Wooden Structures and Bridges* (Proceedings). Lahti, Finland, 2001.

Kim, Jai B., and Robert H. Kim. "Effectively Spliced Large Timbers and Full-Scale Tests on Spliced Timbers" in *5th World Conference on Timber Engineering* (Proceedings). Montreux, Switzerland, 1998.

Kim, Jai B., and Robert H. Kim. "Non-Destructive Evaluation of Strength Properties of In-Place Wood Members," in *IAWPS International Conference on Forest Products* (Proceedings). Deajeon, Korea, 2003.

Kim, Jai B., and Robert H. Kim. "Non-Destructive Evaluation of the Stiffness and Strength of In-Situ Timber Structural Members" in *1st RILEM Symposium on Timber Engineering* (Proceedings). Stockholm, Sweden: RILEM Publications, 1999.

Nomenclature

Unless defined otherwise in the text, the following symbols are used in this book. The words "reference" and "tabulated" are used interchangeably for "unadjusted" values. The same is true for the words "adjusted" and "allowable." Also, the phrase "nominal" is replaced with the phrase "referenced or tabulated."

symbol	definition (units)
A	area (in^2, ft^2)
A_m	gross cross-sectional area of main wood member (in^2)
A_n	net cross-sectional area of a tension or compression member at a connection (in^2)
A_s	sum of gross cross-sectional areas of side member(s) (in^2)
b	width or thickness of rectangular beam cross section (in)
c	buckling and crushing interaction factor for columns (–)
c	distance between neutral axis and extreme fiber (in, ft)
C_b	bearing area factor (–)
C_d	penetration depth factor for connections (–)
C_D	load duration factor (ASD only) (–)
C_{di}	diaphragm factor for nail connections (–)
C_{eg}	end grain factor for connections (–)
C_F	size factor for sawn lumber (–)
C_{fu}	flat use factor for bending stress (–)
C_g	group action factor for connections with $D \geq 1/4$ in (–)
C_i	incising factor for dimension lumber (–)
C_L	beam stability factor (–)
C_M	wet service factor (–)
C_P	column stability factor
C_r	repetitive member factor (bending stress) for dimension lumber (–)
C_{st}	metal side plate factor for 4 in shear plate connections (–)
C_t	temperature factor (–)
C_T	buckling stiffness factor for 2 × 4 and smaller dimension lumber in trusses (–)
C_{tn}	toe-nail factor for nail connections (–)
C_V	volume factor (–)
C_Δ	geometry factor for connections with fasteners $D \geq 1/4$ in (–)
d	depth of rectangular beam cross section for bending member (in)
d	pennyweight of nail or spike (troy)
d_e	effective depth of member at a connection (in)
d_n	depth of member remaining at a notch (in)
d_x	width of rectangular column parallel to y-axis, used to calculate column slenderness ratio about x-axis (in)

symbol	definition (units)
d_y	width of rectangular column parallel to x-axis, used to calculate column slenderness ratio about y-axis (in)
d_1, d_2	cross-sectional dimensions of rectangular compression member in planes of lateral support (in)
D	dead load (lbf)
D	diameter (in)
e	eccentricity (in)
E, E'	tabulated (reference) and allowable modulus of elasticity (lbf/in^2)
E_{axial}	modulus of elasticity of glulam for axial deformation calculation (lbf/in^2)
E_m	modulus of elasticity of main member (lbf/in^2)
E_{min}, E'_{min}	reference and adjusted modulus of elasticity for ASD beam stability and column stability calculations (lbf/in^2)
$E_{\text{min,LRFD}}$, $E'_{\text{min,LRFD}}$	LRFD adjusted and allowable modulus of elasticity for stability analysis (lbf/in)
E_s	modulus of elasticity of side member (lbf/in^2)
E_x	modulus of elasticity about x-axis (lbf/in^2)
E_y	modulus of elasticity about y-axis (lbf/in^2)
f_b	actual (calculated) bending stress (lbf/in^2)
f_{bx}, f_{b1}	actual bending stress about strong (x) axis (lbf/in^2)
f_{by}, f_{b2}	actual bending stress about weak (y) axis (lbf/in^2)
f_c	actual compression stress parallel to grain (lbf/in^2)
$f_{c\perp}$	actual compression stress perpendicular to grain (lbf/in^2)
$f_{c,\text{LRFD}}$	LRFD actual compress stress parallel to grain (lbf/in^2)
f'_c	concrete compressive strength (lbf/in^2)
f_g	actual bearing stress parallel to grain (lbf/in^2)
f_s	actual plywood rolling shear stress (lbf/in^2)
f_t	actual tension stress in a member parallel to grain (lbf/in^2)
f_v	actual shear stress parallel to grain (horizontal shear) in a beam (lbf/in^2)
f_θ	actual bearing stress at angle, θ, to grain (lbf/in^2)
F_b, F'_b	tabulated (reference) and allowable ASD bending stress (lbf/in^2)
F^*_b	tabulated (reference ASD) bending stress multiplied by all applicable adjustment factors except C_L (lbf/in^2)
F^{**}_b	tabulated (reference ASD) bending stress multiplied by all applicable adjustment factors except C_V (lbf/in^2)
$F_{b,\text{LRFD}}$, $F'_{b,\text{LRFD}}$	LRFD reference and allowable (adjusted) bending design stress (lbf/in^2, kips/in^2)
$F^*_{b,\text{LRFD}}$	LRFD adjusted F^*_b value (lbf/in^2)
F_{bE}	critical ASD buckling (Euler) value for bending member (lbf/in^2)
$F_{bE,\text{LRFD}}$	LRFD adjusted F_{bE} (lbf/in^2)
F_{bx}, F'_{bx}	tabulated (reference) and allowable bending stress about strong (x) axis (lbf/in^2)
$F_{bx,c/t}, F^-_{bx}$	tabulated bending stress about the x-axis with compression zone stressed in tension (lbf/in^2)
$F'_{bx,c/t}, F'^-_{bx}$	allowable ASD bending stress about the x-axis with compression laminations stressed in tension (lbf/in^2)
$F_{bx,t/t}, F^+_{bx}$	tabulated bending stress about the x-axis tension zone stressed in tension (lbf/in^2)
$F'_{bx,t/t}, F'^+_{bx}$	allowable ASD bending stress about the x-axis with high-quality tension laminations stressed in tension (lbf/in^2)
F_{by}, F'_{by}	tabulated and allowable ASD bending stress about weak (y) axis (lbf/in^2)
F_c, F'_c	tabulated and allowable ASD compression stress parallel to grain (lbf/in^2)
F^*_c	tabulated ASD compression stress parallel to grain multiplied by all applicable adjustment factors except C_P (lbf/in^2)

symbol	definition (units)
F_{cE}	critical ASD buckling (Euler) value for compression member (lbf/in^2)
$F_{c,\text{LRFD}}$, $F'_{c,\text{LRFD}}$	LRFD reference and allowable compression design stress parallel to grain (lbf/in^2)
$F^*_{c,\text{LRFD}}$	LRFD adjusted F^*_c value (lbf/in^2)
$F_{cE,\text{LRFD}}$	LRFD adjusted F_{cE} (lbf/in^2)
$F_{c\perp}$, $F'_{c\perp}$	tabulated and allowable ASD compression stress perpendicular to grain (lbf/in^2)
$F_{C\perp,\text{LRFD}}$, $F'_{C\perp,\text{LRFD}}$	LRFD adjusted and allowable compression design stress perpendicular to grain (lbf/in^2)
F_e	dowel bearing strength (lbf/in^2)
$F_{e\parallel}$	dowel bearing strength parallel to grain for bolt or lag bolt connection (lbf/in^2)
$F_{e\perp}$	dowel bearing strength perpendicular to grain for bolt or lag bolt connection (lbf/in^2)
F_{em}	dowel bearing strength of main member (lbf/in^2)
F_{es}	dowel bearing strength of side member (lbf/in^2)
$F_{e\theta}$	dowel bearing strength at angle to grain θ for bolt or lag bolt connection (lbf/in^2)
F_g, F'_g	tabulated and allowable bearing stress parallel to grain (lbf/in^2)
F_s, F'_s	tabulated and allowable plywood rolling shear stress (lbf/in^2)
F_t, F'_t	tabulated and allowable ASD tension stress parallel to grain (lbf/in^2)
$F_{t,\text{LRFD}}$, $F'_{t,\text{LRFD}}$	LRFD reference and allowable tension stress parallel to grain (lbf/in^2)
F_u	ultimate tensile strength for steel (lbf/in^2)
F_v, F'_v	reference and allowable ASD shear stress parallel to grain (horizontal shear) in a beam (lbf/in^2)
$F_{v,\text{LRFD}}$, $F'_{v,\text{LRFD}}$	LRFD reference and allowable (adjusted) shear design stress (lbf/in^2)
F_{yb}	bending yield strength of fastener (lbf/in^2)
F'_θ	allowable bearing stress at angle to grain θ (lbf/in^2)
g	gauge of screw (–)
G	specific gravity (–)
h	width or depth of rectangular beam cross section (in)
I	moment of inertia (in^4)
K	correction factor for effective section moduli for plywood (–)
K_{bE}	Euler buckling coefficient for beams (–)
K_{cE}	Euler buckling coefficient for columns (–)
K_{cr}	creep factor (–)
K_D	diameter coefficient for nail and spike connections with $D < 0.25$ in (–)
K_e	effective length factor for column end conditions (buckling length coefficient for columns) (–)
K_F	format conversion factor (–)
K_L	loading condition coefficient for evaluating volume effect factor C_V for glulam beams (–)
K_M	moisture content coefficient for sawn lumber truss compression chords (–)
KS	effective section moduli for plywood rather than I/C (in^3)
K_T	truss compression chord coefficient for sawn lumber (–)
K_x	spaced column fixity coefficient (–)
K_θ	angle to grain coefficient for bolt and lag bolt connections with $D < 0.25$ in (–)
ℓ	length of beam (in)

symbol	definition (units)
ℓ	length of column (in)
ℓ	span length of bending member (in)
ℓ_b	bearing length (in)
ℓ_c	clear span (in)
ℓ_e	effective length of column (in)
ℓ_e	effective length of compression side of beam (in)
ℓ_e/d	slenderness ratio of column (–)
$(\ell_e/d)_x$	slenderness ratio of column for buckling about strong (x) axis (–)
$(\ell_e/d)_y$	slenderness ratio of column for buckling about weak (y) axis (–)
ℓ_m	length of bolt in wood main member (in)
ℓ_s	total length of bolt in wood side member(s) (in)
ℓ_u	laterally unbraced length of compression side of beam or unbraced length of column (in)
ℓ_x, ℓ_1	unbraced length of column considering buckling about strong (x) axis (in)
ℓ_y, ℓ_2	unbraced length of column considering buckling about weak (y) axis (in)
L	distance between points of lateral support of compression member (ft)
L	length of beam between points of zero moment (ft)
L	span length of beam (in)
L	nail length (in)
L_c	cantilever length in cantilever beam system (ft)
M	bending moment (in-kips, ft-kips)
M_b'	adjusted (allowable ASD) bending moment (in-lbf)
MC	moisture content based on oven-dry weight of wood (%)
M_{LRFD}'	LRFD moment capacity (resistance) (in-lbf, ft-kips, in-kips)
M_r, M_r'	reference and adjusted design moment (in-lbf)
M_u	factored (ultimate) bending moment (in-lbf, ft-kips) for LRFD
n, N	number of fasteners in a row (–)
N, N'	reference (tabulated) and allowable lateral design value at angle to grain θ for a single split ring or shear plate connector (lbf)
p	depth of fastener penetration into wood members (in)
P	total applied concentrated load or total axial load (lbf)
P	wind pressure (lbf/ft^2)
P, P'	nominal and allowable lateral design value parallel to grain for a single split ring connector unit or shear plate connector unit (lbf)
P_{allow}	total allowable concentrated load or total allowable axial load (lbf)
P_r	parallel to grain reference rivet capacity (lbf)
P_u	factored concentrated load (lbf/ft)
P_w	parallel to grain reference wood capacity for timber rivets (lbf)
Q	statical moment of an area about the neutral axis (in^3)
Q, Q'	reference (tabulated) and allowable lateral design value perpendicular to grain for a single split ring connector unit or shear plate connector unit (lbf)
Q_{allow}	total allowable concentrated load or total allowable axial load (lbf)
R	reaction force (lbf)
R_B	slenderness ratio of bending member (–)
R_d	reduction term for dowel-type fastener connections (–)
s	fastener spacing (in)
s_p	spacing between rivets parallel to grain (in)
s_q	spacing between rivets perpendicular to grain (in)
S	section modulus (in^3)
t	exposure time (hr)
t	thickness (in)
t_m	thickness of main (thicker) member (in)
t_s	effective thickness for shear (in)
t_s	thickness of side (thinner) member (in)

symbol	definition (units)
$t_{\parallel}$	effective thickness of parallel plies for plywood (in)
T	temperature (degree)
T	tension force (lbf)
v	unit shear stress (lbf/ft, lbf/in^2)
V	shear force (lbf)
V_h	horizontal shear force (plywood) (lbf)
V'_{LRFD}	LRFD shear capacity (resistance) (lbf, kips)
V_s	rolling shear force (plywood) (lbf)
V_u	factored shear force for LRFD (lbf)
w	uniformly distributed load (lbf/ft, lbf/ft^2)
w_{allow}	allowable uniform load (lbf/ft)
w_D	dead load (lbf/ft)
w_E	earthquake load (lbf/ft)
w_F	flood load (lbf/ft)
w_H	load due to lateral earth pressure, ground water pressure, or pressure of bulk materials (lbf/ft)
w_L	live load (lbf/ft)
$w_{L,\text{roof}}$	roof live load (lbf/ft)
w_R	rain load (lbf/ft)
w'_{LRFD}	LRFD allowable design value for w' in ASD (lbf/ft)
w_S	snow load (lbf/ft)
w_u	LRFD factored load (lbf/ft)
w_W	wind load (lbf)
x	variable distance (ft, in)
x	volume factor (–)
Z	reference lateral design value for a single fastener connection (lbf)
Z'	allowable lateral design value for a single fastener connection (lbf)
Z_{LRFD}, Z'_{LRFD}	LRFD adjusted and allowable lateral design stress (lbf)
$Z_{m\perp}$	reference lateral design value for single bolt or lag bolt in wood-to-wood connection with main member loaded perpendicular to grain and side member loaded parallel to grain (lbf)
$Z_{s\perp}$	reference lateral design value for single bolt or lag bolt in wood-to-wood connection with main member loaded parallel to grain and side member loaded perpendicular to grain (lbf)
$Z_{\parallel}$	reference lateral design value for a single bolt or lag bolt in connection with all wood members loaded parallel to grain (lbf)
$Z_{\perp}$	reference lateral design value for a single bolt or lag screw wood-to-wood, wood-to-metal, or wood-to-concrete connection with all wood members loaded perpendicular to grain (lbf)
β_{eff}	effective char rate adjusted for exposure time (in/hr)
β_n	nominal char rate, linear char rate based on 1-hour exposure (in/hr)
γ	load/slip modulus for a connection (lbf/in)
Δ	actual deflection (in)
Δ_{allow}	maximum allowable deflection (in)
θ	angle between direction of load and direction of wood grain (usually longitudinal axis of member) (degree)
λ	time effect factor (LRFD only) (–)
ϕ	resistance factor (LRFD only) (–)
ϕ_b	resistance factor for bending (LRFD only) (–)
ϕ_c	resistance factor for compression (LRFD only) (–)
ϕ_t	resistance factor for tension (LRFD only) (–)
ϕ_v	resistance factor for shear (LRFD only) (–)
ϕ_z	resistance factor for connections (LRFD only) (–)

1

Structural and Physical Properties of Wood

1. Structure of Wood

The cross section of the tree trunk shown in Fig. 1.1 shows the bark on the outside, the wood on the inside, and the pith at the central core. Wood is composed of thin tubular cells called *fibers*. New wood cells are formed on the inside of new bark cells that are grown under the existing bark. Thus, growth in a tree trunk results from the formation of these new cells.

Figure 1.1 Cross Section of Tree Trunk

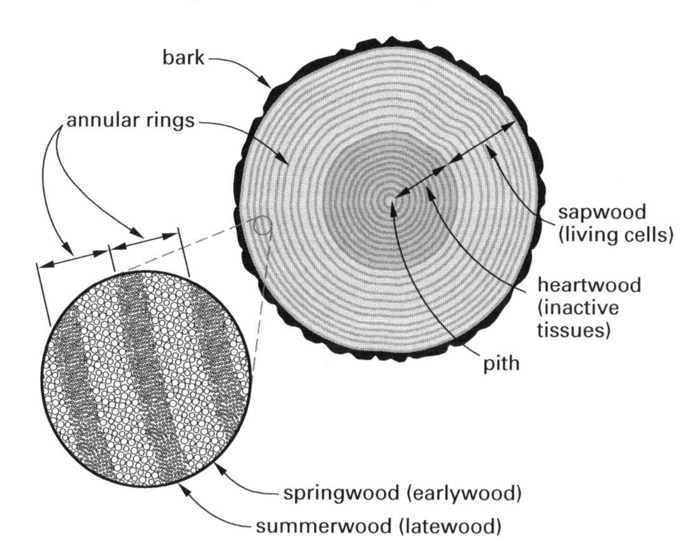

The wood cell walls are made up of cellulose, and the cells are bound together by a substance known as *lignin*. The wood formed late in the growing season differs from wood formed earlier, producing distinct annual growth rings. Each ring consists of two layers: *springwood* (also called *earlywood*) and *summerwood* (also called *latewood*). Summerwood is relatively heavy, so it has greater density and strength. The wood rings nearest the bark, called *sapwood*, are usually light-colored, and the wood inside, called *heartwood*, is relatively darker. There is no difference in mechanical properties between sapwood and heartwood. However, the sapwood is less resistant to decay.

2. Hardwoods and Softwoods

The *hardwoods* (oak, maple, ash, birch, yellow poplar, and so on) have broad leaves that are lost in the fall or winter. The *softwoods* (pines, firs, redwoods, and so on) have narrow needle-like leaves that remain on the tree throughout the year; thus, these species are called *evergreen*. The terms *hardwood* and *softwood* are unrelated to the actual hardness or strength of the wood.

3. Wood Defects During Growth

A. Knots

Knots are portions of branches that have been enveloped into the trunk of the tree during growth. They reduce the wood strength because they interrupt the fiber directions in a wood member.

B. Reaction Wood

Reaction wood is abnormal wood that forms on the compression or tension sides of leaning and crooked trees as a response to the tree's own dead weight. Many properties of reaction wood differ from those of straight wood. Reaction wood should not be used in structural members.

C. Cross Grain

Cross grain is a generic term describing wood fibers (cell walls) that are not aligned with the member's major axis. Cross grain can occur during growth or because of taper cuts for lumber.

D. Shakes

Shakes are cracks that are parallel to the annual growth ring and they can develop in a standing tree.

4. Sawn Lumber

Wood members that have been manufactured by cutting a piece from a tree log are called *sawn lumber.*

5. Moisture Content

The *moisture content*, MC, of wood varies among species, and even in a single species it varies depending on the location in the tree trunk. The moisture content of wood is a comparison of the weight of the moisture in the wood to the weight of the dry wood, known as *oven dry weight.* The moisture content in greenwood (i.e., fresh cut lumber) based on the oven-dry weight can vary from 30% to more than 200%.

A. Fiber Saturation Point

Moisture in green wood that exists in the cell cavities is called *free water.* Moisture that exists in the cell walls is called *bound water.* There is generally about five times as much free water as there is bound water in a typical tree. The moisture content when the free water has dissipated is called the *fiber saturation point*, FSP. The FSP varies with species, but it is typically about 30%. When the moisture content is below the FSP, the wood starts to shrink.

B. Equilibrium Moisture Content

After reaching the FSP, wood loses bound water (it dries) until the moisture in the wood is the same as that in the surrounding atmosphere. The moisture content at this point is known as the *equilibrium moisture content*, EMC. The EMC in the United States ranges from 5% to 25%, with 10% to 15% being the most common range. The EMC can be affected by humidity and air temperature. For example, if the relative humidity is 30%, the EMC will likely be less than 5%. The EMC can also be reduced by an air temperature greater than 100°F. According to the *Wood Handbook*, the EMC could also be greater than 25% if the humidity is 98% with an air temperature lower than 100°F.

For most buildings, the MC at the time of construction is higher than the EMC and the MC reaches the EMC gradually during service.

C. Green (Wet), Dry, and Seasoning

The term *greenwood* refers to the fresh-cut state of lumber. It is often defined as wood in which the cell walls are saturated; however, greenwood usually contains additional water in the cavities. The terms *green* and *dry* are also used in the NDS tables for design values. The term *dry* in the tables refers to areas where moisture content in the wood will not exceed 19% in sawn lumber and 16% in glued laminated timber for an extended time period.

The term *seasoning* usually refers to a controlled drying process by air or kiln drying.

6. Seasoning Defects from Lumber Shrinkage

As lumber dries below the FSP, shrinkage decreases the size of the cross section and can cause cracks and warping in the lumber (see Figs. 1.2 and 1.3). However, most structural strength properties increase as the MC in the wood decreases to about 15%.

Figure 1.2 Common Defects in Lumber

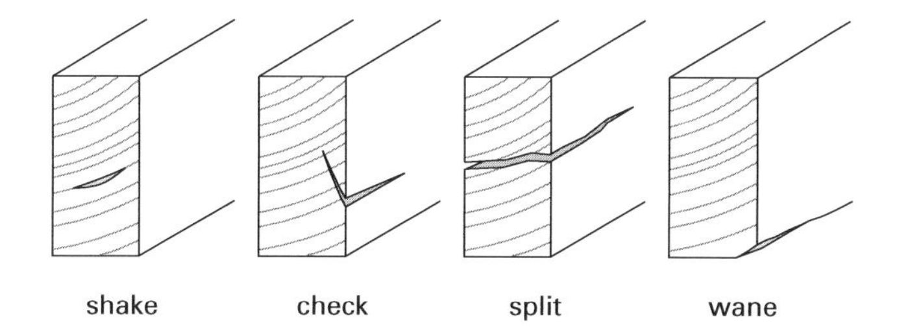

Figure 1.3 Warping for 2 × 10 Lumber

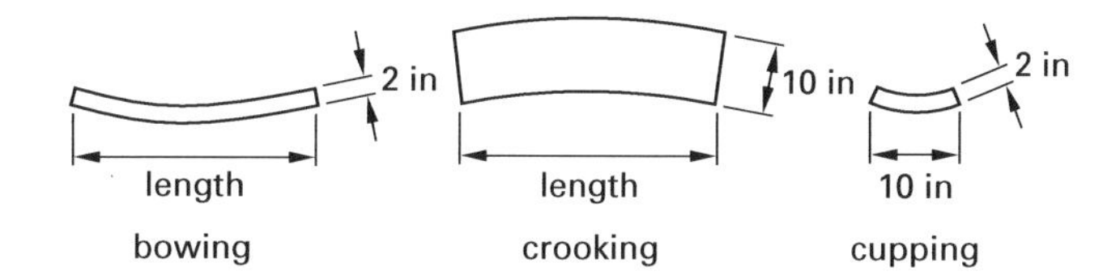

A. Longitudinal Side Checks

The wood near the surface dries faster than the wood at the inner core. This nonuniform drying process causes longitudinal cracks known as *seasoning checks*, *side checks*, or *surface checks* to form near the middle of the wide dimension of lumber lengths.

B. End Checks (Radial Cracks)

The dimensional changes in the wood cross section caused by drying are not uniform. For example, shrinkage parallel to the annual growth ring, called *tangential shrinkage*, is greater than perpendicular shrinkage, called *radial shrinkage.*

Tangential shrinkage causes end checks in the radial direction and is about twice the amount of radial shrinkage.

C. Splits

A *split* is a check occurring across the entire cross section, parallel to the lumber length.

D. Wane

Wane is bark or a lack of wood at the corner(s) of the lumber cross section.

E. Warping

Warping is bowing, crooking, and cupping caused by a difference in tangential and radial shrinkage.

F. Longitudinal Shrinkage

Any *longitudinal shrinkage* is usually negligible and ignored.

G. Shrinkage and Swelling Calculations

The amount of shrinkage or swelling that may occur in a structure in service can be estimated by using the data in Tables 3-5 and 3-6 in the *Wood Handbook*, along with Eq. 3-4 from the *Wood Handbook*. This equation, which gives the shrinkage from the green condition to any moisture content, S_m, is given as follows.

$$S_m = S_0 \left(\frac{30\% - \text{MC}}{30\%} \right)$$

The moisture content, MC (below 30%), and the total shrinkage, S_0 (from the *Wood Handbook* Tables 3-5 and 3-6), are given as percentages.

Greenwood may have a moisture content greater than 100% based on oven-dried weight. The *Wood Handbook* Table 3-3 contains average values for the moisture content of green wood by species. Heartwood's moisture content ranges from 31% to 162% and sapwood's moisture content ranges from 52% to 249%.

7. Thermal Properties

The thermal conductivity of wood, often given in units of Btu/hr-ft-in-°F, is much less than that of many other materials. For example, it is only about $1/16$ that of sand and gravel, $1/6$ that of clay brick, and $1/400$ that of steel. Therefore, wood is a good insulator.

Since wood is a good insulator, the temperature inside a large timber during a fire is elevated only slightly. Consequently, charring on the surface protects the inside, which aids in the prevention of a complete collapse of heavy timber structures.

The thermal coefficient of expansion for wood (parallel to grain) is approximately one-half that of steel. Since the dimensional changes caused by variations in moisture content are much greater than those caused by temperature changes, it is customary to neglect thermal expansion and contraction under most normal conditions.

8. Electrical Properties

Electrical conductivity of wood increases greatly and resistivity decreases as MC increases. The electrical resistivity of wood is about 10×10^{12} Ω·m at the oven-dry state and 5×10^3 Ω·m at the FSP. By comparison, steel's electrical resistivity is 20×10^{-8} Ω·m.

2

Mechanical Properties of Lumber and National Design Specification

Fibers (cells) are arranged in concentric cylinders in the tree trunk. As a result, wood has different characteristics in different directions; that is, wood is *anisotropic*. For simplification, wood is considered *orthotropic* with only three directions: longitudinal (parallel to the length of the fibers), radial (perpendicular to the length of the fibers), and tangential (in a direction tangential to the annual growth ring). (See Fig. 2.1.)

Figure 2.1 Orientation of Axes with Respect to Grain Direction

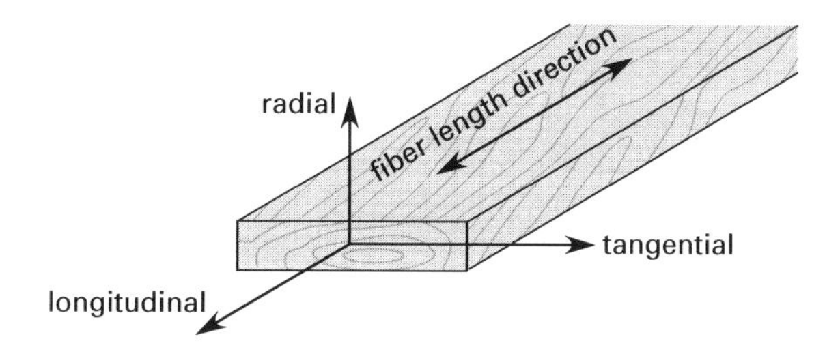

To further simplify these directional options for practical use, the radial and tangential directions are lumped into a perpendicular-to-grain category. The other category is parallel-to-grain in the longitudinal direction. Strength, modulus of elasticity, and other characteristics, such as shrinkage, swelling, and thermal coefficients, differ in the two directions.

1. Lumber Grading

Lumber is sawn from a tree log and is quite variable in mechanical properties. For simplicity and economy for users, lumbers of similar mechanical properties are considered to be of the same *grade*. Lumber is graded in terms of strength and utility in accordance with the grading rules of an approved grading agency.

A. Visual Grading and Design Values

The majority of sawn lumber is graded by visual inspection in accordance with established grading rules. The grade stamp in lumber includes the commercial grade, the species or species group, and other pertinent information, such as the MC range. For example, S-GRN (surfaced green) indicates that the MC is greater than 19% at the time of manufacture, and S-DRY (surfaced dry) indicates that the MC is 19% or less at the time of manufacture.

The *NDS Supplement* includes NDS design values for strength properties of lumber with a moisture content of 19% (16% for glulam) or less for an extended time period.

The design values for most species and grades of visually graded dimension lumber (those with a nominal thickness of 2 in to 4 in) are determined using the full-sized lumber in-grade test program. Design values of visually graded timbers (those with a 5 in minimum nominal dimension) and decking (2 in to 4 in thick and loaded on the wide face) are determined using the clear wood testing program. Clear wood strengths are then reduced to account for density, slope of wood grain, knots, shakes, checks, splits, size, test load duration, test moisture content, variability, and so on, to arrive at the NDS reference design values.

Lumber grading rules also provide limits on the slope of grain allowed in different stress grades. The slope of grain is measured from a line that is parallel to the longitudinal axis of a lumber length and it is expressed as a ratio. Because wood is relatively weaker and more variable in tension perpendicular to grain, designs that include lumber members with steep slopes of grain should be avoided (see Table 2.1).

Table 2.1 Effects of Grain Slopes

maximum slope of grain in member	modulus of rupture	impact bending	compression parallel to grain
	percentage reduction		
straight-grained	100	100	100
1 to 25	96	95	100
1 to 20	93	90	100
1 to 15	89	81	100
1 to 10	81	62	99
1 to 5	55	36	93

Source: *Wood Handbook: Wood as an Engineering Material*, Gen. Tech. Rep. FPL-GTR-113, U.S. Department of Agriculture, Forest Service, Forest Products Laboratory, Madison, WI, 1999, Table 4-12.

B. Machine Stress-Rated Grading

Machine stress-rated (MSR) grading (or E-rated grading) is based on a stiffness and deflection relationship. MSR grading machines measure the flatwise bending stiffness in a span of about 4 ft. MSR lumber is stamped with a grade stamp and has less variability in mechanical properties than visually graded lumber. The thickness of MSR lumber must be 2 in or less. *NDS Supplement* Table 4C lists design values for MSR lumber.

C. Machine Evaluated Lumber

Machine evaluated lumber (MEL) is nondestructively evaluated by approved mechanical grading equipment. This equipment also evaluates the lumber strength value. This lumber has greater variations in F_b/E combinations than MSR lumber. Tabulated design values are given in *NDS Supplement* Table 4C.

2. National Design Specification (NDS)

The *National Design Specification for Wood Construction* (NDS) requires that timber design meet the provisions for allowable stress design (ASD) or load and resistance factor design (LRFD), as applicable. The *NDS Supplement* provides reference design values for structural sawn lumber and structural glued laminated timber. These values are given for normal load duration (cumulative 10 years of full design load) under dry service conditions.

In ASD, adjusted (allowable) design values are determined using applicable ASD adjustment factors to the reference (tabulated) and the adjusted (allowable) stresses of a material are compared to calculated working stress based on the service loads. A factor of safety has already been applied to the reference (tabulated) design values.

In LRFD, adjusted design values are determined using applicable LRFD factors in NDS App. N and the adjusted capacities (resistance) are compared to stress based on the effects of factored loads.

Many of the LRFD provisions used in this book are provided in Table 3.2 (NDS Table 4.3.1), Table 3.3 (NDS Table 5.3.1), NDS Table 10.3.1, and NDS App. N.

In these tables, ASD uses the load duration factor, C_D. LRFD replaces C_D with the format conversion factor, K_F, as given in Table 2.2 (NDS App. Table N1), the resistance factor, ϕ, as given in Table 2.3 (NDS App. Table N2), and the time effect factor, λ, as given in Table 2.4 (NDS Table N3). Therefore, where C_D would be used in ASD, and $K_F\phi\lambda$ would be used in LRFD.

For example, in NDS Table 4.3.1 for bending, the adjusted (allowable) ending stresses F'_b are

ASD method $F'_b = F_bC_DC_MC_tC_LC_FC_{fu}C_iC_r$

LRFD method $F'_b = F_b(K_F\phi_b\lambda)C_MC_tC_LC_FC_{fu}C_iC_r$

The format conversion factor, K_F, is used for converting the *NDS Supplement* design values to LRFD nominal design values. The reference design values in the *NDS Supplement* include safety adjustment factors appropriate for ASD. These ASD safety adjustments are removed to obtain nominal design values for LRFD. The nominal design value for bending for LRFD is F_bK_F. The format conversion factor, K_F, also adjusts the reference (tabulated) design value for short duration (10 min) loading versus the normal duration (10 yr) loading used in ASD.

Table 2.2 Format Conversion Factor, K_F (LRFD Only) (NDS Appendix Table N1)

application	property	K_F
member	F_b, F_t, F_v, F_c, F_{rt}, F_s	$2.16/\phi$
	$F_{c\perp}$	$1.875/\phi$
	$E_{\min}$	$1.5/\phi$
connections	all connections in the NDS	$2.16/\phi$

Reproduced from *National Design Specification for Wood Construction,* 2005 Edition, courtesy, American Forest & Paper Association, Washington, D.C.

The resistance factor, ϕ, is determined using Table 2.3. ϕ is a strength reduction factor to allow for the possibility that the member capacity may actually be less than theoretically calculated. According to Table 2.3, the resistance factor for bending, ϕ_b, is 0.85 and the resistance factor for shear, ϕ_v, is 0.75.

Table 2.3 Resistance Factor, ϕ (LRFD Only) (NDS Appendix Table N2)

application	property	symbol	value
member	F_b	ϕ_b	0.85
	F_t	ϕ_t	0.80
	F_v, F_{rt}, F_s	ϕ_v	0.75
	F_c, $F_{c\perp}$	ϕ_c	0.90
	E_{min}	ϕ_s	0.85
connections	all connections in the NDS	ϕ_z	0.65

Reproduced from *National Design Specification for Wood Construction,* 2005 Edition, courtesy, American Forest & Paper Association, Washington, D.C.

In LRFD, the load duration effect is accounted for using the time effect factor, λ. Wood can support higher stresses if the load is applied for a short period of time. Therefore, λ values range for 0.6 for load combination dominated by dead loads, to 1.0 for load combinations dominated by short-duration loads such as wind forces.

The time effect factors are specified in accordance with load combinations required by the applicable building code. In the absence of a governing building code, the load combinations are specified in ASCE/SEI7-05 and are summarized in Table 2.4.

Table 2.4 Time Effect Factor, λ (LRFD Only) (NDS Appendix Table N3)

load combination[a]	λ
$1.4(D+F)$	0.6
$1.2(D+F)+1.6H+0.5(w_{L,\text{roof}}$ or S or $R)$	0.6
$1.2(D+F)+1.6(L+H)+0.5(w_{L,\text{roof}}$ or S or $R)$	0.7 when L is from storage 0.8 when L is from occupancy 1.25 when L is from impact[b]
$1.2D+1.6(w_{L,\text{roof}}$ or S or $R)+L$ or $0.8w$	0.8
$1.2D+1.6w+L+0.5(w_{L,\text{roof}}$ or S or $R)$	1.0
$1.2D+1.0E+L+0.2S$	1.0
$0.9D+1.6w+1.6H$	1.0
$0.9D+1.0E+1.6H$	1.0

[a]Load combinations and load factors consistent with ASCE/SEI7-05 (Article 2.3, Combining Factored Loads Using Strength Design) are listed for ease of reference. Nominal loads shall be in accordance with NDS App. N.1.2.

[b]Time effect factors, λ, greater than 1.0 shall not apply to connections or to structural members pressure-treated with water-borne preservatives or fire retardant chemicals.

Reproduced from *National Design Specification for Wood Construction,* 2005 Edition, courtesy, American Forest & Paper Association, Washington, D.C.

To prevent confusion with other variables similar to those used in the NDS and ASCE/SEI7-05, the variable w is used to represent loads throughout this book's explanations and examples. All loads are represented as follows. For example, the commonly used dead load and snow load combination that NDS represents as $1.2D + 1.6S$ will be presented as $1.2w_D + 1.6w_S$ throughout this book's remaining chapters.

w_D dead load (D in NDS and ASCE/SEI7-05)
w_E earthquake load (E in NDS and ASCE/SEI7-05)
w_F flood load (F_a in NDS and ASCE/SEI7-05)

w_H load due to lateral earth pressure, ground water pressure, or pressure of bulk materials (H in NDS and ASCE/SEI7-05)
w_L live load (L in NDS and ASCE/SEI7-05)
$w_{L,\text{roof}}$ roof live load (L_r in NDS and ASCE/SEI7-05)
w_R rain load (R in NDS and ASCE/SEI7-05)
w_S snow load (S in NDS and ASCE/SEI7-05)

3. NDS Design Values for Lumber and Glued Laminated Timbers

A. Design Values in the *NDS Supplement*

Reference (tabulated) design values for visually graded structural lumber in the *NDS Supplement* have undergone major changes from the earlier NDS editions. New adjustment factors for determining allowable design values from reference design values have also been included in the *NDS Supplement*.

B. Allowable Design Values

Reference design values in the *NDS Supplement* are reduced or increased, usually by applying adjustment factors, to arrive at allowable design values.

$$\text{allowable design value} = (\text{reference design value})(\text{adjustment factors})$$

C. Frequently Used Table Adjustment Factors

[Tbl 3.2, 3.3]

Adjustment factors in design values include the following.

load duration (ASD only)	C_D
wet service condition	C_M
temperature	C_t
beam stability	C_L
size (sawn lumber only)	C_F
volume (glulam beams only)	C_V
curvature (glulam only)	C_c
flat use	C_{fu}
incising (sawn lumber only)	C_i
repetitive member (sawn lumber only)	C_r
form	C_f
column stability	C_p
buckling stiffness (sawn lumber only)	C_T
bearing area	C_b

Table 2.5 List of Commonly Used Design Properties

design properties	symbol for reference (tabulated) design value	symbol for allowable (adjusted) design value
bending stress	F_b	F'_b
tension stress parallel to grain	F_t	F'_t
shear stress parallel to grain	F_v	F'_v
compression stress perpendicular to grain	$F_{c\perp}$	$F'_{c\perp}$
compression stress parallel to grain	F_c	F'_c
modulus of elasticity	E	E'
modulus of elasticity for beam stability and column stability calculations	E_{min}	E'_{min}

4. NDS Design Values for Mechanical Connections

[NDS Tbls 10.3.1, 10.3.3]

Nominal design values for mechanical connections in the NDS are multiplied by all applicable adjustment factors to obtain the allowable design values.

Adjustment factors for determining allowable design values of mechanical connections include the following.

load duration (ASD only)	C_D
wet service condition	C_M
temperature	C_t
group action	C_g
geometry	C_Δ
penetration depth	C_d
end grain	C_{eg}
metal side plate	C_{st}
toe-nail	C_{tn}
diaphragm	C_{di}

These factors are also used in Tables 7.1 and 7.2 (NDS Tables 10.3.1 and 10.3.3).

3

Lumber Size Categories and Allowable Design Stress

Resistances for wood members are based on various adjustment factors, such as species, size, conditions of use, moisture content, load time duration, and so on. Table design values for visually graded and mechanically graded lumber and glulam are specified in *NDS Supplement* Tables 4A, 4B, 4C, 4D, 4E, 4F, 5A, 5B, 5C, and 5D.

Figure 3.1 Use and Size Categories

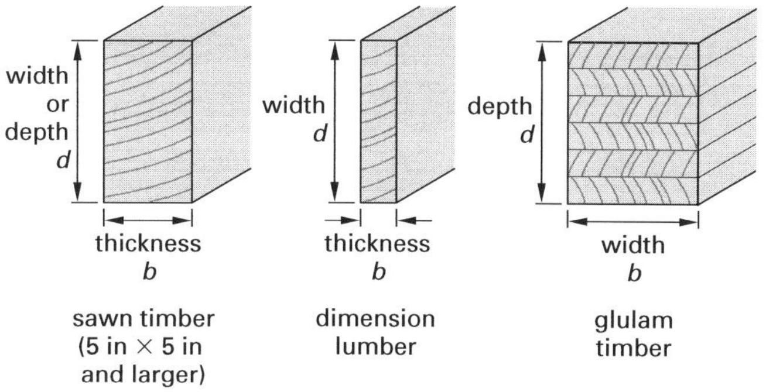

1. Size Categories

A. Size Classifications for Sawn Lumber

Dimension Lumber

Dimension lumber has a nominal thickness ranging from 2 in to 4 in and a width of 2 in or more. The load is applied to either the narrow or wide face (decking is loaded on the wide face only).

Types of dimension lumber include light framing (2 in to 4 in wide), structural joists and planks (5 in or wider), studs (2 in to 6 in wide and 10 ft maximum length), and decking (4 in or wider and loaded about the minor axis).

Timbers

Timbers have a minimum nominal dimension of 5 in or more.

Beams and stringers are sawn into a rectangular cross section, nominal 5 in or thicker, with a width of more than 2 in greater than the thickness and loaded on the narrow face during bending. Examples are 6 × 10, 7 × 10, and 10 × 14.

Posts and timbers are sawn into square or approximately square cross sections of 5 in × 5 in nominal dimensions or larger with a width no greater than 2 in more than the thickness. These can be used as posts or columns. Strength in bending is not especially important. The lateral load can be applied to either the narrow face or the wide face. Examples are 6 × 6, 6 × 8, 8 × 10, and 10 × 12.

Boards are 1 in to 1½ in thick and 2 in or wider.

B. Sizes for Structural Classifications

Lumber dimensions can be specified in one of three ways: dressed, rough sawn, and full sawn. Dressed lumber is dressed on a planing machine to create smooth surfaces and uniform sizes and it has net (actual) dimensions that are less than the nominal dimensions, which are usually specified on construction plans. Rough sawn lumber is approximately ⅛ in larger than dressed lumber. (Large timbers are commonly rough sawn to the desired dimensions.) Full-sawn lumber size is specified by using the actual dimensions and is generally not available (see Table 3.1).

Table 3.1 Examples of Nominal and Net Dimensions

nominal size b (thickness) × d (width)	dressed size (S4S) $b \times d$
timbers	
6 × 6	5½ in × 5½ in
8 × 8	7½ in × 7½ in
dimension lumber	
2 × 6	1½ in × 5½ in
2 × 8	1½ in × 7¼ in
3 × 6	2½ in × 5½ in
3 × 10	2½ in × 9¼ in
4 × 6	3½ in × 5½ in
4 × 8	3½ in × 7¼ in

Structural Calculations

Structural calculations are based on net dimensions, not on nominal sizes, for the anticipated use conditions. Shrinkage may have to be accounted for when detailing connections, but standard dimensions are accepted for stress calculations. Dimensions stated in construction plans are nominal dimensions. Net dimensions for dressed lumber thickness are generally taken as ½ in less than nominal. However, the net width of dimension lumber exceeding a width of 6 in is taken as ¾ in less than nominal.

For 4 in (nominal) and thinner lumber, net *dry* sizes are used in structural calculations regardless of the moisture content at the time of manufacture or use.

For 5 in (nominal) and thicker lumber, net *green* dressed sizes are used in structural calculations regardless of the moisture content at the time of manufacture or use.

C. Glued Laminated Timber

Structural calculations are based on either the net finished dimensions as specified in *NDS Supplement* Tables 1C and 1D Section Properties, or the manufacturer's data.

2. Reference Design Values Specified

[NDS Sec 4.2; NDS Supp Tbls 4, 5]

A. Visually Graded and Mechanically Graded Lumber

NDS Supplement Tables 4 and 5 give reference design values for visually graded and mechanically graded sawn lumber for a 10-year load (in which a member is fully stressed to its allowable design value by the application of the full design load for a cumulative duration of 10 years; see NDS Sec. 2.3.2), and for dry service conditions (in which moisture content is less than or equal to 19% for sawn lumbers).

For dimension grade, reference bending design values apply to x-axis bending only (in which the load is applied on the narrow face, b). For y-axis bending (in which the load is applied to the wide face) the reference design values are multiplied by the flat use factors. (See *NDS Supplement* Table 4 adjustment factors.)

For decking grade, the reference bending design values are already adjusted for flat wise use (in which the load is applied to the wide face).

For post and timber grade, reference bending design values apply to members with the load applied to either the narrow or wide face.

For beam and stringer grade, the reference bending design values apply to members with the load applied to the narrow face only.

B. Glued Laminated Timber

Reference design values are valid for load durations of 10 years and for dry service conditions (in which moisture content is less than or equal to 16%). Reference design values for bending are applicable for loading perpendicular to the wide face of laminations. For bending with loading parallel to the wide face of laminations, the reference design values are multiplied by the flat use factors. (See *NDS Supplement* Table 5A adjustment factors.)

3. Adjustment of Reference (Tabulated) Design Values

[NDS Sec 2.3; NDS Tbls 2.3.2, 2.3.3, 3.3.3, 3.10.4, 4.3.1, 5.3.1; NDS Supp Tbls 4, 5 Adj Fac]

NDS Supplement reference design values (F_b, F_t, F_v, $F_{c\perp}$, E, etc.) are multiplied by all applicable adjustment factors to determine allowable (adjusted) design values (F_b', F_t', F_v', $F_{c\perp}'$, E, etc.) for actual conditions of use in accordance with the *NDS Supplement* Tables 4 and 5 adjustment factors.

$$\text{allowable (adjusted) design value} = (\text{reference design value})(\text{adjustment factors})$$

4. Adjustment Factors

[NDS Secs 2.3, 4.3, 5.3; NDS Supp Tbls 4, 5 Adj Fac]

This section identifies some of the adjustment factors frequently used in the NDS. The applicability of these factors is shown in Tables 3.2 and 3.3. Some of the footnotes for the tables discussed in this section will be explained in subsequent sections.

A. Load Duration Factor, C_D

[NDS Tbl 2.3.2; NDS App B]

Wood can sustain greater maximum loads for short-load durations than for long-load durations (see Table 3.4). When the full maximum load is applied either cumulatively or continuously for periods less than 10 years, NDS reference design stresses and mechanical fastenings are multiplied by the load duration factor, C_D.

Table 3.2 Applicability of Adjustment Factors (Sawn Lumber) (NDS Table 4.3.1)

	ASD only	ASD and LRFD										LRFD only		
	load duration factor	wet service factor	temperature factor	beam stability factor	size factor	flat use factor	incising factor	repetitive member factor	column stability factor	buckling stiffness factor	bearing area factor	format conversion factor	resistance factor	time effect factor
$F'_b = F_b \times$	C_D	C_M	C_t	C_L	C_F	C_{fu}	C_i	C_r	–	–	–	K_F	ϕ_b	λ
$F'_t = F_t \times$	C_D	C_M	C_t	–	C_F	–	C_i	–	–	–	–	K_F	ϕ_t	λ
$F'_v = F_v \times$	C_D	C_M	C_t	–	–	–	C_i	–	–	–	–	K_F	ϕ_v	λ
$F'_{c\perp} = F_{c\perp} \times$	–	C_M	C_t	–	–	–	C_i	–	–	–	C_b	K_F	ϕ_c	λ
$F'_c = F_c \times$	C_D	C_M	C_t	–	C_F	–	C_i	–	C_P	–	–	K_F	ϕ_c	λ
$E' = E \times$	–	C_M	C_t	–	–	–	C_i	–	–	–	–	–	–	–
$E'_{min} = E_{min} \times$	–	C_M	C_t	–	–	–	C_i	–	–	C_T	–	K_F	ϕ_s	–

Reproduced from *National Design Specification for Wood Construction*, 2005 Edition, courtesy, American Forest & Paper Association, Washington, D.C.

Table 3.3 Applicability of Adjustment Factors (Glulam) (NDS Table 5.3.1)

	ASD only	ASD and LRFD								LRFD only		
	load duration factor	wet service factor	temperature factor	beam stability factor*	volume factor*	flat use factor	curvature factor	column stability factor	bearing area factor	format conversion factor	resistance factor	time effect factor
$F'_b = F_b \times$	C_D	C_M	C_t	C_L	C_V	C_{fu}	C_c	–	–	K_F	ϕ_b	λ
$F'_t = F_t \times$	C_D	C_M	C_t	–	–	–	–	–	–	K_F	ϕ_t	λ
$F'_v = F_v \times$	C_D	C_M	C_t	–	–	–	–	–	–	K_F	ϕ_v	λ
$F'_{c\perp} = F_{c\perp} \times$	–	C_M	C_t	–	–	–	–	–	C_b	K_F	ϕ_c	λ
$F'_c = F_c \times$	C_D	C_M	C_t	–	–	–	–	C_P	–	K_F	ϕ_c	λ
$F'_{rt} = F_{rt} \times$	C_D	C_M	C_t	–	–	–	–	–	–	K_F	ϕ_v	λ
$E' = E \times$	–	C_M	C_t	–	–	–	–	–	–	–	–	–
$E'_{min} = E_{min} \times$	–	C_M	C_t	–	–	–	–	–	–	K_F	ϕ_s	–

*The beam stability factor, C_L, shall not apply simultaneously with the volume factor, C_V, for structural glued laminated timber bending members (see NDS Sec. 5.3.6). Therefore, the lesser of these adjustment factors shall apply.

Reproduced from *National Design Specification for Wood Construction*, 2005 Edition, courtesy, American Forest & Paper Association, Washington, D.C.

Table 3.4 Frequently Used Load Duration Factors, C_D[a] (NDS Table 2.3.2)

load duration	C_D	typical design loads
permanent	0.9	dead load
ten years	1.0	occupancy live load
two months	1.15	snow load
seven days	1.25	construction load
ten minutes	1.6	wind/earthquake load
impact[b]	2.0	impact load

[a]Load duration factors shall not apply to modulus of elasticity, E, and the modulus of elasticity for beam and column stability, E_{min}, nor to compression perpendicular to grain design values, $F_{c\perp}$, based on a deformation limit.

[b]Load duration factors greater than 1.6 shall not apply to structural members pressure-treated with water-borne preservatives, or fire retardant chemicals. The impact load duration factor shall not apply to connections.

B. Wet Service Factor, C_M

[NDS Secs 4.1.4, 5.1.5, 7.1.4, 8.1.4, 9.3.3; NDS Supp Tbls 4A, 4B, 4C, 4D, 4E, 4F, 5A, 5B, 5C, 5D]

When moisture content at installation or in service is expected to exceed 19% for sawn lumber and 16% for glulam, reference design values are reduced by the wet service factor, C_M, given in *NDS Supplement* Tables 4 and 5.

C. Size Factor, C_F

[NDS Sec 4.3.6; NDS Supp Tbl 4 Adj Fac]

For dimension lumber (lumber that is 2 in to 4 in thick) for bending, tension, and compression parallel to the grain, the size factor, C_F, is specified in *NDS Supplement* Tables 4A, 4B, 4D, 4E, and 4F adjustment factors.

When the depth, d, of a beam, stringer, post, or timber exceeds 12 in, C_F applies for bending only and is given in *NDS Supplement* Table 4D adjustment factors as

$$C_F = \left(\frac{12}{d}\right)^{1/9} \leq 1.0$$

NDS Supplement Table 4D also includes a table of adjustment factors for beams and stringers subjected to loads applied to the wide face.

Bending design values for decking are based on 4 in thick decking. With the exception of redwood, when 2 in or 3 in thick decking is used, C_F for bending is given in *NDS Supplement* Table 4E adjustment factors.

D. Flat Use Factor, C_{fu}

[NDS Sec 4.3.7; NDS Supp Tbls 4, 5 Adj Fac]

When dimension lumber is loaded in bending on the wide face, C_{fu} is given in *NDS Supplement* Tables 4A, 4B, 4C, and 4F adjustment factors. Table 5A adjustment factors apply to glulam bending design values.

Reference (tabulated) bending design values for decking have already been adjusted for flatwise usage (with the load applied to the wide face) in *NDS Supplement* Table 4E.

E. Bearing Area Factor, C_b

[NDS Sec 2.3.10]

Reference (tabulated) compression design values perpendicular to grain, $F_{c\perp}$, apply to bearings of any length at the ends of a member and to all bearings 6 in or more in length at any other location. For bearings less than 6 in in length and not nearer than 3 in to the end of a member (see NDS Table 3.10.4).

F. Incising Factor for Structural Sawn Lumber, C_i

[NDS Sec 4.3.8]

Reference design values shall be multiplied by an incising factor, C_i (NDS Table 4.3.8), when structural sawn lumber is incised to increase penetration of preservatives. Incisions are cut parallel to grain a maximum depth of 0.4 in, a maximum length of $^3/_8$ in, and a maximum density of up to 1100 incisions per square foot.

G. Repetitive Member Factor, C_r

[NDS Sec 4.3.9]

Bending design values, F_b, in *NDS Supplement* Tables 4A, 4B, 4C, and 4F for dimension lumber are multiplied by the repetitive member factor, C_r. C_r is 1.15 when such members are used as joists, truss chords, rafters, studs, planks, decking, or similar members that are in contact or spaced not more than 24 in on centers, are not less than three in number, and are joined by floor, roof, or other load distributing elements adequate to support the design load.

H. Other Factors

Other adjustment factors, such as temperature factor, C_t (equals 1.0 for 100°F or less), fire-retardant treatment, and so on are described in NDS Sec. 2.3 and in other relevant sections of the NDS. These other adjustment factors will be covered as they are needed.

Example 3.1
Design of Dimension Lumber Beam

A 2 × 8 no. 1 douglas fir-larch is loaded on its narrow face. It is adequately braced to prevent lateral torsional buckling. The loads consist of dead load, w_D, plus snow load, w_S. The wet service condition prevails with a moisture content greater than 19%. Assume normal temperature and deck beams spaced at 15 in.

w_D	dead load	14 lbf/ft^2
w_S	snow load	20 lbf/ft^2
MC	moisture content	> 19%
L	simple span length	14 ft

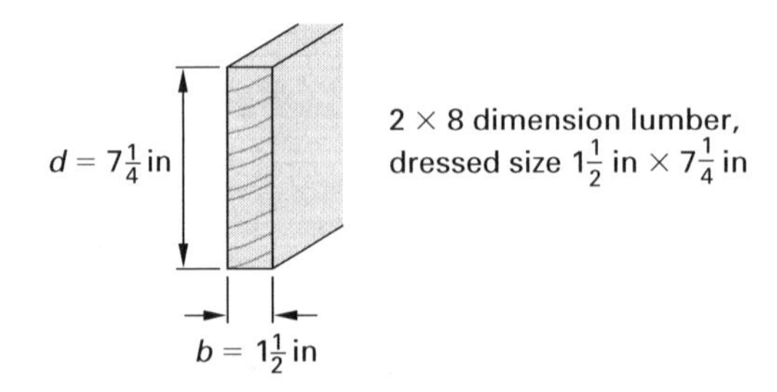

Find the allowable (adjusted) design values and allowable (adjusted) shear and moment values in accordance with the NDS.

REFERENCE

Solution (ASD Method)

2 × 8 is a visually graded dimension lumber.

NDS Tbl 2.3.2

For snow load, the load duration factor, C_D, is 1.15.

NDS Tbls 2.3.2, 4.3.1; NDS Supp Tbl 4A, 4A Adj Fac

The adjusted bending stress is

$$\begin{aligned} F'_b &= F_b C_D C_M C_t C_L C_F C_{fu} C_r \\ &= \left(1000\ \frac{\text{lbf}}{\text{in}^2}\right)(1.15)(0.85)(1.0)(1.0)(1.2)(1.0)(1.0) \\ &= 1173\ \text{lbf/in}^2 \end{aligned}$$

The adjusted tension value parallel to the grain is

$$\begin{aligned} F'_t &= F_t C_D C_M C_t C_F = \left(675\ \frac{\text{lbf}}{\text{in}^2}\right)(1.15)(1.0)(1.0)(1.2) \\ &= 931.5\ \text{lbf/in}^2 \end{aligned}$$

The adjusted shear design value parallel to the grain is

$$\begin{aligned} F'_v &= F_v C_D C_M C_t = \left(180\ \frac{\text{lbf}}{\text{in}^2}\right)(1.15)(0.97)(1.0) \\ &= 200.79\ \text{lbf/in}^2 \end{aligned}$$

The adjusted modulus of elasticity is

$$\begin{aligned} E' &= E C_M C_t = \left(1{,}700{,}000\ \frac{\text{lbf}}{\text{in}^2}\right)(0.9)(1.0) \\ &= 1{,}530{,}000\ \text{lbf/in}^2 \end{aligned}$$

The adjusted modulus of elasticity for beam stability is

$$\begin{aligned} E'_m &= E_{\text{min}} C_M C_t \\ E'_{\text{min}} &= \left(620{,}000\ \frac{\text{lbf}}{\text{in}^2}\right)(0.9)(1.0) \\ &= 558{,}000\ \text{lbf/in}^2 \end{aligned}$$

The section modulus, S_x, is 13.14 in^3.

The adjusted bending moment capacity (or resistance) is

$$\begin{aligned} M'_b &= F'_b S_x = \left(1173\ \frac{\text{lbf}}{\text{in}^2}\right)(13.14\ \text{in}^3) \\ &= 15{,}413.2\ \text{in-lbf} \end{aligned}$$

The total design load is

$$\begin{aligned} w_{\text{TL}} &= w_D + w_S \\ &= \left(14\ \frac{\text{lbf}}{\text{ft}^2}\right)(15\ \text{in})\left(\frac{1\ \text{ft}}{12\ \text{in}}\right) + \left(20\ \frac{\text{lbf}}{\text{ft}^2}\right)(15\ \text{in})\left(\frac{1\ \text{ft}}{12\ \text{in}}\right) \\ &= 42.5\ \text{lbf/ft} \end{aligned}$$

The actual bending moment is

$$M = \frac{w_{\text{TL}}L^2}{8} = \frac{\left(42.5\ \frac{\text{lbf}}{\text{ft}}\right)(14\ \text{ft})^2\left(12\ \frac{\text{in}}{\text{ft}}\right)}{8}$$

$$= 12{,}495\ \text{in-lbf} < M_b' = 15{,}413.2\ \text{in-lbf} \quad [\text{OK}]$$

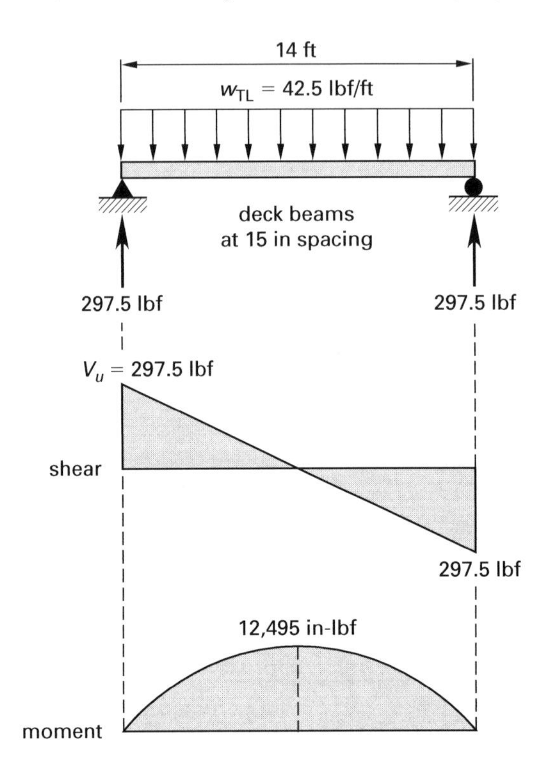

The actual shear value parallel to the grain is

$$f_v \approx \frac{1.5V}{A} = \frac{(1.5)(297.5\ \text{lbf})}{10.88\ \text{in}^2}$$

$$= 41.01\ \text{lbf/in}^2 < F_v' = 200.29\ \text{lbf/in}^2 \quad [\text{OK}]$$

REFERENCE

Solution (LRFD Method)

NDS App Tbls N1, N2

For bending, the LRFD reference bending stress is

$$F_{b,\text{LRFD}} = F_b K_F \phi_b \lambda$$

K_F converts the reference design value for ASD 10-year load duration (normal duration) loading to a nominal design value for LRFD short duration loading (10-minute duration).

$$K_F = \frac{2.16}{\phi_b} = \frac{2.16}{0.85}$$

$$= 2.54$$

Ex. 3.1

2×8 no. 1 douglas fir-larch $(w_D + w_S)$ and MC > 19%. Assume that C_L is 1.0.

NDS Supp Tbl 4A, 4A Adj Fac

reference values	C_M	C_F
$F_b = 1000\ \frac{\text{lbf}}{\text{in}^2}$	0.85	1.2
$F_t = 675\ \frac{\text{lbf}}{\text{in}^2}$	1.0	1.2
$F_v = 180\ \frac{\text{lbf}}{\text{in}^2}$	0.97	–
$E - 1{,}700{,}000\ \frac{\text{lbf}}{\text{in}^2}$	0.90	–
$E_{\text{min}} = 620{,}000\ \frac{\text{lbf}}{\text{in}^2}$	0.90	–

NDS App Tbls N1, N2, N3

$F_b,\ F_t,\ F_v$	$K_F = \dfrac{2.16}{\phi}$
E_{min}	$K_F = \dfrac{1.5}{\phi}$
F_b	$\phi_b = 0.85$
F_t	$\phi_t = 0.80$
F_v	$\phi_v = 0.75$
E_{min}	$\phi_s = 0.85$
E	–

For load combination of dead load and snow load, $1.2w_D + 1.6w_S$. The time effect factor, λ, is 0.8.

Note that C_D used in ASD is replaced with $K_F\phi_b\lambda$ for LRFD bending.

NDS Tbl 4.3.1

The LRFD adjusted bending design value is

$$\begin{aligned}
F'_{b,\text{LRFD}} &= F_{b,\text{LRFD}}C_MC_tC_LC_F = F_b(K_F\phi_b\lambda)C_MC_tC_LC_F \\
&= \left(1000\ \frac{\text{lbf}}{\text{in}^2}\right)\left(\left(\frac{2.16}{\phi_b}\right)\phi_b(0.8)\right)(0.85)(1.0)(1.0)(1.2) \\
&= 1762.6\ \text{lbf/in}^2
\end{aligned}$$

The LRFD adjusted tension design value is

$$\begin{aligned}
F'_{t,\text{LRFD}} &= F_t(K_F\phi_t\lambda)C_MC_tC_F \\
&= \left(675\ \frac{\text{lbf}}{\text{in}^2}\right)\left(\frac{2.16}{\phi_t}\right)\phi_t(0.8)(1.0)(1.0)(1.2) \\
&= 1399.7\ \text{lbf/in}^2
\end{aligned}$$

The LRFD adjusted shear design value is

$$\begin{aligned}
F'_{v,\text{LRFD}} &= F_{v,\text{LRFD}}C_MC_t = F_v(K_F\phi_v\lambda)C_MC_t \\
&= \left(180\ \frac{\text{lbf}}{\text{in}^2}\right)\left(\frac{2.16}{\phi_v}\right)\phi_v(0.8)(0.97)(1.0) \\
&= 301.71\ \text{lbf/in}^2
\end{aligned}$$

The LRFD adjusted modulus of elasticity is

$$\begin{aligned} E'_{\text{LRFD}} &= EC_MC_t \\ &= \left(1{,}700{,}000\ \frac{\text{lbf}}{\text{in}^2}\right)(0.9)(1.0) \\ &= 1{,}530{,}000\ \text{lbf/in}^2 \end{aligned}$$

The LRFD adjusted modulus of elasticity for beam stability is

$$\begin{aligned} E'_{\text{min,LRFD}} &= E_{\text{min}}K_F\phi_sC_MC_t \\ &= \left(620{,}000\ \frac{\text{lbf}}{\text{in}^2}\right)\left(\frac{1.5}{\phi_s}\right)\phi_s(0.9)(1.0) \\ &= 837{,}000\ \text{lbf/in}^2 \end{aligned}$$

The total (factored) LRFD designed load is

$$\begin{aligned} w_{\text{TL,LRFD}} &= 1.2w_D + 1.6w_S \\ &= (1.2)\left(17.5\ \frac{\text{lbf}}{\text{ft}}\right) + (1.6)\left(25.0\ \frac{\text{lbf}}{\text{ft}}\right) \\ &= 61\ \text{lbf/ft} \end{aligned}$$

Calculate the LRFD moment resistance value.

$$\begin{aligned} M'_{\text{LRFD}} &= F'_{b,\text{LRFD}}S_x \\ &= \left(1.762\ \frac{\text{kips}}{\text{in}^2}\right)(13.14\ \text{in}^3) \\ &= 23.15\ \text{in-kips} \end{aligned}$$

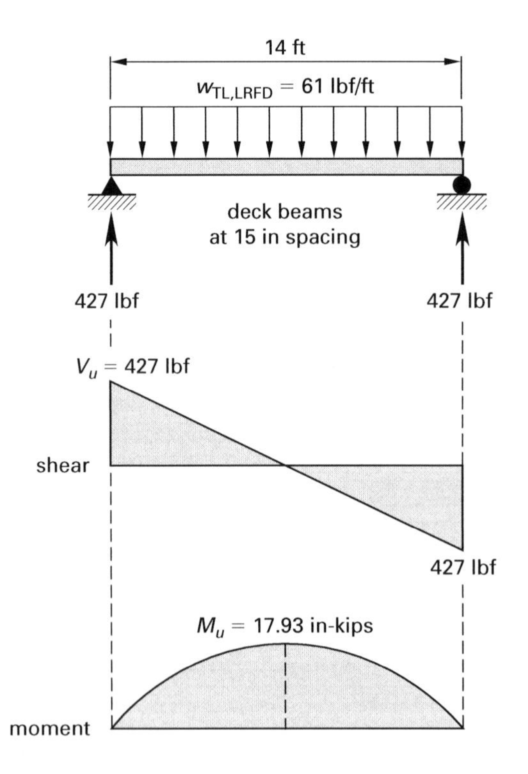

NDS App Tbl N3

Calculate the actual factored moment.

$$M_u = \frac{w_{\text{TL,LRFD}}L^2}{8} = \left(\frac{\left(61\ \frac{\text{lbf}}{\text{ft}}\right)(14\ \text{ft})^2}{8}\right)\left(12\ \frac{\text{in}}{\text{ft}}\right)\left(\frac{1\ \text{kip}}{1000\ \text{lbf}}\right)$$

$$= 17.93\ \text{in-kips} < M'_{\text{LRFD}} = 23.15\ \text{in-kips} \quad [\text{OK}]$$

The factored shear is

$$V_u = w_{\text{TL,LRFD}}(14\ \text{ft})\left(\tfrac{1}{2}\right) = \left(61\ \frac{\text{lbf}}{\text{ft}}\right)(14\ \text{ft})\left(\tfrac{1}{2}\right) = 427\ \text{lbf}$$

The adjusted shear design value, $F'_{v,\text{LRFD}}$, was previously calculated as 301.71 lbf/in^2. Rearrange the following equation to find the adjusted LRFD shear resistance.

$$F'_{v,\text{LRFD}} = \frac{\frac{3}{2}V'_{\text{LRFD}}}{A}$$

The LRFD shear resistance is

$$V'_{\text{LRFD}} = \tfrac{2}{3}F'_{v,\text{LRFD}}A = \left(\tfrac{2}{3}\right)\left(301.71\ \frac{\text{lbf}}{\text{in}^2}\right)(10.88\ \text{in}^2)$$

$$= 2188.40\ \text{lbf} > V_u = 427\ \text{lbf} \quad [\text{OK}]$$

Example 3.2
Allowable Design Values: Beams and Stringers

A full-sawn 6×10 no. 2 red oak in bending is subjected to a combination loading of dead load, w_D, live load, w_L, snow load, w_S, and wind load, w_W. Assume use conditions to be dry service, normal temperature, and adequate lateral bracing.

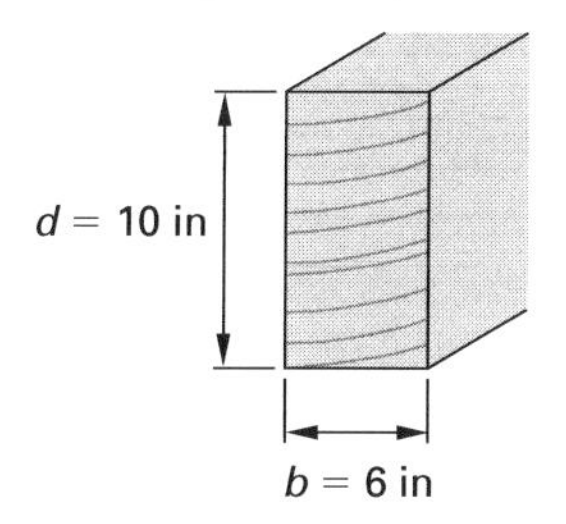

Find the relevant allowable design values in accordance with the NDS.

REFERENCE

Solution (ASD Method)

The size classification is
thickness = 6 in > 5 in
width = 10 in > thickness + 2 in

Therefore, the classification is beams and stringers, visually graded.

NDS Tbl 2.3.2

For wind load, the load duration factor, C_D, is 1.6.

NDS Tbl 4.3.1;
NDS Sec 2.3;
NDS Supp Tbl 4D,
4D Adj Fac

The adjusted bending design value is

$$\begin{aligned}F'_b &= F_b C_D C_M C_t C_L C_F C_{fu} C_r \\ &= \left(725\ \frac{\text{lbf}}{\text{in}^2}\right)(1.6)(1.0)(1.0)(1.0)(1.0)(1.0)(1.0) \\ &= 1160\ \text{lbf/in}^2\end{aligned}$$

The adjusted tension design label is

$$\begin{aligned}F'_t &= F_t C_D C_M C_t \\ &= \left(375\ \frac{\text{lbf}}{\text{in}^2}\right)(1.6)(1.0)(1.0) \\ &= 600\ \text{lbf/in}^2\end{aligned}$$

C_F is not applicable to tension in timbers (i.e., 5×5 and larger).

The adjusted shear design value is

$$\begin{aligned}F'_v &= F_v C_D C_M C_t \\ &= \left(155\ \frac{\text{lbf}}{\text{in}^2}\right)(1.6)(1.0)(1.0)(1.0) \\ &= 248\ \text{lbf/in}^2\end{aligned}$$

The adjusted compression design value perpendicular to the grain is

$$\begin{aligned}F'_{c\perp} &= F_{c\perp} C_M C_t C_b \\ &= \left(820\ \frac{\text{lbf}}{\text{in}^2}\right)(1.0)(1.0)(1.0) \\ &= 820\ \text{lbf/in}^2\end{aligned}$$

4

Beam Design: Sawn Lumber

The design of rectangular sawn wood beams includes bending (with lateral stability), shear, deflection, and bearing at supports and load points.

1. Bending

[NDS Secs 3.2, 3.3, 3.4, 3.5, 4.3.1, 4.3.13]

Allowable bending design values, F_b', will be greater than the actual (calculated) bending stress, f_b.

c	distance between neutral axis to extreme fiber	in
F_b'	allowable bending design stress	lbf/in^2
f_b	actual (calculated) bending stress	lbf/in^2
I	moment of inertia	in^4
M	actual (calculated) moment	in-lbf
MC	moisture content	%
S	section modulus	in^3

$$f_b = \frac{Mc}{I} = \frac{M}{S} \le F_b' \quad \text{[NDS Eq. 3.3-1]}$$

$$S = \frac{I}{c} \quad \text{[NDS Eq. 3.3-4]}$$

A. Section Modulus and Beam Span Definition

[NDS Sec 3.2.1]

The section modulus is for the net, dressed section. The span for calculating bending moment, M, is taken as the distance from face to face, plus one-half the required bearing length at each end.

Figure 4.1 End Support of Typical Beam Span

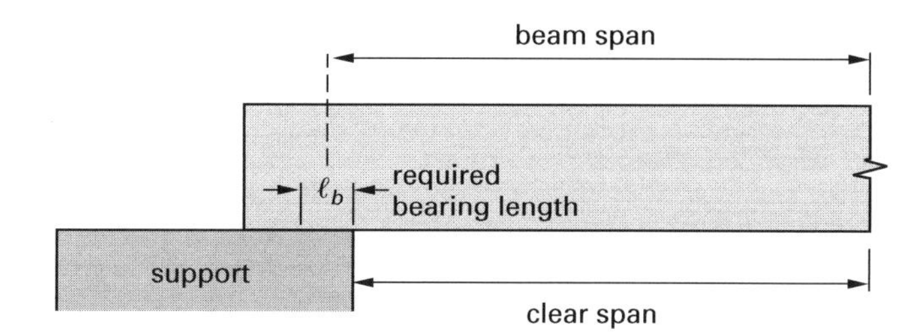

NDS Sec. 3.2.3.1 states that notches are permitted by NDS Secs. 4.4.3, 5.4.4, 7.4.4, and 8.4.1. On the compression face (see NDS Secs. 3.2.3 and 3.4.3.2) shear force, V, shall be determined by principles of engineering mechanics (except those given in NDS Sec. 3.4.3.1). According to NDS Sec. 3.2.3.2, the stiffness of a bending member is almost completely unaffected by a notch with a depth that is less than or equal to one-sixth the beam depth. The same is true if the notch length is less than or equal to one-third the beam depth.

For notched bending members with rectangular cross sections and notches on the tension face and on the compression face (see NDS Secs. 3.2.3 and 3.4.3.2), shear force, V, shall be determined by principles of engineering mechanics (except those given in NDS Sec. 3.4.3.1).

For bending members with rectangular cross sections and notches on the tension face, calculate the adjusted design shear, V_r', as follows.

b	breadth of rectangular bending member	in
d	depth of unnotched bending member	in
d_n	depth of member remaining at a notch	in
F_v'	adjusted shear design value parallel to grain	lbf/in^2

$$V_r' = \tfrac{2}{3}F_v' b d_n \left(\frac{d_n}{d}\right)^2 \quad \text{[NDS Eq. 3.4-3]}$$

When a bending member is notched on the compression face at the end as shown in Fig. 4.2, calculate the adjusted design shear, V_r', as follows.

e	the distance the notch extends inside the inner edge of the support must be less than or equal to the depth remaining at the notch, $e \leq d_n$. If $e > d_n$, use d_n to calculate f_v using NDS Eq. 3.4-2.	in
d_n	depth of member remaining at a notch meeting the provisions of NDS Sec. 3.2.3. If the end of the beam is beveled, as shown by the dashed line in Fig. 4.2, measure d_n from the inner edge of the support.	in

$$V_r' = \tfrac{2}{3}F_v' b \left(d - \left(\frac{d - d_n}{d_n}\right) e\right) \quad \text{[NDS Eq. 3.4-5]}$$

Figure 4.2 Bending Member End-Notched on Compression Face (NDS Figure 3D)

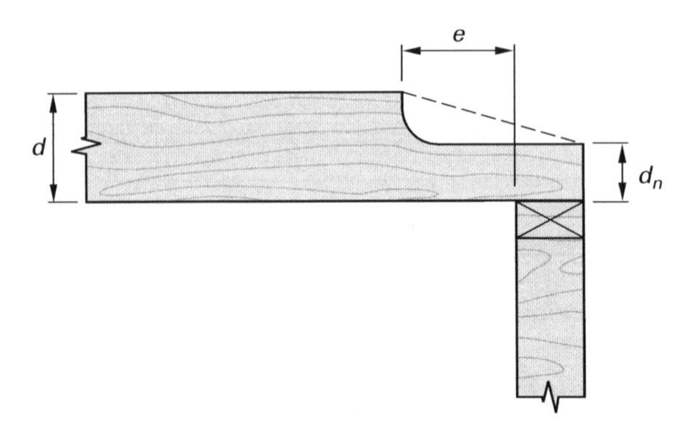

B. Allowable Bending Design Values

[NDS Supp Tbl 4; NDS Tbl 4.3.1]

This section summarizes the allowable bending stresses for sawn lumber beams of a rectangular cross section. Some obvious adjustment factors can be assumed to be 1.0 and can be used to obtain allowable bending stresses from the NDS table values, as listed in Table 3.2 (NDS Table 4.3.1). As a result, some of these factors are often omitted. For example, for covered structures with a normal temperature, C_M for the wet service factor and C_t for the temperature factor are 1.0 and do not have to be included in the calculations.

Figure 4.3 Sawn Lumber Beam Bending About the Strong Axis

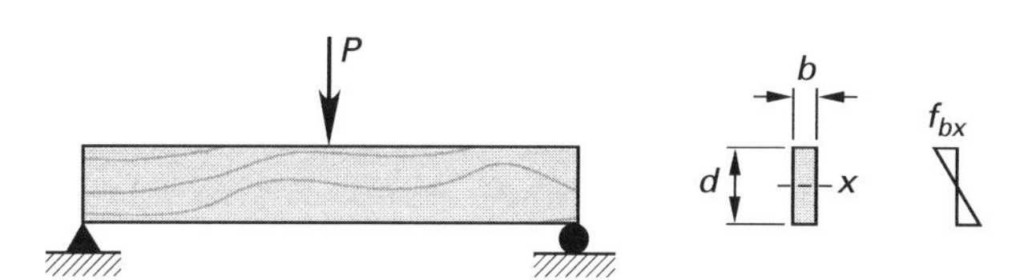

The allowable bending stress for the strong axis (x-axis) is given using the following.

F'_{bx}	allowable bending stress for the strong axis	lbf/in^2
F_{bx}, F_b	reference bending stress (For sawn lumber, reference design values of bending stress apply to the x-axis (except decking). Values are listed in *NDS Supplement* Tables 4A, 4B, 4C, and 4F for dimension lumber and in Table 4D for timbers.)	lbf/in^2
C_D	load duration factor (See NDS Table 2.3.2; NDS App. B.)	–
C_M	wet service factor (When MC $\leq$ 19%, as in most covered structures, C_M is 1.0. See *NDS Supplement* Tables 4A, 4B, 4C, 4D, 4E, and 4F.)	–
C_t	temperature factor (For normal temperature conditions, C_t is 1.00. See NDS Table 2.3.3.)	–
C_L	beam stability factor (For continuous lateral support of the compression face of the beam, C_L is 1.0. For other conditions, use NDS Sec. 3.3.3.)	–
C_F	size factor for sawn lumber (Obtain the values from the adjustment factors section of *NDS Supplement* Tables 4A, 4B, 4E, and 4F for dimension lumber and in Table 4D for timbers. See NDS Sec. 4.3.6.)	–
C_r	repetitive member factor for dimension lumber, prefabricated wood I-joists, and structural composite lumber. (For dimension lumber applications that meet the definition of a repetitive member, C_r is 1.15. See NDS Sec. 4.3.9 and also *NDS Supplement* Tables 4A, 4B, 4C, and 4F adjustment factors. C_r is 1.0 for all other conditions.)	–
C_i	incising factor (See NDS Sec. 4.3.8 and NDS Table 4.3.8. C_i is 1.0 for all other conditions.)	–

$$F'_{bx} = F_{bx}C_DC_MC_tC_LC_FC_r$$

Figure 4.4 Sawn Lumber Beam Bending About the Weak Axis

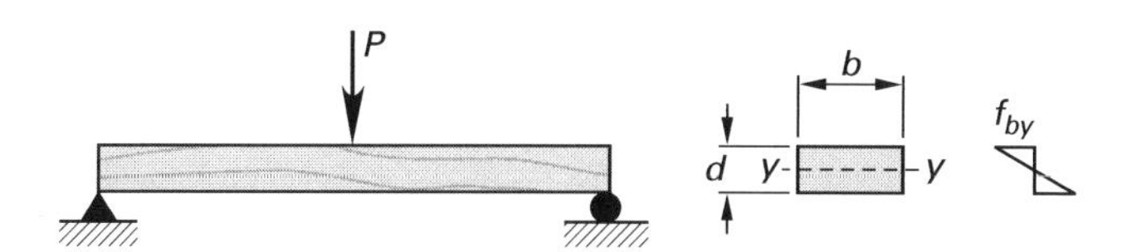

The allowable bending stress for the weak axis (y-axis) is given using the following.

F_{by}, F_b	bending stress given in NDS Tables, which are based on edgewise use (load applied to narrow face). (Values of F_b are listed in *NDS Supplement* Tables 4A, 4B, 4C, and 4F for dimension lumber and in Table 4D for timbers. The load can also be applied to the y-axis with flat use factors for all sizes of sawn lumber except beams and stringers. See NDS Sec. 4.2.5.4.)	lbf/in^2
C_{fu}	flat use factor (Obtain values from the adjustment factors section of *NDS Supplement* Tables 4A, 4B, 4C, and 4F for dimension lumber loaded on the wide face. Decking values listed in *NDS Supplement* Table 4E already have C_{fu} applied to them. C_{fu} is 1.0 for posts and timbers; that is, F_b for posts and timbers in NDS Table 4D are for either x-axis or y-axis bending. See NDS Sec. 4.2.5.3.)	–

$$F'_{by} = F_{bx}C_D C_M C_t C_F C_{fu} C_i$$

C. Load Duration Factor, C_D, for ASD Only

[NDS Sec 2.3.2; NDS App B]

The term *duration of load* refers to the total accumulated length of time that a load is applied during the life of a structure. With regard to the duration, the full design load (not the length of time over a portion of the load) is applied. The factor C_D is associated with the shortest duration load in a given load combination. When both wind and earthquake loads are possible, there is no need to assume that they act simultaneously for normal design situations.

D. Beam Stability Factor, C_L

[NDS Secs 3.3.3, 4.4.1]

If the compression zone of the beam is not braced to prevent lateral torsional buckling, the beam may buckle at a bending stress that is less than the allowable design stress when buckling is prevented.

Sawn lumber bending members shall be designed in accordance with the lateral stability calculations in NDS Sec. 3.3.3 or shall meet the lateral support requirements in NDS Secs. 4.4.1.2 and 4.4.1.3.

A stability requirement in NDS Sec. 3.3.3 shall meet the lateral support requirements in NDS Secs. 4.4.1.2 and 4.4.1.3. As an alternative to the NDS Sec. 4.4.1 requirements, rectangular sawn lumber bending members can be designed in accordance with the following methods to provide restraint against rotation or lateral displacement of the bending members.

Rule of Thumb Method

[NDS Sec 4.4.1]

C_L is 1.0 if d/b (based on nominal dimensions) is

- 2; no lateral support is required.
- 3 or 4; the beam ends are held in position.
- 5; the compression edge of a beam is laterally supported throughout its length.
- 6; bridging or full-depth solid blocking is provided at 8 ft intervals or less.
- 7; both edges are held in line for their entire length.

Analytic Method

[NDS Sec 3.3.3.4]

When $d > b$,

- lateral support will be provided at bearing points to prevent rotation and/or lateral displacement at bearing points
- the unsupported beam length, ℓ_u, is the distance between points of end bearings or between points of intermediate lateral support preventing rotation and/or lateral displacement (see NDS Table 3.3.3)

F_{bE}	critical buckling (Euler) value for a bending member	lbf/in^2
F_{bx}^*, F_{bx}	reference bending stress for x-axis multiplied by all adjustment factors except C_{fu}, C_V, and C_L	lbf/in^2
E'_{min}	adjusted modulus of elasticity for beam and column stability calculations	lbf/in^2
E_y, E'_y	modulus and adjusted modulus of elasticity associated with lateral torsional buckling; adjusted modulus of elasticity about the y-axis (C_D does not apply to E)	lbf/in^2
R_B	slenderness ratio of bending member	–
ℓ_e	effective length for bending members (see NDS Sec. 3.3.3)	in

The beam stability factor is given by

$$C_L = \frac{1 + \frac{F_{bE}}{F_{bx}^*}}{1.9} - \sqrt{\left(\frac{1 + \frac{F_{bE}}{F_{bx}^*}}{1.9}\right)^2 - \frac{\frac{F_{bE}}{F_{bx}^*}}{0.95}} \quad \text{[NDS Eq. 3.3-6]}$$

The critical buckling (Euler) value for bending a member is

$$F_{bE} = \frac{1.20E'_{\text{min}}}{R_B^2}$$

The adjusted modulus of elasticity associated with lateral torsional buckling is

$$E'_y = E_y C_M C_t$$

For sawn lumber, $E_y = E_x$. For glulam, E_x and E_y may be different.

The slenderness ratio of unbraced beam length is

$$R_B = \sqrt{\frac{\ell_e d}{b^2}} \quad \text{[NDS Eq. 3.3-5]}$$

Table 4.1 Effective Unbraced Length, ℓ_e, for Bending Members (NDS Table 3.3.3)

cantilever[a]	when $\ell_u/d < 7$	when $\ell_u/d \geq 7$
uniformly distributed load	$\ell_e = 1.33\ell_u$	$\ell_e = 0.90\ell_u + 3d$
concentrated load at unsupported end	$\ell_e = 1.87\ell_u$	$\ell_e = 1.44\ell_u + 3d$

single-span beam[a,b]	when $\ell_u/d < 7$	when $\ell_u/d \geq 7$
uniformly distributed load	$\ell_e = 2.06\ell_u$	$\ell_e = 1.63\ell_u + 3d$
concentrated load at center with no intermediate lateral support	$\ell_e = 1.80\ell_u$	$\ell_e = 1.37\ell_u + 3d$
concentrated load at center with lateral support at center	$\ell_e = 1.11\ell_u$	
two equal concentrated loads at one-third points with lateral support at one-third points	$\ell_e = 1.68\ell_u$	
three equal concentrated loads at one-fourth points with lateral support at one-fourth points	$\ell_e = 1.54\ell_u$	
four equal concentrated loads at one-fifth points with lateral support at one-fifth points	$\ell_e = 1.68\ell_u$	
five equal concentrated loads at one-sixth points with lateral support at one-sixth points	$\ell_e = 1.73\ell_u$	
six equal concentrated loads at one-seventh points with lateral support at one-seventh points	$\ell_e = 1.78\ell_u$	
seven or more equal concentrated loads, evenly spaced, with lateral support at points of load application	$\ell_e = 1.84\ell_u$	
equal end moments	$\ell_e = 1.84\ell_u$	

[a]For single-span or cantilever bending members with loading conditions not specified in Table 4.1:

$\ell_e = 2.06\ell_u$ when $\ell_u/d < 7$

$\ell_e = 1.63\ell_u + 3d$ when $7 \leq \ell_u/d \leq 14.3$

$\ell_e = 1.84\ell_u$ when $\ell_u/d > 14.3$

[b]Multiple span applications shall be based on table values or engineering analysis.

2. Shear Parallel to Grain (Horizontal Shear)

[NDS Sec 3.4; NDS Tbl 4.3.1]

The actual shear stress parallel to grain, f_v, shall not be greater than the allowable shear stress, F'_v.

F_v reference shear design value parallel to grain (horizontal shear; see *NDS Supplement* Tables 4A, 4B, 4C Footnotes, 4D, and 4F) lbf/in^2

For a rectangular cross section,

$$f_v = \frac{VQ}{Ib} = \frac{3V}{2bd}$$

$$F'_v = F_v C_D C_M C_t C_i$$

Figure 4.5 Shear at Supports

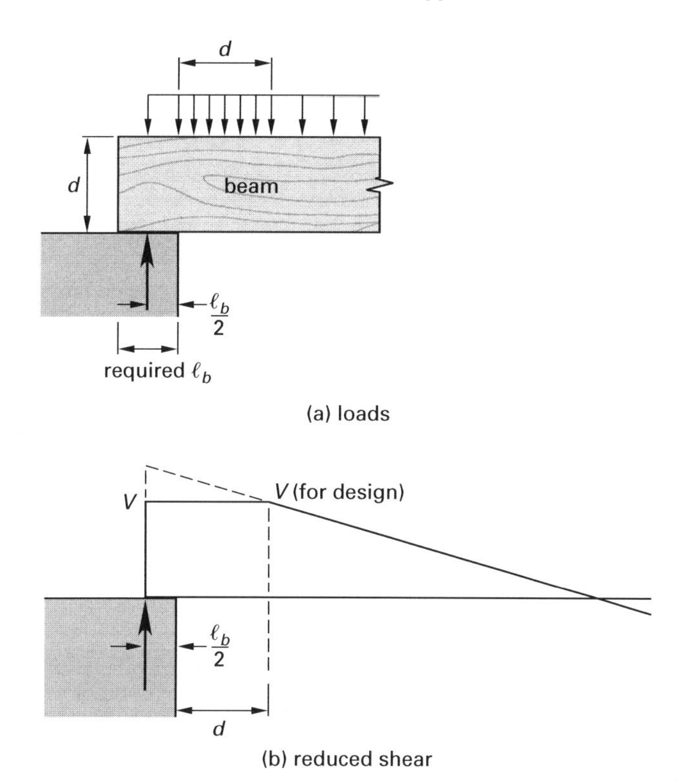

When calculating the maximum shear force, V, in bending members, the following rules apply. (See NDS Sec. 3.4.3.)

For beams supported by full bearing on one surface and loads applied to the opposite surface, uniformly distributed loads within a distance from the face of supports equal to the depth, d, of the bending member are ignored. Concentrated loads within a distance, d, from the support end shall be multiplied by x/d where x is the distance from the load to the support end.

When there is a single moving load or one moving load that is considerably greater than the others, place that load at a distance from each support equal to d; place any other loads in their normal relation.

When there are two or more moving loads of about equal weight and in proximity, place the loads in the position that produces the maximum shear force, V, neglecting any loads within a distance d from each support. Loads within d are ignored for shear force determination.

3. Deflection

[NDS Sec 3.5]

For long-term loading (see NDS Sec. 3.5.2), deflection shall be calculated with a creep factor, K_{cr}.

For seasoned lumber, glulam, or prefabricated wood I-joists used in dry service conditions, K_{cr} is 1.5 (see NDS Secs. 4.1.4, 5.1.5, and 7.1.4, respectively).

For glulam used in wet service conditions, K_{cr} is 2.0.

For wood structural panels used in dry service conditions, K_{cr} is also 2.0 (see NDS Sec. 9.1.4).

The immediate deflection due to the long term component of the design load, Δ_{LT}, is given in inches.

The deflection due to the short term or normal component of the design load, Δ_{ST}, is given in inches.

Deflection is calculated as follows.

$$\Delta = f\left(\frac{P, w, L}{E'I}\right)$$

The adjusted modulus of elasticity is found as follows.

$$E' = EC_M C_t C_i \quad \text{[NDS Table 4.3.1]}$$

The total deflection is found as follows.

$$\Delta_T = K_{cr}\Delta_{LT} + \Delta_{ST} \quad \text{[NDS Sec. 3.5.2]}$$

4. Bearing

A. Bearing Perpendicular to Grain

[NDS Sec 3.10]

Bearing stress perpendicular to the grain of wood occurs at either the beam supports or where loads from one wood member are transferred to another member.

Figure 4.6 Compression (Bearing) Perpendicular to Grain

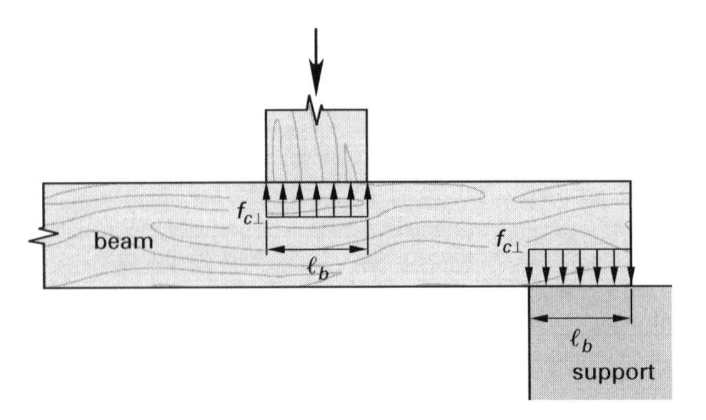

The actual compressive stress perpendicular to the grain, $f_{c\perp}$, will not exceed the allowable compressive design value perpendicular to grain. The reference design value for compression perpendicular to grain, $F_{c\perp}$, is determined using *NDS Supplement* Tables 4A, 4B, 4C, 4D, and 4F.

The actual compressive stress perpendicular to the grain is given by

$$f_{c\perp} = \frac{P}{A} \leq F'_{c\perp}$$

Using Table 3.2 (NDS Table 4.3.1), the allowable compressive (bearing) design value is given by

$$F'_{c\perp} = F_{c\perp} C_M C_t C_i C_b$$

The bearing area is

$$A = \ell_b b \quad \text{[NDS Sec. 3.10]}$$

When ℓ_b is less than 6 in but at least 3 in from a member end, the bearing area factor is

$$C_b = \frac{\ell_b + \frac{3}{8}\text{ in}}{\ell_b}$$

When ℓ_b is greater than 6 in (see NDS Table 3.10.4) and at least 3 in from a member end, C_b is 1.0.

B. Bearing Parallel to Grain

[NDS Sec 3.10.1]

This type of bearing stress occurs when two wood members bear end to end with each other, as well as when a wood end bears on other surfaces.

The actual bearing stress shall not exceed the allowable design value given by

$$f_c \leq F_c^*$$

F_c^* is the reference compression design value multiplied by all applicable factors except C_P. (See NDS Sec. 4.3.10 and NDS Table 4.3.1.)

When $f_c > 0.75F_c^*$, bearing shall be a metal plate or equivalent material. (See NDS Sec. 3.10.1.3.)

C. Bearing Stress Occurring at an Angle Other than 0° or 90° with Respect to Grain Direction

[NDS Sec 3.10.3 and NDS App. J]

Bearing at some angle to grain other than 0° or 90° is checked by the Hankinson formula and NDS App. J.

$$F'_\theta = \frac{F_c^* F'_{c\perp}}{F_c^* \sin^2\theta + F'_{c\perp}\cos^2\theta} \quad \text{[NDS Eq. 3.10-1]}$$

The actual compressive stress on the member at angle θ, f_θ, must be less than or equal to F'_θ ($f_\theta \leq F'_\theta$).

Figure 4.7 Bearing at an Angle to Grain

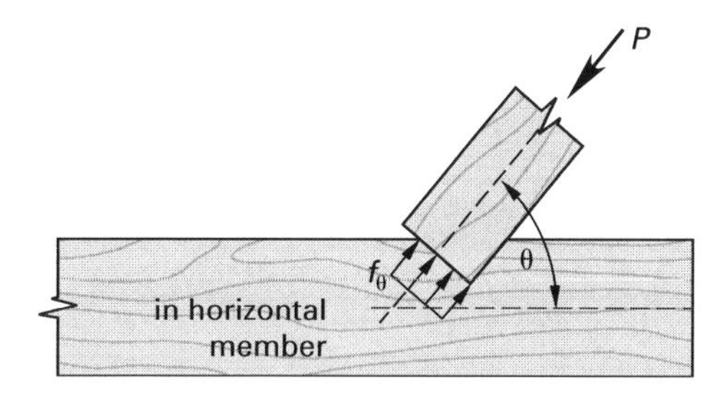

Example 4.1
Sawn Lumber Beam Analysis

The select structural southern pine 4 in × 16 in beam of 20 ft span supports a hoist located at the center of the span. Assume normal load duration (10 years) and dry service conditions. Assume wood density is 37.5 lbf/ft^3. Lateral support is provided. The hoist load, P, is 3000 lbf. The bearing length at each beam support is 4 in. The allowable live load deflection criterion is $L/360$, where L is the beam span length.

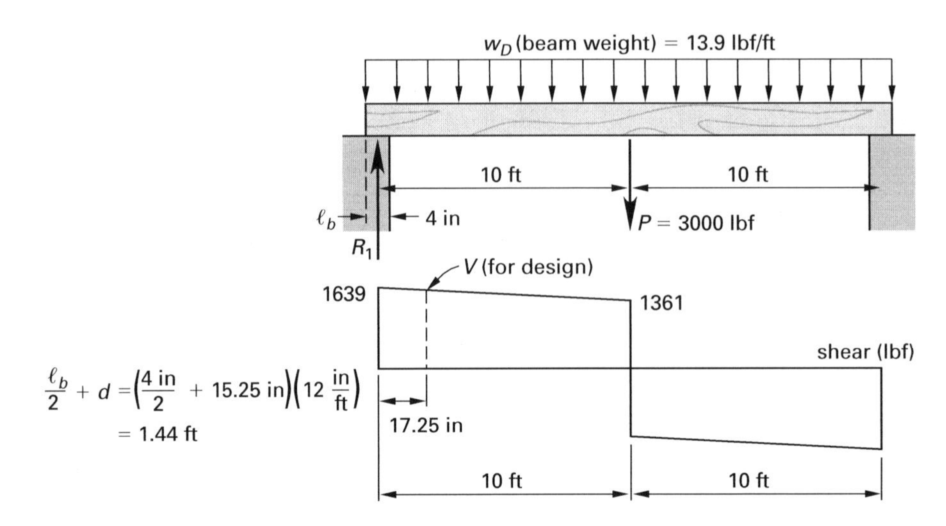

$$\frac{\ell_b}{2} + d = \left(\frac{4\text{ in}}{2} + 15.25\text{ in}\right)\left(12\ \frac{\text{in}}{\text{ft}}\right) = 1.44\text{ ft}$$

Review the beam.

REFERENCE

Solution (ASD Method)

For 4 in × 16 in (3.5 in × 15.25 in actual) dimension lumber,

I_x	moment of inertia about x	1034.4 in^4
A	area	53.4 in^2
S_x	section modulus about x	135.7 in^3

1. Bending

Use the following to calculate the allowable bending design stress.

NDS Tbl 4.3.1
NDS Supp Tbl 4B Adj Fac

$$F_b' = F_bC_DC_MC_tC_LC_FC_{fu}C_iC_r = F_b^*C_L$$

The reference bending design value multiplied by all applicable adjustment factors except C_L is

$$\begin{aligned} F_b^* &= F_bC_DC_MC_tC_FC_{fu}C_iC_r \\ &= \left(1900\ \frac{\text{lbf}}{\text{in}^2}\right)(1.0)(1.0)(1.0)(0.9)(1.0)(1.0)(1.0) \\ &= 1710\ \text{lbf/in}^2 \end{aligned}$$

The adjusted modulus of elasticity for deflection calculations is

$$\begin{aligned} E' = E_y' = E_x' &= EC_MC_tC_i \\ &= \left(1{,}800{,}000\ \frac{\text{lbf}}{\text{in}^2}\right)(1.0)(1.0)(1.0) \\ &= 1{,}800{,}000\ \text{lbf/in}^2 \end{aligned}$$

Calculate the adjusted modulus of elasticity for beam stability and column stability calculations.

$$
\begin{aligned}
E'_{\text{min}} &= E_{\text{min}} C_M C_t C_i C_T \\
&= \left(660{,}000\ \frac{\text{lbf}}{\text{in}^2}\right)(1.0)(1.0)(1.0)(1.0) \\
&= 660{,}000\ \text{lbf/in}^2
\end{aligned}
$$

NDS Sec 3.3.3

Find the beam stability factor, C_L.

The unbraced length is

$$
\ell_u = (20\ \text{ft})\left(12\ \frac{\text{in}}{\text{ft}}\right) = 240\ \text{in}
$$

$$
\frac{\ell_u}{d} = \frac{240\ \text{in}}{15.25\ \text{in}} = 15.7 > 7
$$

NDS Tbl 3.3.3

The effective unbraced length is

$$
\begin{aligned}
\ell_e &= 1.37\ell_u + 3d \\
&= (1.37)(240\ \text{in}) + (3)(15.25\ \text{in}) = 374.6\ \text{in}
\end{aligned}
$$

NDS Eq 3.3-5; NDS Sec 3.3.3.7

The slenderness ratio is

$$
\begin{aligned}
R_B &= \sqrt{\frac{\ell_e d}{b^2}} = \sqrt{\frac{(374.6\ \text{in})(15.25\ \text{in})}{(3.5\ \text{in})^2}} \\
&= 21.6 < 50 \quad [\text{OK}]
\end{aligned}
$$

NDS Sec 3.3.3.6

The critical buckling design value is

$$
\begin{aligned}
F_{bE} &= \frac{1.20 E'_{\text{min}}}{R_B^2} = \frac{(1.20)\left(660{,}000\ \frac{\text{lbf}}{\text{in}^2}\right)}{(21.6)^2} \\
&= 1697.53\ \text{lbf/in}^2
\end{aligned}
$$

NDS Eq 3.3-6

The beam stability factor is

$$
\begin{aligned}
C_L &= \frac{1 + \frac{F_{bE}}{F^*_{bx}}}{1.9} - \sqrt{\left(\frac{1 + \frac{F_{bE}}{F^*_{bx}}}{1.9}\right)^2 - \frac{\frac{F_{bE}}{F^*_{bx}}}{0.95}} \\
&= \frac{1 + \frac{1697.53\ \frac{\text{lbf}}{\text{in}^2}}{1710\ \frac{\text{lbf}}{\text{in}^2}}}{1.9} - \sqrt{\left(\frac{1 + \frac{1697.53\ \frac{\text{lbf}}{\text{in}^2}}{1710\ \frac{\text{lbf}}{\text{in}^2}}}{1.9}\right)^2 - \frac{\frac{1697.53\ \frac{\text{lbf}}{\text{in}^2}}{1710\ \frac{\text{lbf}}{\text{in}^2}}}{0.95}} \\
&= 0.80
\end{aligned}
$$

The adjusted bending design value is

$$
\begin{aligned}
F'_b &= F^*_b C_L = \left(1710\ \frac{\text{lbf}}{\text{in}^2}\right)(0.80) \\
&= 1368\ \text{lbf/in}^2
\end{aligned}
$$

The maximum bending moment is

$$M = \frac{w_D L^2}{8} + \frac{PL}{4}$$
$$= \frac{\left(13.9\ \frac{\text{lbf}}{\text{ft}}\right)(20\ \text{ft})^2\left(12\ \frac{\text{in}}{\text{ft}}\right)}{8} + \frac{(3000\ \text{lbf})(20\ \text{ft})\left(12\ \frac{\text{in}}{\text{ft}}\right)}{4}$$
$$= 188{,}340\ \text{in-lbf}$$

The actual bending stress is

$$f_b = \frac{M}{S_x} = \frac{188{,}340\ \text{in-lbf}}{135.7\ \text{in}^3} = 1387.9\ \text{lbf/in}^2$$

The adjusted bending stress is

$$F'_b = 1368.0\ \frac{\text{lbf}}{\text{in}^2} \approx f_b = 1387.9\ \text{lbf/in}^2 \quad \text{[consider OK]}$$

2. Shear

NDS Tbl 4.3.1
NDS Supp Tbl 4B

The adjusted shear design value parallel to the grain is

$$F'_v = F_v C_D C_M C_t$$
$$= \left(175\ \frac{\text{lbf}}{\text{in}^2}\right)(1.0)(1.0)(1.0)$$
$$= 175\ \text{lbf/in}^2$$

The beam reaction is

$$R_1 = \frac{w_D L}{2} + \frac{P}{2}$$
$$= \left(13.9\ \frac{\text{lbf}}{\text{ft}}\right)(20\ \text{ft})\left(\frac{1}{2}\right) + (3000\ \text{lbf})\left(\frac{1}{2}\right)$$
$$= 1639\ \text{lbf}$$

NDS Secs 3.4.2, 3.4.3

The shear force is

$$V = 1639\ \text{lbf} - \left(13.9\ \frac{\text{lbf}}{\text{ft}}\right)(1.44\ \text{ft})$$
$$= 1619\ \text{lbf}$$

The actual shear stress parallel to grain is

$$f_v = 1.5\left(\frac{V}{bd}\right) = (1.5)\left(\frac{1619\ \text{lbf}}{53.38\ \text{in}^2}\right)$$
$$= 45.5\ \text{lbf/in}^2$$
$$F'_v = 175\ \text{lbf/in}^2 > 45.5\ \text{lbf/in}^2 \quad \text{[OK]}$$

3. Deflection (live load only)

The deflection is

$$\Delta = \frac{PL^3}{48E'I} = \frac{(3000\ \text{lbf})\left((20\ \text{ft})\left(12\ \frac{\text{in}}{\text{ft}}\right)\right)^3}{(48)\left(1{,}800{,}000\ \frac{\text{lbf}}{\text{in}^2}\right)(1034.4\ \text{in}^4)} = 0.464\ \text{in}$$

Check that the live load deflection criterion is met.

$$\frac{L}{360} = \frac{(20\ \text{ft})\left(12\ \frac{\text{in}}{\text{ft}}\right)}{360} = 0.667\ \text{in} > 0.464\ \text{in} \quad [\text{OK}]$$

NDS Sec 3.10.2

4. Support bearing

Calculate the support reaction.

$$\frac{w_D L}{2} + \frac{P}{2} = \left(13.9\ \frac{\text{lbf}}{\text{ft}}\right)(20\ \text{ft})\left(\frac{1}{2}\right) + \frac{3000\ \text{lbf}}{2} = 1639\ \text{lbf}$$

The actual compression stress perpendicular to the grain is

$$f_{c\perp} = \frac{R_1}{\ell_b b} = \frac{1639\ \text{lbf}}{(4\ \text{in})(3.5\ \text{in})} = 117.1\ \text{lbf/in}^2$$

NDS Eq 3.10-2

When ℓ_b is 4 in, the bearing area factor, C_b, is 1.10.

NDS Tbl 4.3.1; Supp Tbl 4B Adj Fac

The adjusted compression design value perpendicular to the grain is

$$F'_{c\perp} = F_{c\perp} C_M C_t C_i C_b = \left(565\ \frac{\text{lbf}}{\text{in}^2}\right)(1.0)(1.0)(1.0)(1.10) = 621.5\ \text{lbf/in}^2 > 117.1\ \text{lbf/in}^2 \quad [\text{OK}]$$

REFERENCE

Solution (LRFD Method)

The load combination is $1.2w_D + 1.6w_L$.

NDS App Tbl N3

$$w_u = 1.2w_D = (1.2)\left(13.9\ \frac{\text{lbf}}{\text{ft}}\right) = 16.68\ \text{lbf/ft}$$

$$P_u = 1.6w_L = (1.6)(3000\ \text{lbf}) = 4800\ \text{lbf}$$

1. Bending

The factored moment is

$$M_u = \frac{w_u L^2}{8} + \frac{P_u L}{4}$$

$$M_u = \frac{\left(16.68\ \frac{\text{lbf}}{\text{ft}}\right)(20\ \text{ft})^2\left(12\ \frac{\text{in}}{\text{ft}}\right)}{8} + \frac{(4800\ \text{lbf})(20\ \text{ft})\left(12\ \frac{\text{in}}{\text{ft}}\right)}{4} = 298{,}008\ \text{in-lbf}$$

The factored shear is

$$V_u = \frac{w_u L}{2} + \frac{P_u}{2} = \frac{\left(16.68\ \frac{\text{lbf}}{\text{ft}}\right)(20\ \text{ft})}{2} + \frac{4800\ \text{lbf}}{2}$$
$$= 2566.8\ \text{lbf}$$

NDS App Tbls N1, N2, N3

For dead load and live load combination, $1.2w_D + 1.6w_L$, the time effect factor, λ, is 0.80.

NDS Supp Tbl 4B; Adj Fac

For 4 × 16 select structural southern pine,

C_F	size factor for sawn lumber	0.9
C_M	wet service factor	1.0
C_t	temperature factor	1.0
F_b	reference bending stress	$1900\ \frac{\text{lbf}}{\text{in}^2}$
F_v	reference shear design	$175\ \frac{\text{lbf}}{\text{in}^2}$
$F_{c\perp}$	reference compression design value perpendicular to grain	$565\ \frac{\text{lbf}}{\text{in}^2}$
E_{min}	reference modulus of elasticity for beam stability and column stability calculations	$660{,}000\ \frac{\text{lbf}}{\text{in}^2}$

NDS Tbl 4.3.1; NDS App Tbls N1, N2, N3

Use the following to calculate the allowable bending design stress.

$$\begin{aligned} F'_{b,\text{LRFD}} &= F_b(K_F\phi_b\lambda)C_M C_t C_L C_F C_{fu} \\ &= F_b(K_F\phi_b\lambda)C_M C_t C_F C_{fu} C_L \\ &= F^*_{b,\text{LRFD}} C_L \end{aligned}$$

The reference bending design value multiplied by all applicable adjustment factors except C_L is

$$\begin{aligned} F^*_{b,\text{LRFD}} &= F_b(K_F\phi_b\lambda)C_M C_t C_F C_{fu} \\ &= \left(1900\ \frac{\text{lbf}}{\text{in}^2}\right)\left(\left(\frac{2.16}{\phi_b}\right)\phi_b\,(0.8)\right)(1.0)(1.0)(0.9)(1.0) \\ &= 2954.9\ \text{lbf/in}^2 \end{aligned}$$

Calculate the adjusted modulus of elasticity for beam stability and column stability calculations.

NDS Tbl 4.3.1

$$\begin{aligned} E'_{\text{min,LRFD}} &= (E_{\text{min}} K_F \phi_s)C_M C_t \\ &= E_{\text{min,LRFD}} C_M C_t \\ &= \left(660{,}000\ \frac{\text{lbf}}{\text{in}^2}\right)\left(\frac{1.5}{\phi_s}\right)\phi_s(1.0)(1.0) \\ &= 990{,}000\ \text{lbf/in}^2 \end{aligned}$$

Find the beam stability factor, C_L.

The unbraced length, ℓ_u, for LRFD is the same as for ASD. The effective unbraced length, ℓ_e, and the slenderness ratio, R_B, are also the same.

$$\frac{\ell_u}{d} = 15.7 < 7$$

$$\ell_e = 374.6\ \text{in};\ R_B = 21.6$$

The critical buckling design value for bending members is

NDS Sec 3.3.3.8; NDS Eq 3.3-6

$$F_{bE,\text{LRFD}} = \frac{1.20E'_{\text{min,LRFD}}}{R_B^2}$$

$$= \frac{(1.20)\left(990{,}000\ \frac{\text{lbf}}{\text{in}^2}\right)}{(21.6)^2}$$

$$= 2546.3\ \text{lbf/in}^2$$

The beam stability factor, C_L, is

$$C_L = \frac{1+\left(\frac{F_{bE,\text{LRFD}}}{F^*_{b,\text{LRFD}}}\right)}{1.9} - \sqrt{\left(\frac{1+\frac{F_{bE,\text{LRFD}}}{F^*_{b,\text{LRFD}}}}{1.9}\right)^2 - \left(\frac{\frac{F_{bE,\text{LRFD}}}{F^*_{b,\text{LRFD}}}}{0.95}\right)}$$

$$= \frac{1+\left(\frac{2546.3\ \frac{\text{lbf}}{\text{in}^2}}{2954.9\ \frac{\text{lbf}}{\text{in}^2}}\right)}{1.9} - \sqrt{\left(\frac{1+\frac{2546.3\ \frac{\text{lbf}}{\text{in}^3}}{2954.9\ \frac{\text{lbf}}{\text{in}^3}}}{1.9}\right)^2 - \frac{\left(\frac{2546.3\ \frac{\text{lbf}}{\text{in}^2}}{2954.9\ \frac{\text{lbf}}{\text{in}^2}}\right)}{0.95}}$$

$$= 0.749$$

The allowable bending design stress is

$$F'_{b,\text{LRFD}} = F^*_{b,\text{LRFD}}C_L = \left(2954.9\ \frac{\text{lbf}}{\text{in}^2}\right)(0.749) = 2213.2\ \text{lbf/in}^2$$

The actual bending stress is

$$f_b = \frac{M_u}{S_x} = \frac{298{,}008\ \text{in-lbf}}{135.7\ \text{in}^3}$$

$$= 2196.1\ \text{lbf/in}^2$$

$$F'_{b,\text{LRFD}} = 2213.2\ \text{lbf/in}^2 > 2196.1\ \text{lbf/in}^2 \quad \text{[OK]}$$

Or, calculate the moment capacity (resistance).

$$F'_{b,\text{LRFD}}S_x = \left(2213.2\ \frac{\text{lbf}}{\text{in}^2}\right)(135.7\ \text{in})$$

$$= 300{,}331\ \text{in-lbf} > M_u = 298{,}008\ \text{in-lbf} \quad \text{[OK]}$$

2. Shear

NDS Tbl 4.3.1

The adjusted shear design value parallel to the grain is

$$\begin{aligned} F'_{v,\text{LRFD}} &= F_v(K_F\phi_v\lambda)C_M C_t \\ &= \left(175\ \frac{\text{lbf}}{\text{in}^2}\right)\left(\left(\frac{2.16}{\phi_v}\right)\phi_v(0.80)\right)(1.0)(1.0) \\ &= 302.4\ \text{lbf/in}^2 \end{aligned}$$

The factored shear (approximately equal to the reaction at beam ends, R_1) is 2566.8 lbf.

Calculate the actual shear stress parallel to grain,

$$\begin{aligned} f_v &= \tfrac{3}{2}\left(\frac{V_u}{bd}\right) = \left(\frac{3}{2}\right)\left(\frac{2566.8\ \text{lbf}}{53.4\ \text{in}^2}\right) \\ &= 72.1\ \text{lbf/in}^2 \end{aligned}$$

$$F'_{v,\text{LRFD}} = 302.4\ \text{lbf/in}^2 > f_v = 72.1\ \text{lbf/in}^2 \quad [\text{OK}]$$

Or, calculate the shear capacity by setting f_v equal to $F_{v,\text{LRFD}}$.

$$\begin{aligned} V'_{\text{LRFD}} &= \tfrac{2}{3}F'_{v,\text{LRFD}}A = \left(\frac{2}{3}\right)\left(302.4\ \frac{\text{lbf}}{\text{in}^2}\right)(53.4\ \text{in}^2) \\ &= 10{,}765.4\ \text{lbf} > V_u = 2566.8\ \text{lbf} \quad [\text{OK}] \end{aligned}$$

3. Deflection (live load only)

Same as ASD method.

4. Support bearing

The factored support reaction, V_u, is 2566.8 lbf. The actual compression stress perpendicular to the grain is

$$\begin{aligned} f_{c\perp} &= \frac{R_1}{\ell_b b} = \frac{2566.8\ \text{lbf}}{(4\ \text{in})(3.5\ \text{in})} \\ &= 183.34\ \text{lbf/in}^2 \end{aligned}$$

NDS Eq 3.10-2

The bearing area factor is

$$\begin{aligned} C_b &= \frac{\ell_b + \frac{3}{8}\ \text{in}}{\ell_b} = \frac{4\ \text{in} + \frac{3}{8}\ \text{in}}{4\ \text{in}} \\ &= 1.1 \end{aligned}$$

The adjusted compression design value perpendicular to the grain is

NDS Tbl 4.3.1; NDS App N

$$\begin{aligned} F_{c\perp,\text{LRFD}} &= F_{c\perp}(K_F\phi_c\lambda)C_M C_t C_b \\ &= \left(565\ \frac{\text{lbf}}{\text{in}^2}\right)\left(\left(\frac{1.875}{\phi_c}\right)\phi_c(0.80)\right)(1.0)(1.0)(1.1) \\ &= 932.3\ \text{lbf/in}^2 > f_{c\perp} = 183.34\ \text{lbf/in}^2 \quad \text{[OK]} \end{aligned}$$

Example 4.2
Review of 2 × 8 Roof Rafters for Bending

Determine required section modulus, S_x, for 2×8 douglas fir-larch no. 1 grade 14 ft roof rafters spaced at 16 in. Assume that the wet service factor, C_M, and the temperature factor, C_t, are 1.0. Dressed size for 2×8 is $1^1/_2 \times 7^1/_4$.

C_D	load duration factor (seven days for roof live load or construction; *NDS Supplement* Table 4A and adjustment factors)	1.25
C_F	size factor for sawn lumber	1.20
C_M, C_t	wet service factor, temperature factor	1.0
F_b	reference bending design value (NDS Table 2.3.2)	1000 lbf/in^2
C_r	repetitive member factor for dimension lumber	1.15
S_x	section modulus	13.14 in^3
w_D	dead load	14 lbf/ft^2
$w_{L,\text{roof}}$	roof live load	20 lbf/ft^2

REFERENCE

Solution (ASD Method)

The allowable bending design stress is

NDS Tbl 4.3.1

$$\begin{aligned} F_b' &= F_b C_D C_M C_t C_F C_r \\ &= \left(1000\ \frac{\text{lbf}}{\text{in}^2}\right)(1.25)(1.0)(1.0)(1.2)(1.15) \\ &= 1725\ \text{lbf/in}^2 \end{aligned}$$

The total design load is

$$\begin{aligned} w_{\text{TL}} &= w_L + w_D = \left(14\ \frac{\text{lbf}}{\text{ft}^2}\right)\left(\frac{16\text{ in}}{12\text{ in}}\right) + \left(20\ \frac{\text{lbf}}{\text{ft}^2}\right)\left(\frac{16\text{ in}}{12\text{ in}}\right) \\ &= 45.34\ \text{lbf/ft}^2 \end{aligned}$$

$$\begin{aligned} M_{\max} &= \frac{w_{\text{TL}}L^2}{8} = \frac{\left(45.34\ \frac{\text{lbf}}{\text{ft}}\right)(14\text{ ft})^2\left(12\ \frac{\text{in}}{\text{ft}}\right)}{8} \\ &= 13{,}330\text{ in-lbf} \end{aligned}$$

The section modulus is

$$\begin{aligned} S_{x,\text{required}} &= \frac{M_{\max}}{F_b'} = \frac{13{,}330\text{ in-lbf}}{1725\ \frac{\text{lbf}}{\text{in}^2}} \\ &= 7.73\ \text{in}^3 < S_x = 13.14\ \text{in}^3 \quad \text{[OK]} \end{aligned}$$

REFERENCE

Solution (LRFD Method)

NDS App Tbl N3

For dead load alone, the load combination is

$$1.4w_D = (1.4)\left(18.67\ \frac{\text{lbf}}{\text{ft}}\right) = 26.14\ \text{lbf/ft}$$

For dead load and roof live load, the load combination is

$$w_D = 1.2w_D + 1.6w_{L,\text{roof}} = (1.2)\left(18.67\ \frac{\text{lbf}}{\text{ft}}\right) + (1.6)\left(20\ \frac{\text{lbf}}{\text{ft}^2}\right)\left(\frac{16\ \text{in}}{12\ \frac{\text{in}}{\text{ft}}}\right)$$

$$= 65.08\ \text{lbf/ft}\quad \text{[controls]}$$

For this load combination, the time effect factor, λ, is 0.8.

Calculate the factored moment.

$$M_u = \frac{w_u L^2}{8} = \frac{\left(65.08\ \frac{\text{lbf}}{\text{ft}}\right)(14\ \text{ft})^2\left(12\ \frac{\text{in}}{\text{ft}}\right)}{8} = 19{,}133\ \text{in-lbf}$$

NDS Tbl 4.3.1; App N

The allowable bending design stress is

$$F'_{b,\text{LRFD}} = F_b(K_F\phi_b\lambda)C_M C_t C_F C_r$$

$$= \left(1000\ \frac{\text{lbf}}{\text{in}^2}\right)\left(\left(\frac{2.16}{\phi_b}\right)\phi_b(0.8)\right)(1.0)(1.0)(1.2)(1.15)$$

$$= 2384.6\ \text{lbf/in}^2$$

The section modulus is

$$S_{x,\text{required}} = \frac{M_u}{F'_{b,\text{LRFD}}} = \frac{19{,}133\ \text{in-lbf}}{2384.6\ \frac{\text{lbf}}{\text{in}^2}}$$

$$= 8.02\ \text{in}^3\quad [7.73\ \text{in}^3\ \text{by ASD}] < S_x = 13.14\ \text{in}^3\quad \text{[OK]}$$

5

Beam Design: Glued Laminated Timber

Glued laminated timber, or glulam, is made up of wood laminations, or lams, that are bonded together. The grain of laminations runs parallel to the length of the member. Individual lams are $1^3/_8$ in thick for southern pine and $1^1/_2$ in thick for western species. The net widths for glulam members typically range from $2^1/_2$ to $10^3/_4$ in. *NDS Supplement* Tables 5A to 5D provide two member types of reference design values: bending combination members (stressed primarily in bending) and axial combination members (stressed primarily in axial tension or compression).

Bending combination members may be fabricated as unbalanced or balanced members. The most critical zone with respect to controlling strength is the outermost tension zone. In unbalanced beams, the quality of lumber used on the tension side is higher than the lumber used on the compression side. Therefore, unbalanced beams have different bending stresses assigned to the compression and tension zones. Unbalanced beams are mainly intended for simple-span beams. In *NDS Supplement* Table 5A, members with the combination symbol 24F-V4 and outer/core species of DF/DF are an example of this type of unbalanced beam. 24F-E4 DF/DF is another example from the table. "V" indicates that the lams are visually graded lumber and "E" is for mechanically graded lumber. The number "4" further indicates a specific combination of lumber used for a horizontal shear, a modulus of elasticity, and so on. Balanced members are symmetrical in lumber quality about the mid-height, and they are used in applications such as in cantilevers or continuous beams for more efficient use. Examples are 24F-V8 DF/DF and 24F-E4 SP/SP.

In axial combination members, the lumber quality is uniform for all lams in the member. In *NDS Supplement* Table 5B, a member with an identification number 3, species DF, and grade L2D for visually graded lumbers is an example. For mechanically graded lumber, 27 DF 1.9E2 is an example.

1. Bending Combinations

[NDS Secs 5.3.1 through 5.3.10; NDS Supp Tbls 5A, 5C]

Members are stressed principally in bending but can be loaded axially, if necessary. However, since the bending combinations have higher-quality laminating wood at the outer fibers, they are more efficient as beams.

A combination symbol and the species of the laminating wood define the bending designation. For example, a combination symbol 24F-V8 indicates a combination with a tabulated bending stresses of 2400 lbf/in^2 for a normal duration of loading (10 years) and dry service conditions (16% or less moisture content). The letter "V" indicates that the lumber laminations have been visually graded. In addition, the species of wood is listed DF/DF, DF/HF, and so on, for outer lams/inner lams, with the outer lams a higher-quality lam wood and the inner lams a lower-quality lam wood.

The other combination is for laminating stock that has been mechanically graded, or E-rated, 24-E13 DF/DF.

Actual bending stresses are determined by

$$f_b = \frac{Mc}{I} = \frac{M}{S}$$

M is the bending moment that uses the clear distance between the support faces plus one-half the required bearing length at each support for the span length.

A. Volume Factor, C_V

[NDS Sec 5.3.6]

When glued laminated timber is loaded perpendicular to the wide face of the laminations, reference bending design values for loading perpendicular to the wide faces of the laminations, F_{bxx}, shall be multiplied by the volume factor, C_V.

The reference design values for bending about the x-axis, F_{bxx}, are based on a simple-span member 5¹/₈ in wide, 12 in deep, and 21 ft long, that is loaded with a uniform load on the 5¹/₈ in face. When a different member size is used or a different loading condition exists, reference value is to be multiplied by the the volume factor, C_V.

Figure 5.1 Base Dimensions for Bending Stress in Glulam

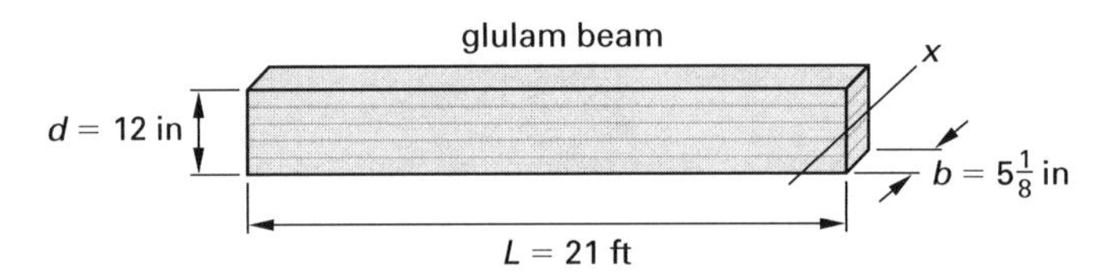

For structural glued laminated bending members, the volume factor and its components are given as follows.

L	length of beam between points of zero moment	ft
d	depth of the beam	in
b	width of the beam (For laminations that consist of more than one piece, b is the width of the widest piece in layup; thus, $b \leq 10.75$ in.)	in
$x = x_{\text{sp}}$	volume factor for southern pine	20
$x = x_{\text{ot}}$	volume factor for other species	10

The volume factor is

$$C_V = \left(\frac{21}{L}\right)^{1/x} \left(\frac{12}{d}\right)^{1/x} \left(\frac{5.125}{b}\right)^{1/x} \leq 1.0 \quad \text{[NDS Eq. 5.3-1]}$$

B. Beam Stability Factor, C_L

[NDS Sec 3.3.3]

NDS Eq. 3.3-6, used in Ch. 4 for sawn lumber beams, is also applicable to glulam beams.

C. Allowable Bending Stress for Strong Axis (bending about x-axis)

[NDS Tbl 5.3.1]

A glulam beam bending combination stressed about the x-axis is usually with the tension laminations stressed in tension. The notation F_{bx} refers to this loading situation.

Figure 5.2 Glulam Beam Bending About the Strong Axis

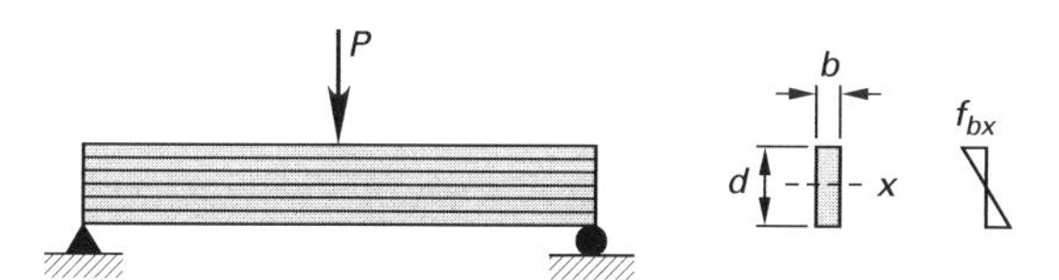

C_D	load duration factor (See NDS Table 2.3.2; NDS App. B.)	–
C_M	wet service factor (When MC < 16%, C_M is 1.0.)	–
C_t	temperature factor (For normal temperature conditions, C_t is 1.0; see NDS Table 2.3.4.)	–
C_L	beam stability factor (For continuous lateral support of compression face of beam, C_L is 1.0. For other conditions, use NDS Sec. 3.3.3.)	
C_V	volume factor (See NDS Sec. 5.3.6 and NDS Eq. 5.3-1.)	–
F_{bx}	reference bending stress about the x-axis tension zone stressed in tension (See *NDS Supplement* Table 5A.)	lbf/in^2
F'^{-}_{bx}	allowable bending stress about the x-axis with compression zone stressed in tension (negative bending)	lbf/in^2
F^{-}_{bx}	reference bending stress about the x-axis with compression zone stressed in tension.	lbf/in^2
F^{+}_{bx}	reference bending stress about the x-axis tension zone stressed in tension. For balanced layup, same as F^{-}_{bx}. (See *NDS Supplement* Table 5A).	lbf/in^2
F'_{bx}, F'^{+}_{bx}	allowable bending stress about the x-axis with tension zone stressed in tension (positive bending)	lbf/in^2

The allowable bending stress, when tension zone is stressed in tension, is taken as the smaller of the following values.

With lateral stability,

$$F'_{bx} = F'^{+}_{bx,t/t} = F^{+}_{bx} C_D C_M C_t C_L C_{fu} \quad \text{[NDS Table 5.3.1]}$$

With volume effect,

$$F'_{bx} = F'^{+}_{bx,t/t} = F^{+}_{bx,t} C_D C_M C_t C_V C_{fu}$$

Glulam beams are sometimes loaded in bending about the x-axis with the compression laminations stressed in tension. The typical application of this case is in a beam with a relatively short cantilever. The allowable bending stress is taken as the smaller of the following values.

With lateral stability,

$$F'^{-}_{bx,c/t} = F^{-}_{bx} C_D C_M C_t C_L C_{fu}$$

With volume effect,

$$F'^{-}_{bx,c/t} = F^{-}_{bx} C_D C_M C_t C_V C_{fu}$$

D. Allowable Bending Stress for Weak Axis (bending about y-axis)

[NDS Sec 5.3.7]

Figure 5.3 Glulam Beam Bending About the Weak Axis

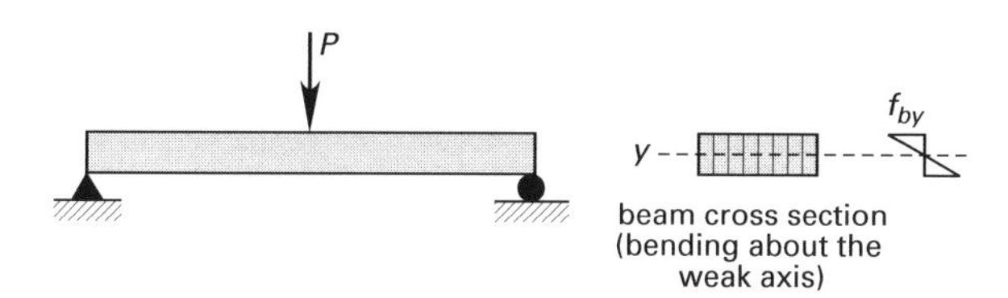

F_{by}	reference bending stress about the y-axis. (See NDS Table 5A–5D and adjustment factors.)	lbf/in^2
C_{fu}	flat use factor (See NDS Sec. 5.3.7. Obtain values from the adjustment factors section of *NDS Supplement* Table 5A. Conservatively, C_{fu} can be called 1.0.)	–

C_L and C_V apply to bending about the x-axis (strong) only. (See NDS Secs. 3.3.3 and 5.3.6.)

The allowable bending stress about the y-axis is given by

$$F'_{by} = F'_{by} C_D C_M C_t C_{fu}$$

2. Axial Combinations

[NDS Supp Tbl 5B]

Members are stressed primarily in axial tension or compression. The combination symbols are numbered 1, 2, 3, and so on, followed by the wood species and grade, such as "2DF L2." The combination member will be more efficient as an axial member because the distribution of laminating wood grades is uniform across the member cross section.

The use of axial combination members as beams is allowable, although it is not as efficient as the bending combination beams. However, the same adjustment factors for the bending combination are applicable.

Example 5.1
Glulam Beam Analysis

Analyze the beam shown.

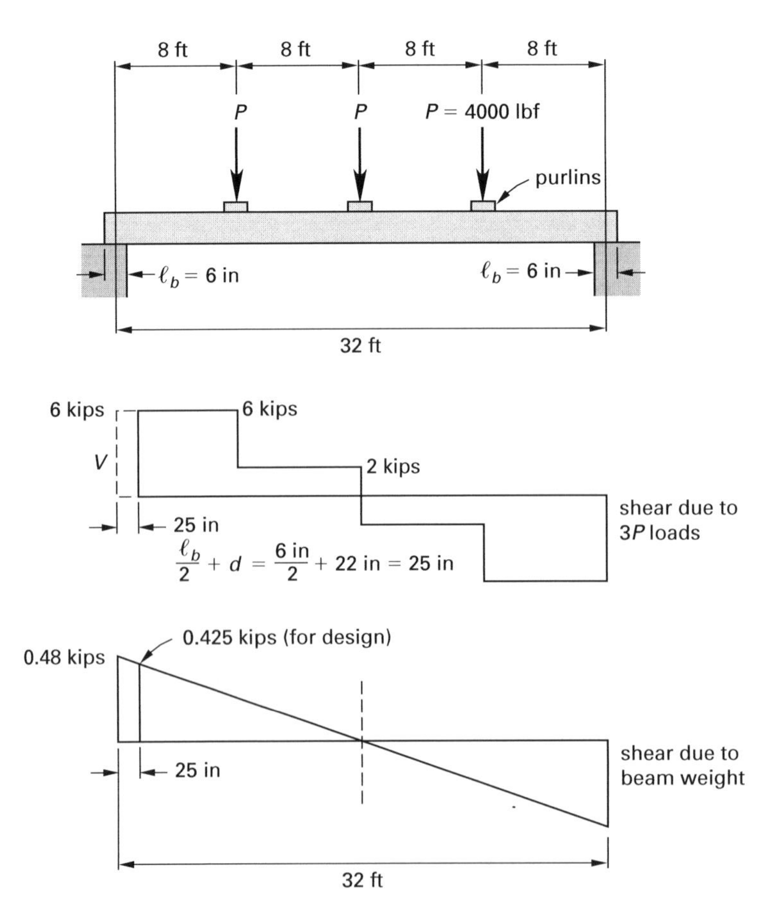

A 32 ft glulam beam (5 in × 22 in, 24F-V1 SP/SP, southern pine bending combination) is laterally supported at 8 ft intervals (purlin points). Assume conditions of wet service and normal temperature (i.e., $C_t = 1.0$). The three loads ($P = 4000$ lbf each) consist of dead load and snow load ($w_D + w_S$; mostly snow load). The beam weight is 30 lbf/ft, as illustrated. The total deflection may not exceed $L/240$.

REFERENCE *Solution (ASD Method)*

For 24F-V1 SP/SP glulam,

NDS Supp Tbl 5A

	C_M
$F_{bx}^{+} = 2400\ \text{lbf/in}^2$	0.80
$F_{vx} = 300\ \text{lbf/in}^2$	0.875
$F_{c\perp x} = 740\ \text{lbf/in}^2$ (tension force)	0.53
$F_{c\perp x} = 650\ \text{lbf/in}^2$ (compression face)	0.53
$E_x = 1{,}700{,}000\ \text{lbf/in}^2$	0.833
$E_{x,\min} = 0.88 \times 10^6\ \text{lbf/in}^2$	0.833
$E_y = 1{,}500{,}000\ \text{lbf/in}^2$	0.833
$E_{y,\min} = 0.78 \times 10^6\ \text{lbf/in}^2$	0.833

NDS Tbl 2.3.2

For snow load, the load duration factor, C_D, is 1.15.

NDS Supp Tbl 1D

For 5 × 22 glulam,

$$A = 110.0\ \text{in}^2$$

$$S_x = 403.3\ \text{in}^3$$

$$I_x = 4437.0\ \text{in}^4$$

1. Bending

NDS Tbl 5.3.1

Use the following to calculate the allowable bending stress. The smaller of the following values controls.

$$F'_{bx},\ F'^{+}_{bx} = \begin{cases} F_{bx}^{+} C_D C_M C_t C_L = F_{bx}^{*+} C_L \\ F_{bx}^{+} C_D C_M C_t C_V = F_{bx}^{*+} C_V \end{cases}$$

NDS Sec 3.3.3; NDS Tbl 3.3.3

Find the beam stability factor, C_L.

The unbraced length is

$$\ell_u = (8\ \text{ft})\left(12\ \frac{\text{in}}{\text{ft}}\right) = 96\ \text{in}$$

The effective length is

$$\begin{aligned} \ell_e &= 1.54\ell_u = (1.54)(96\ \text{in}) \\ &= 147.8\ \text{in} \end{aligned}$$

NDS Eq 3.3-5

The slenderness ratio is

$$\begin{aligned} R_B &= \sqrt{\frac{\ell_e d}{b^2}} = \sqrt{\frac{(147.8\ \text{in})(22\ \text{in})}{(5\ \text{in})^2}} \\ &= 11.4 \end{aligned}$$

The adjusted modulus of elasticity for the y-axis is

$$\begin{aligned} E'_{y,\text{min}} &= E_{y,\text{min}} C_M C_t \\ &= \left(0.78 \times 10^6 \ \frac{\text{lbf}}{\text{in}^2}\right)(0.833)(1.0) \\ &= 0.649 \times 10^6 \ \text{lbf/in}^2 \end{aligned}$$

NDS Sec 3.3.3.8

The critical buckling design value is

$$\begin{aligned} F_{bE} &= \frac{1.20 E'_{\text{min}}}{R_B^2} = \frac{(1.20)\left(0.649 \times 10^6 \ \frac{\text{lbf}}{\text{in}^2}\right)}{(11.4)^2} \\ &= 5992.6 \ \text{lbf/in}^2 \end{aligned}$$

The reference bending stress with all adjustments except C_V and C_L is

$$\begin{aligned} F_{bx}^{*+} &= F_{bx}^{+} C_D C_M C_t C_{fu} \\ &= \left(2400 \ \frac{\text{lbf}}{\text{in}^2}\right)(1.15)(0.80)(1.0)(1.0) \\ &= 2208.0 \ \text{lbf/in}^2 \end{aligned}$$

Let $F_b^* = F_{bx}^{*+}$.

NDS Eq 3.3-6

The beam stability factor is

$$\begin{aligned} C_L &= \frac{1 + \frac{F_{bE}}{F_b^*}}{1.9} - \sqrt{\left(\frac{1 + \frac{F_{bE}}{F_b^*}}{1.9}\right)^2 - \frac{\frac{F_{bE}}{F_b^*}}{0.95}} \\ &= \frac{1 + \frac{5992.6 \ \frac{\text{lbf}}{\text{in}^2}}{2208 \ \frac{\text{lbf}}{\text{in}^2}}}{1.9} - \sqrt{\left(\frac{1 + \frac{5992.6 \ \frac{\text{lbf}}{\text{in}^2}}{2208 \ \frac{\text{lbf}}{\text{in}^2}}}{1.9}\right)^2 - \frac{\frac{5992.6 \ \frac{\text{lbf}}{\text{in}^2}}{2208 \ \frac{\text{lbf}}{\text{in}^2}}}{0.95}} \\ &= 0.973 \end{aligned}$$

Find the volume factor of glulam, C_V.

For southern pine, $x = 20$.

$$\begin{aligned} C_V &= \left(\frac{21}{L}\right)^{1/x} \left(\frac{12}{d}\right)^{1/x} \left(\frac{5.125}{b}\right)^{1/x} \leq 1.0 \\ &= \left(\frac{21}{31 \ \text{ft}}\right)^{1/20} \left(\frac{12}{22 \ \text{in}}\right)^{1/20} \left(\frac{5.125}{5 \ \text{in}}\right)^{1/20} \\ &= 0.953 \leq 1.0 \end{aligned}$$

Since $C_L = 0.973 > C_V = 0.953$, use C_V.

Calculate the bending stress with all adjustments except C_L.

$$\begin{aligned} F'_{bx} &= F_{bx}^{*+} C_V = \left(2208.0 \ \frac{\text{lbf}}{\text{in}^2}\right)(0.953) \\ &= 2104.2 \ \text{lbf/in}^2 \end{aligned}$$

Calculate the maximum bending moment.

$$M = M_{\text{beam}} + M_L \quad \text{[from the shear diagram]}$$

$$= \frac{\left(30 \ \frac{\text{lbf}}{\text{ft}}\right)(32 \text{ ft})^2 \left(12 \ \frac{\text{in}}{\text{ft}}\right)}{8} + \big((6000 \text{ lbf})(8 \text{ ft}) + (2000 \text{ lbf})(8 \text{ ft})\big)\left(12 \ \frac{\text{in}}{\text{ft}}\right)$$

$$= 814{,}080 \text{ in-lbf}$$

Calculate the actual bending stress.

$$f_{bx} = \frac{M}{S_x} = \frac{814{,}080 \text{ in-lbf}}{403.3 \text{ in}^3}$$

$$= 2018.6 \text{ lbf/in}^2 < F'_{bx} = 2104.2 \text{ lbf/in}^2 \quad \text{[OK]}$$

2. Shear

NDS Secs 3.4.3, 3.4.3.1

Account for the allowed disregard of loads at a distance from beam support equal to beam depth for beam weight.

Calculate the shear force.

$$V_L = 6000 \text{ lbf} + \left(30 \ \frac{\text{lbf}}{\text{ft}}\right)\left(\frac{32 \text{ ft} - \frac{(2)(22 \text{ in})}{12 \ \frac{\text{in}}{\text{ft}}}}{2}\right)$$

$$= 6425 \text{ lbf}$$

NDS Eq 3.4-2

Calculate the actual shear stress parallel to grain.

$$f_v = 1.5\left(\frac{V}{bd}\right) = (1.5)\left(\frac{6425 \text{ lbf}}{(5 \text{ in})(22 \text{ in})}\right)$$

$$= 87.6 \text{ lbf/in}^2$$

NDS Tbl 5.3.1

Calculate the adjusted shear design value parallel to the grain.

$$F'_{vx} = F_{vx}C_DC_MC_t$$

$$= \left(300 \ \frac{\text{lbf}}{\text{in}^2}\right)(1.15)(0.875)(1.0)$$

$$= 301.9 \text{ lbf/in}^2 > 87.6 \text{ lbf/in}^2 \quad \text{[OK]}$$

3. Deflection

NDS Tbl 5.3.1

Calculate the adjusted modulus of elasticity.

$$E'_x = E_xC_MC_t = \left(1{,}700{,}000 \ \frac{\text{lbf}}{\text{in}^2}\right)(0.833)(1.0)$$

$$= 1{,}416{,}100 \text{ lbf/in}^2$$

Calculate the deflection.

$$\Delta = \frac{19PL^3}{384E'_xI_x} + \frac{5wL^4}{384E'_xI_x}$$

$$= \frac{(19)(4000\text{ lbf})\left((32\text{ ft})\left(12\ \frac{\text{in}}{\text{ft}}\right)\right)^3 + (5)\left(30\ \frac{\text{lbf}}{\text{ft}}\right)\left((32\text{ ft})\left(12\ \frac{\text{in}}{\text{ft}}\right)\right)^4}{(384)\left(1{,}416{,}100\ \frac{\text{lbf}}{\text{in}^2}\right)(4436.7\text{ in}^4)}$$

$$= 3.14\text{ in}$$

Check that the deflection criterion is met.

$$\frac{L}{240} = \frac{(32\text{ ft})\left(12\ \frac{\text{in}}{\text{ft}}\right)}{240}$$

$$= 1.6\text{ in} < 3.14\text{ in} \quad \text{[no good]}$$

4. Support bearing

Calculate the support reaction.

$$R = R_{\text{beam}} + P + \frac{P}{2}$$

$$= \left(30\ \frac{\text{lbf}}{\text{ft}}\right)(16\text{ ft}) + 4000\text{ lbf} + \frac{4000\text{ lbf}}{2}$$

$$= 6480\text{ lbf}$$

NDS Tbl 3.10.4

When ℓ_b is 6 in or more, the bearing area factor, C_b, is 1.0.

NDS Tbl 5.3.1

Calculate the adjusted compression stress perpendicular to the grain.

$$f'_{c\perp} = \frac{P}{\ell_b b} = \frac{6480\text{ lbf}}{(6\text{ in})(5\text{ in})}$$

$$= 216\text{ lbf/in}^2$$

Calculate the adjusted compression design value perpendicular to the grain.

NDS Tbl 4.3.1

$$F'_{c\perp} = F_{c\perp x}C_MC_tC_b$$

$$= \left(740\ \frac{\text{lbf}}{\text{in}^2}\right)(0.53)(1.0)(1.0)$$

$$= 392.2\text{ lbf/in}^2 > f'_{c\perp} = 216\text{ lbf/in}^2 \quad \text{[OK]}$$

REFERENCE

Solution (LRFD Method)

For 24F-V1 SP/SP glulam,

NDS Supp Tbl 5A Adj Fac

	C_M
$F_{bx}^{+} = 2400$ lbf/in^2	0.80
$F_{vx} = 300$ lbf/in^2	0.875
$F_{c\perp x} = 740$ lbf/in^2 (tension force)	0.53
$F_{c\perp x} = 650$ lbf/in^2 (compression face)	0.53
$E_x = 1{,}700{,}000$ lbf/in^2	0.833
$E_{x,\text{min}} = 0.88 \times 10^6$ lbf/in^2	0.833
$E_y = 1{,}500{,}000$ lbf/in^2	0.833
$E_{y,\text{min}} = 0.78 \times 10^6$ lbf/in^2	0.833

NDS Supp Tbl 1D

For 5×22 glulam, $A = 110.0 \text{ in}^2$, $S_x = 403.3 \text{ in}^3$, and $I_x = 4436.7 \text{ in}^4$.

NDS App Tbl N3

For the dead load and snow load combination, $1.2w_D + 1.6w_S$, the time effect factor, λ, is 0.8.

$$w_u = 1.2w_D = (1.2)\left(30\ \frac{\text{lbf}}{\text{ft}}\right) = 36\text{ lbf/ft}$$

$$P_u = (1.6)(4000\text{ lbf}) = 6400\text{ lbf}$$

The factored moment is

$$\begin{aligned} M_u &= \frac{w_u L^2}{8} + (1.6)(\text{area under shear diagram}) \\ &= \frac{\left(36\ \frac{\text{lbf}}{\text{ft}}\right)(32\text{ ft})^2\left(12\ \frac{\text{in}}{\text{ft}}\right)}{8} \\ &\quad + (1.6)(6000\text{ lbf})(8\text{ ft}) + (2000\text{ lbf})(8\text{ ft})\left(12\ \frac{\text{in}}{\text{ft}}\right) \\ &= 1{,}284{,}096\text{ in-lbf} \end{aligned}$$

1. Bending

NDS Tbl 5.3.1

$$\begin{aligned} F'_{bx} &= F_{bx}(K_F\phi_b\lambda)C_M C_t \quad [\text{the lesser of } C_V \text{ and } C_L] \\ &= F^*_{bx,\text{LRFD}} \quad [\text{the lesser of } C_V \text{ and } C_L] \end{aligned}$$

NDS App Tbls N1, N2, N3

$K_F = 2.16/\phi_b$; $\phi_b = 0.85$; $\lambda = 0.80$ for the load combination $(1.2w_D + 1.6w_S)$

$$\begin{aligned} F^*_{bx,\text{LRFD}} &= F_{bx}(K_F\phi_b\lambda)C_M C_t \\ &= \left(2400\ \frac{\text{lbf}}{\text{in}^2}\right)\left(\left(\frac{2.16}{\phi_b}\right)\phi_b(0.8)\right)(0.80)(1.0) \\ &= 3317.75\text{ lbf/in}^2 \end{aligned}$$

The volume factor for LRFD is the same as for ASD. Therefore, $C_V = 0.951$.

Calculate the beam stability factor, C_L.

NDS Tbl 5.3.1

$$E'_{\text{min}} = E_{\text{min}}(K_F\phi_s)C_M C_t$$

NDS App Tbls N1, N2, N3

$$\begin{aligned} E'_{y,\text{min,LRFD}} &= E_{y,\text{min}}\left(\frac{1.5}{\phi_s}\right)\phi_s C_M C_T \\ &= \left(0.78 \times 10^6\ \frac{\text{lbf}}{\text{in}^2}\right)(1.5)(0.833)(1.0) \\ &= 974{,}610\text{ lbf/in}^2 \end{aligned}$$

The slenderness ratio, R_B, for LRFD is the same as for ASD. Therefore, $R_B = 11.4$.

NDS 3.3.3.8

$$F_{bE,\mathrm{LRFD}} = \frac{1.20E'_{y,\min,\mathrm{LRFD}}}{{R_B}^2} = \frac{(1.20)\left(974{,}610\ \frac{\mathrm{lbf}}{\mathrm{in}^2}\right)}{(11.4)^2}$$
$$= 8999.2\ \mathrm{lbf/in}^2$$

$$C_L = \frac{1+\frac{F_{bE,\mathrm{LRFD}}}{F^*_{bx,\mathrm{LRFD}}}}{1.9} - \sqrt{\left(\frac{1+\frac{F_{bE,\mathrm{LRFD}}}{F^*_{bx,\mathrm{LRFD}}}}{1.9}\right)^2 - \left(\frac{\frac{F_{bE,\mathrm{LRFD}}}{F^*_{bx,\mathrm{LRFD}}}}{0.95}\right)}$$

$$= \frac{1+\frac{8999.2\ \frac{\mathrm{lbf}}{\mathrm{in}^2}}{3317.75\ \frac{\mathrm{lbf}}{\mathrm{in}^2}}}{1.9} - \sqrt{\left(\frac{1+\frac{8999.2\ \frac{\mathrm{lbf}}{\mathrm{in}^2}}{3317.75\ \frac{\mathrm{lbf}}{\mathrm{in}^2}}}{1.9}\right)^2 - \frac{\frac{8999.2\ \frac{\mathrm{lbf}}{\mathrm{in}^2}}{3317.75\ \frac{\mathrm{lbf}}{\mathrm{in}^2}}}{0.95}}$$

$$= 0.98 > C_V = 0.951 \quad [C_L\ \text{controls}]$$

Calculate the adjusted bending stress with an adjustment for volume factor.

$$F'_{bx,\mathrm{LRFD}} = F^*_{bx,\mathrm{LRFD}}C_V = \left(3317.75\ \frac{\mathrm{lbf}}{\mathrm{in}^2}\right)(0.951)$$
$$= 3155.2\ \mathrm{lbf/in}^2$$

Calculate the actual bending stress.

$$f_{bx,\mathrm{LRFD}} = \frac{M_u}{S_x} = \frac{1{,}284{,}096\ \text{in-lbf}}{403.3\ \mathrm{in}^3}$$
$$= 3184.0\ \mathrm{lbf/in}^2 \approx F'_{bx,\mathrm{LRFD}} = 3155.2\ \mathrm{lbf/in}^2 \quad [\text{consider OK}]$$

2. Shear

NDS App Tbls N1, N2

For shear,

$$K_F = \frac{2.16}{\phi_v} = \frac{2.16}{0.75}$$
$$= 2.88$$

Calculate the factored shear. (This is approximately equal to the reaction at beam ends, R_u.)

$$R_u \approx V_u = 1.2w_D + 1.6w_S = 1.2\left(\frac{w_D L}{2}\right) + 1.6w_S$$

$$= (1.2)\left(\frac{\left(30\ \frac{\mathrm{lbf}}{\mathrm{ft}}\right)(32\ \mathrm{ft})}{2}\right) + (1.6)(4000\ \mathrm{lbf} + 2000\ \mathrm{lbf})$$

$$= 10{,}176\ \mathrm{lbf}$$

Calculate the adjusted shear design value parallel to grain.

NDS Tbl 5.3.1

$$\begin{aligned} F'_{v,\text{LRFD}} &= F_v(K_F\phi_V\lambda)C_MC_t \\ &= \left(300\ \frac{\text{lbf}}{\text{in}^2}\right)\big((2.88)(0.75)(0.8)\big)(0.875)(1.0) \\ &= 453.6\ \text{lbf/in}^2 \end{aligned}$$

Calculate the shear capacity.

$$\begin{aligned} V'_{\text{LRFD}} &= \tfrac{2}{3}F'_{v,\text{LRFD}}A \\ &= \left(\frac{2}{3}\right)\left(453.6\ \frac{\text{lbf}}{\text{in}^2}\right)(110.0\ \text{in}^2) \\ &= 33{,}264\ \text{lbf} > V_u = 10{,}176\ \text{lbf} \quad \text{[OK]} \end{aligned}$$

3. Deflection

Same as ASD method.

4. Bearing

NDS App Tbls N1, N2

For bearing,

$$\begin{aligned} K_F &= \frac{1.875}{\phi_c} = \frac{1.875}{0.9} \\ &= 2.08 \end{aligned}$$

NDS Supp Tbl 5A

$$F_{c\perp,x} = 740\ \text{lbf/in}^2 \quad \text{[tension face]}$$

NDS Tbl 5.3.1

Calculate the adjusted compression design value perpendicular to the grain.

$$\begin{aligned} F'_{c\perp,\text{LRFD}} &= F_{c\perp x}(K_F\phi_C\lambda)C_MC_t \\ &= \left(740\ \frac{\text{lbf}}{\text{in}^2}\right)\big((2.08)(0.9)(0.8)\big)(0.53)(1.0) \\ &= 587.3\ \text{lbf/in}^2 \end{aligned}$$

The required bearing area is

$$\begin{aligned} \frac{R_u}{F'_{c\perp,\text{LRFD}}} &= \frac{10{,}176\ \text{lbf}}{587.3\ \frac{\text{lbf}}{\text{in}^2}} \\ &= 17.3\ \text{in}^2 \end{aligned}$$

The required bearing length is

$$\ell_b = \frac{\text{required bearing area}}{b} = \frac{17.3\ \text{in}^2}{5\ \text{in}} = 3.46\ \text{in} \quad (3.5\ \text{in})$$

Example 5.2
Glulam Beam Design

A glulam beam has a 24F-V3 SP/SP bending combination (see *NDS Supplement* Table 5A). The design loads are caused by a combined dead load, w_D, and roof live load, $w_{L,\text{roof}}$. The top face of the beam is laterally supported at the load points, and the bottom face in the negative moment area is laterally unsupported except at the support point. Assume wet service conditions at normal temperatures. The beam weight can be neglected for this example.

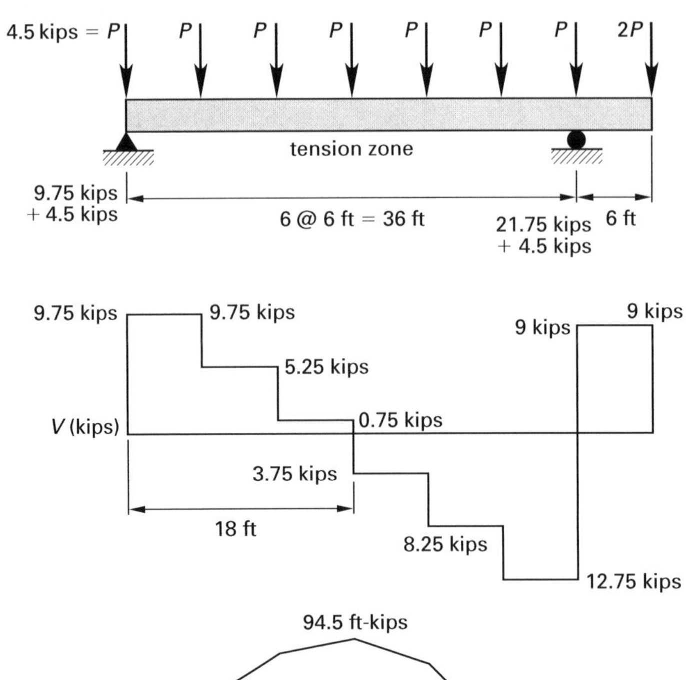

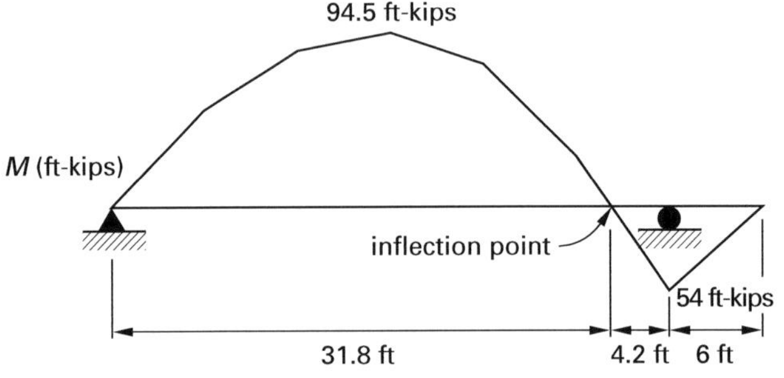

Determine the necessary glulam beam size (considering bending and shear stresses only) using design procedure.

REFERENCE *Solution (ASD Method)*

For 24F-V3 SP/SP,

NDS Supp Tbl 5A

	C_M
$F_{bx}^{+} = 2400\ \text{lbf/in}^2$ $\begin{bmatrix}\text{tension zone}\\ \text{stressed in tension}\end{bmatrix}$	0.80
$F_{bx}^{-} = 1950\ \text{lbf/in}^2$ $\begin{bmatrix}\text{compression zone}\\ \text{stressed in tension}\end{bmatrix}$	0.80
$F_{vx} = 300\ \text{lbf/in}^2$	0.875
$E_x = 1{,}800{,}000\ \text{lbf/in}^2$	0.833
$E_{x,\text{min}} = 0.93 \times 10^6\ \text{lbf/in}^2$	0.833
$E_y = 1{,}600{,}000\ \text{lbf/in}^2$	0.833
$E_{y,\text{min}} = 0.83 \times 10^6\ \text{lbf/in}^2$	0.833

NDS Tbl 2.3.2 For roof live load, C_D is 1.25 (for seven days construction live load).

1. Positive moment (tension zone stressed in tension)

NDS Tbl 5.3.1

Use the following to calculate the allowable bending stress.

$$F'_{bx} = F^{+}_{bx} C_D C_M C_t \quad \text{[the lesser of } C_V \text{ and } C_L\text{]}$$
$$= F^{*+}_{bx} \quad \text{[the lesser of } C_V \text{ and } C_L\text{]}$$

The reference bending stress with all adjustments except C_V and C_L is

$$\begin{aligned} F^{*+}_{bx} &= F^{+}_{bx} C_D C_M C_t \\ &= \left(2400 \ \frac{\text{lbf}}{\text{in}^2}\right)(1.25)(0.80)(1.0) \\ &= 2400 \ \text{lbf/in}^2 \end{aligned}$$

The initial beam size is

$$\begin{aligned} S &= \frac{M}{F^{*+}_{bx}} \\ &= \frac{(94.5 \text{ ft-kips})\left(1000 \ \frac{\text{lbf}}{\text{kip}}\right)\left(12 \ \frac{\text{in}}{\text{ft}}\right)}{2400 \ \frac{\text{lbf}}{\text{in}^2}} \\ &= 472.5 \ \text{in}^3 \end{aligned}$$

NDS Supp Tbl 1D

Try beam size 5 × 24³/₄, southern pine glulam.

$$A = 123.8 \ \text{in}^2$$
$$S_x = 510.5 \ \text{in}^3$$

Find the beam stability factor, C_L.

The unbraced length is

$$\begin{aligned} \ell_u &= (6 \text{ ft})\left(12 \ \frac{\text{in}}{\text{ft}}\right) \\ &= 72 \text{ in} \end{aligned}$$

NDS Tbl 3.3.3, Ftn 1

$$\begin{aligned} \frac{\ell_u}{d} &= \frac{72 \text{ in}}{24.75 \text{ in}} \\ &= 2.9 < 7.0 \end{aligned}$$

The effective length is

NDS Tbl 3.3.3

$$\begin{aligned} \ell_e &= 2.06\ell_u = (2.06)(72 \text{ in}) \\ &= 148.32 \text{ in} \end{aligned}$$

The slenderness ratio is

NDS Eq 3.3-5

$$\begin{aligned} R_B &= \sqrt{\frac{\ell_e d}{b^2}} = \sqrt{\frac{(148.32 \text{ in})(24.75 \text{ in})}{(5 \text{ in})^2}} \\ &= 12.12 \end{aligned}$$

The adjusted modulus of elasticity is

NDS Tbl 5.3.1

$$E'_{y,\min} = E_{y,\min}C_M C_t = \left(0.83 \times 10^6 \ \frac{\text{lbf}}{\text{in}^2}\right)(0.833)(1.0) = 0.69 \times 10^6 \ \text{lbf/in}^2$$

The critical buckling design value is

NDS 3.3.3.8

$$F_{bE} = \frac{1.20E'_{y,\min}}{R_B^2} = \frac{(1.20)\left(0.69 \times 10^6 \ \frac{\text{lbf}}{\text{in}^2}\right)}{(12.12)^2} = 5637 \ \text{lbf/in}^2$$

$$\frac{F_{bE}}{F_{bx}^{*+}} = \frac{5637 \ \frac{\text{lbf}}{\text{in}^2}}{2400 \ \frac{\text{lbf}}{\text{in}^2}} = 2.35$$

The beam stability factor is

$$C_L = \frac{1 + \frac{F_{bE}}{F_{bx}^{*+}}}{1.9} - \sqrt{\left(\frac{1 + \frac{F_{bE}}{F_{bx}^{*+}}}{1.9}\right)^2 - \frac{\frac{F_{bE}}{F_{bx}^{*+}}}{0.95}} = \frac{1 + 2.35}{1.9} - \sqrt{\left(\frac{1 + 2.35}{1.9}\right)^2 - \frac{2.35}{0.95}} = 0.966$$

Find the volume factor of glulam, C_V.

NDS Eq 5.3-1

For southern pine, $x = 20$.

The distance between points of zero moment is $L = 31.8$ ft. (See the moment diagram in the problem statement.)

$$C_V = \left(\frac{21}{L}\right)^{1/x}\left(\frac{12}{d}\right)^{1/x}\left(\frac{5.125}{b}\right)^{1/x} \leq 1.0 = \left(\frac{21}{31.8 \text{ ft}}\right)^{1/20}\left(\frac{12}{24.75 \text{ in}}\right)^{1/20}\left(\frac{5.125}{5 \text{ in}}\right)^{1/20} \leq 1.0 = 0.946$$

Since $C_V = 0.946 < C_L = 0.966$, use C_V.

Calculate the bending stress with all adjustments except C_L.

$$F'_{bx} = F_{bx}^{*+}C_V = \left(2400 \ \frac{\text{lbf}}{\text{in}^2}\right)(0.946) = 2270.4 \ \text{lbf/in}^2$$

Calculate the actual bending stress.

$$f_{bx} = \frac{M}{S_x} = \frac{(94.5 \text{ ft-kips})\left(1000 \dfrac{\text{lbf}}{\text{kip}}\right)\left(12 \dfrac{\text{in}}{\text{ft}}\right)}{510.5 \text{ in}^3}$$
$$= 2221.4 \text{ lbf/in}^2 < F'_{bx} = 2270.4 \text{ lbf/in}^2 \quad \text{[OK]}$$

2. Shear

Calculate the actual shear stress parallel to grain.

$$f_v = 1.5\left(\frac{V}{A}\right) = (1.5)\left(\frac{12{,}750 \text{ lbf}}{123.8 \text{ in}^2}\right)$$
$$= 154.5 \text{ lbf/in}^2$$

Calculate the adjusted shear design value parallel to the grain.

NDS Tbl 5.3.1

$$F'_{vx} = F_{vx} C_D C_M C_t$$
$$= \left(300 \frac{\text{lbf}}{\text{in}^2}\right)(1.25)(0.875)(1.0)$$
$$= 328.13 \text{ lbf/in}^2 > f_v = 154.5 \text{ lbf/in}^2 \quad \text{[OK]}$$

3. Negative moment (compression zone stressed in tension)

Use the following to calculate the allowable bending stress. The smaller of the following values controls.

$$F'_{bx} = F'^{-}_{bx} = \begin{cases} C_D C_M C_t C_L = F^{*-} C_L \\ C_D C_M C_t C_V = F^{*-} C_V \end{cases}$$

$$F^{*-}_{bx} = F^{-}_{bx} C_D C_M C_t$$
$$= \left(1950 \frac{\text{lbf}}{\text{in}^2}\right)(1.25)(0.80)(1.0)$$
$$= 1950 \text{ lbf/in}^2$$

For beam size 5 in × 24¾ in,

$$A = 123.8 \text{ in}^2$$
$$S_x = 510.5 \text{ in}^3$$

NDS Sec 3.3.3

Find the beam stability factor, C_L.

The unbraced length is

$$\ell_u = (6 \text{ ft})\left(12 \frac{\text{in}}{\text{ft}}\right) = 72 \text{ in}$$

(Note: There is 4.2 ft between the support and the inflection point. This is less than 6 ft cantilever length. Therefore, $\ell_u = 6.0$ ft.)

NDS Tbl 3.3.3

$$\frac{\ell_u}{d} = \frac{72 \text{ in}}{24.75 \text{ in}}$$
$$= 2.9 < 7.0$$

The effective length is

$$\ell_e = 1.87\ell_u = (1.87)(72 \text{ in}) = 134.64 \text{ in}$$

NDS Sec 3.3.3.8

The slenderness ratio is

$$R_B = \sqrt{\frac{\ell_e d}{b^2}} = \sqrt{\frac{(134.64 \text{ in})(24.75 \text{ in})}{(5 \text{ in})^2}} = 11.54$$

The adjusted modulus of elasticity is

$$E'_{y,\min} = E_{y,\min} C_M C_t = \left(0.83 \times 10^6 \ \frac{\text{lbf}}{\text{in}^2}\right)(0.833)(1.0) = 0.69 \times 10^6 \text{ lbf/in}^2$$

The critical buckling design value is

$$F_{bE} = \frac{1.20 E_{y,\min}}{R_B^2} = \frac{(1.20)\left(0.69 \times 10^6 \ \frac{\text{lbf}}{\text{in}^2}\right)}{(11.54)^2} = 6217.54 \text{ lbf/in}^2$$

$$\frac{F_{bE}}{F_{bx}^{*+}} = \frac{6217.54 \ \frac{\text{lbf}}{\text{in}^2}}{1950 \ \frac{\text{lbf}}{\text{in}^2}} = 3.19$$

The beam stability factor is

$$C_L = \frac{1 + \frac{F_{bE}}{F_{bx}^*}}{1.9} - \sqrt{\left(\frac{1 + \frac{F_{bE}}{F_{bx}^*}}{1.9}\right)^2 - \frac{\frac{F_{bE}}{F_{bx}^*}}{0.95}}$$

$$= \frac{1 + 3.19}{1.9} - \sqrt{\left(\frac{1 + 3.19}{1.9}\right)^2 - \frac{3.19}{0.95}} = 0.978$$

Find the volume factor of glulam, C_V.

NDS Eq 5.3-1

For southern pine, $x = 20$.

The distance between points of zero moments is $L = 4.2 \text{ ft} + 6 \text{ ft} = 10.2 \text{ ft}$. (See the negative moment diagram in the problem statement.)

$$C_V = \left(\frac{21}{L}\right)^{1/x} \left(\frac{12}{d}\right)^{1/x} \left(\frac{5.125}{b}\right)^{1/x} \leq 1.0$$

$$= \left(\frac{21}{10.2 \text{ ft}}\right)^{1/20} \left(\frac{12}{24.75 \text{ in}}\right)^{1/20} \left(\frac{5.125}{5 \text{ in}}\right)^{1/20} = 1.001$$

Since $C_L = 0.978 < C_V = 1.0$, use C_L.

Calculate the bending stress with all adjustments except C_V.

$$F'_{bx} = F^{*-}_{bx} C_L = \left(1950 \ \frac{\text{lbf}}{\text{in}^2}\right)(0.978)$$
$$= 1907.1 \ \text{lbf/in}^2$$

Calculate the actual bending stress.

$$f_{bx} = \frac{M}{S_x} = \frac{(54 \text{ ft-kips})\left(1000 \ \frac{\text{lbf}}{\text{kip}}\right)\left(12 \ \frac{\text{in}}{\text{ft}}\right)}{510.5 \text{ in}^3}$$
$$= 1269.3 \ \text{lbf/in}^2 < F'_{bx} = 1907.1 \ \text{lbf/in}^2 \quad \text{[OK]}$$

Example 5.3
Cantilevered Glulam Beam: Two Equal Spans

Roof girders consist of $5^1/8$ in by $28^1/2$ in glulam timbers. The combination symbol is 24F-V10 DF/HF (see *NDS Supplement* Table 5A). The combined roof dead and snow load is 600 lbf/ft. The maximum allowable deflection is $L/180$.

Review the girders for bending and shear and determine the deflections at B (midpoint of span AC) and D (hinge point).

REFERENCE

Solution (ASD Method)

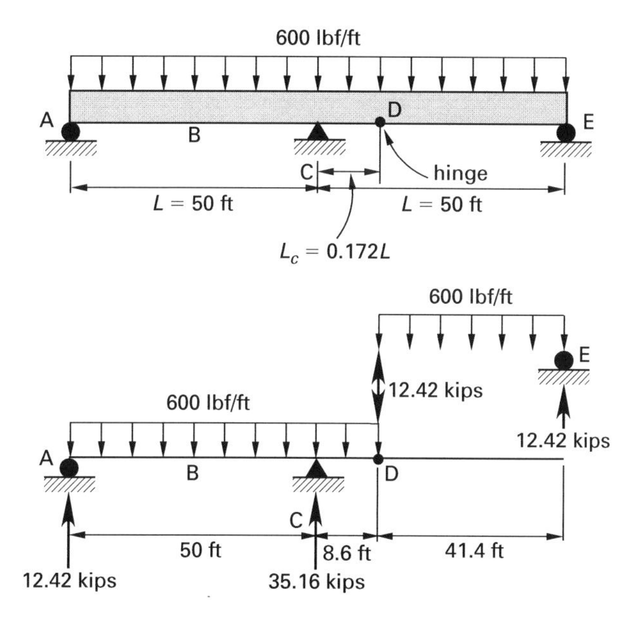

The shear diagram is as follows.

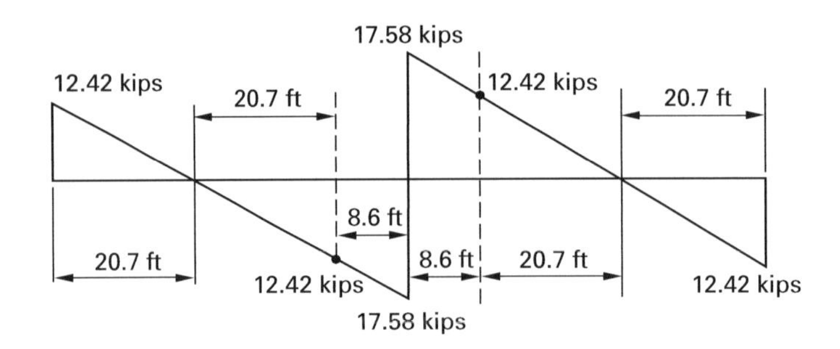

The moment diagram is as follows.

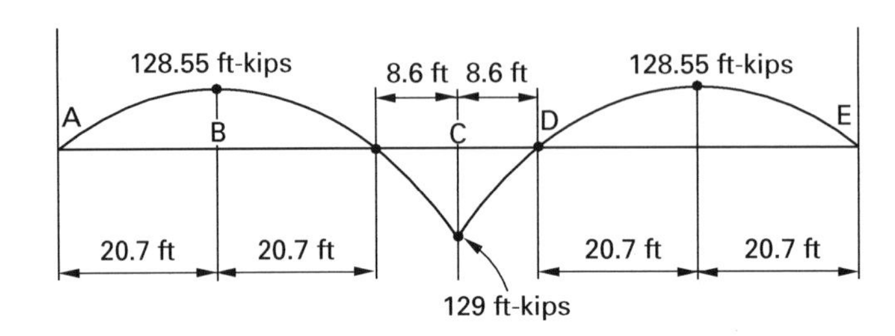

For $5^1/8$ in × $28^1/2$ in 24F-V10 DF/HF glulam,

NDS Supp Tbl 5A

$$C_M = C_t = 1.0$$

This is a balanced glulam section. Thus, $F_{bx} = F_{bx}^+ = F_{bx}^-$.

$$
\begin{aligned}
F_{bx} &= 2400 \text{ lbf/in}^2 \\
F_{vx} &= 215 \text{ lbf/in}^2 \\
E_x &= 1{,}800{,}000 \text{ lbf/in}^2 \\
E_y &= 1{,}500{,}000 \text{ lbf/in}^2 \\
E_{x,\text{min}} &= 0.93 \times 10^6 \text{ lbf/in}^2 \\
E_{y,\text{min}} &= 0.78 \times 10^6 \text{ lbf/in}^2
\end{aligned}
$$

NDS Tbl 2.3.2

For snow load, C_D is 1.15.

NDS Supp Tbl 1C

For $5^1/8$ in × $28^1/2$ in western species glulam,

$$
\begin{aligned}
A &= 146.1 \text{ in}^2 \\
I_x &= 9887.0 \text{ in}^4 \\
S_x &= 693.8 \text{ in}^3
\end{aligned}
$$

Find the allowable bending stresses, F'_{bx}.

For the positive moment region in member ABC, C_L is 1.0 since the roof sheathings and purlins provide lateral stability at the girder topside.

NDS Eq 5.3-1; NDS Supp Tbl 5A Adj Fac

The volume factor is

$$C_V = \left(\frac{21}{L}\right)^{1/10}\left(\frac{12}{d}\right)^{1/10}\left(\frac{5.125}{b}\right)^{1/10} \leq 1.0$$
$$= \left(\frac{21}{20.7\text{ ft} + 20.7\text{ ft}}\right)^{1/10}\left(\frac{12}{28.5\text{ in}}\right)^{1/10}\left(\frac{5.125}{5.125\text{ in}}\right)^{1/10}$$
$$= 0.857$$

The allowable bending stress, F'_{bx}, for member ABC is

$$F'_{bx} = F_{bx}C_DC_MC_tC_V$$
$$= \left(2400\ \frac{\text{lbf}}{\text{in}^2}\right)(1.15)(1.0)(1.0)(0.857)$$
$$= 2365.3\text{ lbf/in}^2$$

For the negative moment region in member BCD, assume the lateral support is provided by bracing the underside at support C.

NDS Sec 3.3.3; NDS Tbl 3.3.3

Find the beam stability factor, C_L.

The unbraced length is

$$\ell_u = (8.6\text{ ft})\left(12\ \frac{\text{in}}{\text{ft}}\right)$$
$$= 103.2\text{ in}$$
$$\frac{\ell_u}{d} = \frac{103.2\text{ in}}{28.5\text{ in}}$$
$$= 3.62 < 7.0$$

NDS Tbl 3.3.3; Ftn 1

The effective span length of the bending member is

$$\ell_e = 2.06\ell_u = (2.06)(103.2\text{ in})$$
$$= 212.6\text{ in}$$

NDS Sec 3.3.3.6

The slenderness ratio is

$$R_B = \sqrt{\frac{\ell_e d}{b^2}} = \sqrt{\frac{(212.6\text{ in})(28.5\text{ in})}{(5.125\text{ in})^2}}$$
$$= 15.2 \leq 50 \quad \text{[OK]}$$

The adjusted modulus of elasticity is

$$E'_{y,\min} = E_{y,\min}C_MC_t$$
$$= \left(0.78 \times 10^6\ \frac{\text{lbf}}{\text{in}^2}\right)(1.0)(1.0)$$
$$= 0.78 \times 10^6\text{ lbf/in}^2$$

NDS Sec 3.3.3.8

The critical buckling design value for glulam is

$$F_{bE} = \frac{1.20E'_{y,\min}}{R_B^2} = \frac{(1.20)\left(0.78 \times 10^6\ \frac{\text{lbf}}{\text{in}^2}\right)}{(15.2)^2}$$
$$= 4051.25\text{ lbf/in}^2$$

The reference bending stress with all adjustments except C_L and C_V is

$$\begin{aligned} F_{bx}^* &= F_{bx}C_DC_MC_t \\ &= \left(2400\ \frac{\text{lbf}}{\text{in}^2}\right)(1.15)(1.0)(1.0) \\ &= 2760\ \text{lbf/in}^2 \end{aligned}$$

The beam stability factor is

$$\begin{aligned} C_L &= \frac{1+\dfrac{F_{bE}}{F_{bx}^*}}{1.9} - \sqrt{\left(\frac{1+\dfrac{F_{bE}}{F_{bx}^*}}{1.9}\right)^2 - \frac{\dfrac{F_{bE}}{F_{bx}^*}}{0.95}} \\ &= \frac{1+\dfrac{4051.25\ \frac{\text{lbf}}{\text{in}^2}}{2760\ \frac{\text{lbf}}{\text{in}^2}}}{1.9} - \sqrt{\left(\frac{1+\dfrac{4051.25\ \frac{\text{lbf}}{\text{in}^2}}{2760\ \frac{\text{lbf}}{\text{in}^2}}}{1.9}\right)^2 - \frac{\dfrac{4051.25\ \frac{\text{lbf}}{\text{in}^2}}{2760\ \frac{\text{lbf}}{\text{in}^2}}}{0.95}} \\ &= 0.922 \end{aligned}$$

Find the volume factor.

NDS Eq 5.3-1

$$\begin{aligned} C_V &= \left(\frac{21}{L}\right)^{1/x}\left(\frac{12}{d}\right)^{1/x}\left(\frac{5.125}{b}\right)^{1/x} \leq 1.0 \\ &= \left(\frac{21}{8.6\ \text{ft} + 8.6\ \text{ft}}\right)^{1/10}\left(\frac{12}{28.5\ \text{in}}\right)^{1/10}\left(\frac{5.125}{5.125\ \text{in}}\right)^{1/10} \\ &= 0.936 \end{aligned}$$

Since $C_L = 0.922 < C_V = 0.936$, C_L controls.

The allowable bending stress, F'_{bx}, for member BCD is

$$\begin{aligned} F'_{bx} &= F_{bx}^*C_V = \left(2760\ \frac{\text{lbf}}{\text{in}^2}\right)(0.922) \\ &= 2544.7\ \text{lbf/in}^2 \end{aligned}$$

For member DE (in the positive moment region), the allowable bending stress is the same as for member ABC, thus the adjusted beam design value, F'_{bx}, is 2365.3 lbf/in^2.

Find the allowable shear stress, F'_v.

For all members, calculate the adjusted shear design value parallel to the grain.

$$\begin{aligned} F'_{vx} &= F_{vx}C_DC_MC_t = \left(215\ \frac{\text{lbf}}{\text{in}^2}\right)(1.15)(1.0)(1.0) \\ &= 247.3\ \text{lbf/in}^2 \end{aligned}$$

Compare the allowable stresses with the calculated stresses.

For members ABC and DE, the actual bending stress is

$$f_b = \frac{M}{S_x} = \frac{(128.55 \text{ ft-kips})\left(1000 \ \frac{\text{lbf}}{\text{kip}}\right)\left(12 \ \frac{\text{in}}{\text{ft}}\right)}{693.8 \text{ in}^3}$$
$$= 2223.4 \text{ lbf/in}^2 < F'_{bx} = 2365.3 \text{ lbf/in}^2 \quad \text{[OK]}$$

The shear stress parallel to the grain is

$$f_v = 1.5\left(\frac{V}{A}\right) = (1.5)\left(\frac{(12.42 \text{ kips})\left(1000 \ \frac{\text{lbf}}{\text{kip}}\right)}{146.1 \text{ in}^2}\right)$$
$$= 127.5 \text{ lbf/in}^2 < F'_v = 247.3 \text{ lbf/in}^2 \quad \text{[OK]}$$

For member BCD (in the negative moment region), the bending stress is

$$f_b = \frac{M}{S_x} = \frac{(129 \text{ ft-kips})\left(1000 \ \frac{\text{lbf}}{\text{kip}}\right)\left(12 \ \frac{\text{in}}{\text{ft}}\right)}{693.8 \text{ in}^3}$$
$$= 2231.2 \text{ lbf/in}^2 < F'_{bx} = 2544.7 \text{ lbf/in}^2 \quad \text{[OK]}$$

The shear stress parallel to the grain is

$$f_v = 1.5\left(\frac{V}{A}\right) = (1.5)\left(\frac{(17.58 \text{ kips})\left(1000 \ \frac{\text{lbf}}{\text{kip}}\right)}{146.1 \text{ in}^2}\right)$$
$$= 180.5 \text{ lbf/in}^2 < F'_v = 247.3 \text{ lbf/in}^2 \quad \text{[OK]}$$

Determine deflections at B (midpoint of span AC) and at hinge point D for camber purposes.

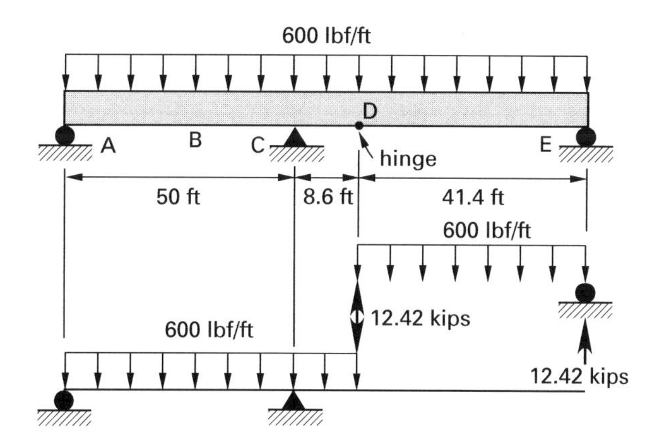

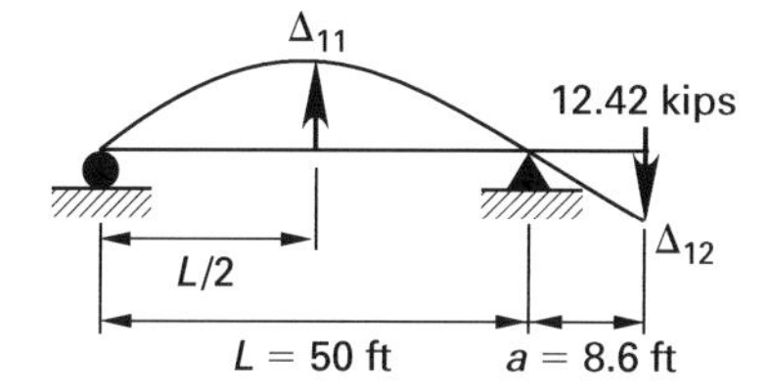

Calculate the adjusted modulus of elasticity.

$$E' = E_x C_M C_t = \left(1.8 \times 10^6 \ \frac{\text{lbf}}{\text{in}^2}\right)(1.0)(1.0)$$
$$= 1.8 \times 10^6 \ \text{lbf/in}^2$$

Calculate the deflection.

Appendix

$$\Delta_{11} = \left(\frac{Pax}{6E'_x IL}\right)(L^2 - x^2)$$
$$= \frac{(12.42 \text{ kips})\left(1000 \ \frac{\text{lbf}}{\text{kip}}\right)(8.6 \text{ ft})(25 \text{ ft})}{(6)\left(1{,}800{,}000 \ \frac{\text{lbf}}{\text{in}^2}\right)(9887 \text{ in}^4)(50 \text{ ft})} \times \left((50 \text{ ft})^2 - (25 \text{ ft})^2\right)\left(12 \ \frac{\text{in}}{\text{ft}}\right)^3$$
$$= 1.62 \text{ in} \quad [\text{upward}]$$
$$= -1.62 \text{ in}$$

$$\Delta_{12} = \left(\frac{Pa^2}{3E'_x I_x}\right)(L + a)$$
$$= \left(\frac{(12.42 \text{ kips})\left(1000 \ \frac{\text{lbf}}{\text{kip}}\right)(8.6 \text{ ft})^2}{(3)\left(1{,}800{,}000 \ \frac{\text{lbf}}{\text{in}^2}\right)(9887 \text{ in}^4)}\right)(50 \text{ ft} + 8.6 \text{ ft})\left(12 \ \frac{\text{in}}{\text{ft}}\right)^3$$
$$= 1.74 \text{ in} \quad [\text{downward}]$$

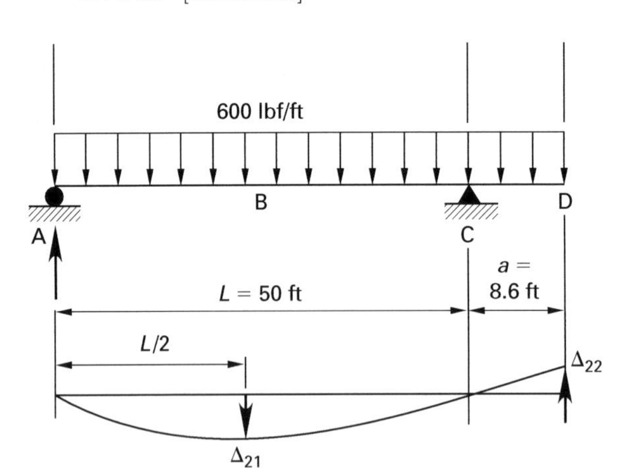

$$\Delta_{22} = \left(\frac{wa}{24E'_x I}\right)(4a^2 L - L^3 + 3a^3)$$
$$= \left(\frac{\left(600 \ \frac{\text{lbf}}{\text{ft}}\right)(8.6 \text{ ft})}{(24)\left(1{,}800{,}000 \ \frac{\text{lbf}}{\text{in}^2}\right)(9887 \text{ in}^4)}\right) \times \left((4)(8.6 \text{ ft})^2(50 \text{ ft}) - (50 \text{ ft})^3 + (3)(8.6 \text{ ft})^3\right)\left(12 \ \frac{\text{in}}{\text{ft}}\right)^3$$
$$= 2.26 \text{ in} \quad [\text{upward}]$$
$$= -2.26 \text{ in}$$

$$\Delta_{21} = \left(\frac{wx}{24E'_xIL}\right)(L^4 - 2L^2x^2 + Lx^3 - 2a^2L^2 + 2a^2x^2)$$

$$= \left(\frac{\left(0.6\ \frac{\text{kips}}{\text{ft}}\right)\left(1000\ \frac{\text{lbf}}{\text{kip}}\right)(25\ \text{ft})}{(24)\left(1{,}800{,}000\ \frac{\text{lbf}}{\text{in}^2}\right)(9887\ \text{in}^4)(50\ \text{ft})}\right)$$

$$\times \left((50\ \text{ft})^4 - (2)(50\ \text{ft})^2(25\ \text{ft})^2(50\ \text{ft})(25\ \text{ft})^3 - (2)(8.6\ \text{ft})^2(50\ \text{ft})^2 + (2)(8.6\ \text{ft})^2(25\ \text{ft})^2\right)\left(12\ \frac{\text{in}}{\text{ft}}\right)^3$$

$$= 4.40\ \text{in} \quad [\text{downward}]$$

The deflection at B (midpoint of AC) is

$$\Delta_{11}\ \text{upward} + \Delta_{21}\ \text{downward} = -1.62\ \text{in} + 4.40\ \text{in} = 2.78\ \text{in} \quad [\text{downward}]$$

The deflection at D (hinge point) is

$$\Delta_{12}\ \text{downward} + \Delta_{22}\ \text{upward} = 1.74\ \text{in} - 2.26\ \text{in} = -0.52\ \text{in} \quad [\text{upward}]$$

6

Axial Members, and Combined Bending and Axial Loading

1. Tension Members

[NDS Tbls 4.3.1, 5.3.1; NDS Sec 3.8; NDS Supp Tbls 4A, 4B, 4C, 4D, 4F, 5A, 5B, 5C, 5D]

The actual tensile stress parallel to grain is based on the net section area and shall not exceed the allowable tensile design value parallel to grain.

P	tensile force	lbf
F_t	reference tension value parallel to grain (See *NDS Supplement* Tables 4A–4F and 5A–5D.)	lbf/in^2
C_D	load duration factor (See NDS Table 2.3.2; NDS App. B.)	–
C_M	wet service factor (When MC > 19% for sawn lumber, or MC > 16% for glulam, $C_M < 1.0$. For dry service conditions, C_M is 1.0. See *NDS Supplement* Tables 4 and 5 adjustment factors.)	–
C_F	size factor for visually graded dimension lumber and timbers (See *NDS Supplement* Tables 4A, 4B, 4D, 4E, and 4F adjustment factors.)	–
C_t	temperature factor (When $T \leq 100°$F, C_t is 1.0. See NDS Table 2.3.3.)	–
C_i	incising factor for sawn lumber (See NDS Sec. 4.3.8.)	–
A_n	net cross-sectional area of member	in^2

For sawn lumber,

$$f_t = \frac{P}{A_n} < F_t' = F_t C_D C_M C_t C_F C_i \quad \text{[see NDS Table 4.3.1]}$$

For glued laminated timber,

$$f_t = \frac{P}{A_n} < F_t' = F_t C_D C_M C_t \quad \text{[see NDS Table 5.3.1]}$$

Example 6.1
Tension Member: Sawn Lumber

Assume no. 2 hem-fir, 2 × 8 tension member with a row of 3/4 in diameter bolts with 1/8 in oversized holes. Moisture content, MC, is greater than 19%. The tensile force, P, is caused by dead load, w_D, and snow load, w_S, and is 6000 lbf.

Calculate the actual tensile stress parallel to grain, f_t, and the allowable tension design value parallel to grain, F_t'.

REFERENCE *Solution (ASD Method)*

The actual tensile stress is

$$f_t = \frac{P}{A_n} = \frac{6000 \text{ lbf}}{(1.5 \text{ in})\big(7.25 \text{ in} - (0.75 \text{ in} + 0.125 \text{ in})\big)} = 627.5 \text{ lbf/in}^2$$

Reference			
NDS Supp Tbl 4A and Adj Fac	F_t	reference tension design value parallel to grain	525 lbf/in^2
	C_F	size factor for sawn lumber	1.2
	C_M	wet service factor	1.0
	C_t	temperature factor	1.0
NDS Tbl 2.3.2	C_D	load duration factor for snow load for two months	1.15

NDS Tbl 4.3.1 — The allowable tension design value parallel to grain is

$$F_t' = F_t C_D C_M C_t C_F = \left(525 \ \frac{\text{lbf}}{\text{in}^2}\right)(1.15)(1.0)(1.0)(1.2) = 724.5 \text{ lbf/in}^2$$

Example 6.2
Tension Member: Glulam

Glulam axial combination 5DF 2½ × 6 (4 lams; A = 15.0 in^2), with a row of ¾ in diameter bolts with ⅛ in over-sized holes. Moisture content, MC, is greater than 19%. The tensile force, P, is caused by dead load, w_D, and snow load, w_S, and is 6000 lbf.

Calculate the actual tensile stress parallel to grain, f_t, and the allowable tension design value parallel to grain, F_t'.

REFERENCE *Solution (ASD Method)*

NDS Supp Tbl 1C — 2½ in × 6 in are net finished dimensions. The actual tensile stress parallel to grain is

$$f_t = \frac{P}{A_n} = \frac{6000 \text{ lbf}}{(2.5 \text{ in})\big(6.0 \text{ in} - (0.75 \text{ in} + 0.125 \text{ in})\big)} = 468 \text{ lbf/in}^2$$

NDS Supp Tbl 5B Adj Fac — For 5DF axial combination,

Reference			
NDS Supp Tbl 5B Adj Fac	F_t	reference tension design value parallel to grain	1600 lbf/in^2
	C_M	wet service factor	0.80
NDS Tbl 2.3.2	C_D	load duration factor	1.15

NDS Tbl 5.3.1 — The allowable tension design value parallel to grain is

$$F_t' = F_t C_D C_M C_t = \left(1600 \ \frac{\text{lbf}}{\text{in}^2}\right)(1.15)(0.80)(1.0) = 1472 \text{ lbf/in}^2$$

2. Combined Tension and Bending

[NDS 3.9.1, Eqs. 3.9-1 and 3.9-2]

Members subjected to a combination of axial tension and bending will satisfy NDS Eqs. 3.9-1 and 3.9-2.

f_b	actual bending tensile stress	lbf/in^2
F_b^*	allowable bending tensile stress without the stability factor, C_L	lbf/in^2
F_b^{**}	allowable bending compressive stress without the volume factor, C_V (see NDS Table 4.3.1)	lbf/in^2

$$\frac{f_t}{F_t'} + \frac{f_b}{F_b^*} \leq 1.0 \quad \text{[NDS Eq. 3.9-1]}$$

To check the buckling on the compression face,

$$\frac{f_b - f_t}{F_b^{**}} \leq 1.0 \quad [\text{if } f_b > f_t] \quad \text{[NDS Eq. 3.9-2]}$$

Example 6.3
Combined Tension and Bending: Sawn Lumber

Determine if a member with the following characteristics meets the requirements of the NDS combined bending and axial loading.

no. 2 hem-fir 2 × 8 member with a row of 3/4 in diameter bolts

MC > 19%

$P = w_D + w_S = 6000 \text{ lbf}$

$M = 10{,}000 \text{ in-lbf}$

REFERENCE

Solution (ASD Method)

The actual tensile stress is

$$f_t = \frac{P}{A_n} = \frac{6000 \text{ lbf}}{(1.5 \text{ in})\big(7.25 \text{ in} - (0.75 \text{ in} + 0.125 \text{ in})\big)} = 627.5 \text{ lbf/in}^2$$

The allowable tensile design value parallel to grain is

Ex 6.1

$$F_t' = F_t C_D C_M C_t C_F = \left(525 \, \frac{\text{lbf}}{\text{in}^2}\right)(1.15)(1.0)(1.0)(1.2) = 724.5 \text{ lbf/in}^2$$

The section modulus is

$$S = \frac{(1.5 \text{ in})(7.25 \text{ in})^2}{6} = 13.14 \text{ in}^3$$

The actual bending stress is

$$f_b = \frac{M}{S} = \frac{10{,}000 \text{ in-lbf}}{13.14 \text{ in}^3} = 761.04 \text{ lbf/in}^2 \quad \text{[tensile or compressive]}$$

NDS Supp Tbl 4A Adj Fac

F_b	reference bending design value	850 lbf/in^2
C_F	size factor for sawn lumber	1.2

Since

$$F_bC_F = (850 \text{ lbf/in}^2)(1.2) = 1020 \text{ lbf/in}^2 < 1150 \text{ lbf/in}^2$$

C_M	wet service factor	1.0
C_D	load duration factor	1.15
C_t	temperature factor	1.0

The reference bending design value multiplied by all applicable adjustment factors is

NDS Tbl 4.3.1

$$F_b^* = F_bC_DC_MC_tC_FC_{fu} = \left(850 \, \frac{\text{lbf}}{\text{in}^2}\right)(1.15)(1.0)(1.0)(1.2)(1.0) = 1173.0 \text{ lbf/in}^2$$

NDS Eq 3.9-1

$$\frac{f_t}{F_t'} + \frac{f_b}{F_b^*} = \frac{627.5 \, \frac{\text{lbf}}{\text{in}^2}}{724.5 \, \frac{\text{lbf}}{\text{in}^2}} + \frac{761.0 \, \frac{\text{lbf}}{\text{in}^2}}{1173.0 \, \frac{\text{lbf}}{\text{in}^2}} = 1.515 > 1.0 \quad \text{[no good]}$$

Assuming the beam stability factor, C_L, is 1.0, calculate the reference bending design value multiplied by all applicable adjustment factors, including C_L.

NDS Eq 3.9-2

$$F_b^{**} = F_bC_DC_MC_tC_LC_FC_{fu} = F_b^*C_L = \left(1173 \, \frac{\text{lbf}}{\text{in}^2}\right)(1.0) = 1173 \text{ lbf/in}^2$$

$$\frac{f_b - f_t}{F_b^{**}} = \frac{761 \, \frac{\text{lbf}}{\text{in}^2} - 627.5 \, \frac{\text{lbf}}{\text{in}^2}}{1173.0 \, \frac{\text{lbf}}{\text{in}^2}} = 0.114 < 1.0 \quad \text{[OK]}$$

Therefore, a 2×8 no. 2 hem-fir member is no good.

Example 6.4
Combined Tension and Bending: Glulam

A 30 ft long glulam tension member, with an axial combination of 3DF (see *NDS Supplement* Table 5B) $5^1/_8 \times 10^1/_2$, is subjected to a combination of 25,000 lbf tension and a bending moment of 80,000 in-lbf about the x-axis (caused by a uniformly distributed load). Assume that MC is more than 16%, the temperature is normal ($C_t = 1.0$), and the load duration is normal ($C_D = 1.0$). Lateral supports are provided at the member ends only.

Check the adequacy of the member.

REFERENCE

Solution (ASD Method)

NDS Supp Tbl 1c

For $5^1/8$ in × $10^1/2$ in western species with seven laminations ($10^1/2$ in/ $1^1/2$ in),

A	area	53.81 in^2
S_x	section modulus	94.17 in^3

NDS Supp Tbl 5B, Adj Fac

For the 3DF axial combination,

	C_M
$E = 1.9 \times 10^6$ lbf/in^2	0.833
$E_{\text{min}} = 0.98 \times 10^6$ lbf/in^2	0.833
$F_t = 1450$ lbf/in^2	0.80
$F_{bx} = 2000$ lbf/in^2	0.80

NDS Sec 3.9.1

1. Axial

The actual tension stress parallel to grain is

$$f_t = \frac{P}{A_n} = \frac{25{,}000 \text{ lbf}}{53.81 \text{ in}^2} = 464.6 \text{ lbf/in}^2 \quad \text{[tensile or compressive]}$$

NDS Table 5.3.1

The adjusted tension design value parallel to grain is

$$F'_t = F_t C_D C_M C_t = \left(1450 \ \frac{\text{lbf}}{\text{in}^2}\right)(1.0)(0.80)(1.0) = 1160 \text{ lbf/in}^2$$

2. Bending

The actual bending design value is

$$f_{bx} = \frac{M}{S_x} = \frac{80{,}000 \text{ in-lbf}}{94.17 \text{ in}^3} = 849.5 \text{ lbf/in}^2$$

NDS Eq 3.9-1

3. Combined axial tension and bending

NDS Eq 5.3-1

For all species other than pine, x is 10. The volume factor is given by

$$C_V = \left(\frac{21}{L}\right)^{1/x}\left(\frac{12}{d}\right)^{1/x}\left(\frac{5.125}{b}\right)^{1/x} \leq 1.0$$

$$= \left(\frac{21}{30 \text{ ft}}\right)^{1/10}\left(\frac{12}{10.5 \text{ in}}\right)^{1/10}\left(\frac{5.125}{5.125 \text{ in}}\right)^{1/10} = 0.978 < 1.0$$

NDS Tbl 5.3.1; NDS Sec 3.9.1

Calculate the reference bending design value multiplied by all applicable adjustment factors except C_L.

$$\begin{aligned} F_b^* &= F_{bx} C_D C_M C_t C_V \\ &= \left(2000\ \frac{\text{lbf}}{\text{in}^2}\right)(1.0)(0.8)(1.0)(0.978) \\ &= 1564.8\ \text{lbf/in}^2 \end{aligned}$$

$$\begin{aligned} \frac{f_t}{F_t'} + \frac{f_{bx}}{F_b^*} &= \frac{464.6\ \frac{\text{lbf}}{\text{in}^2}}{1160\ \frac{\text{lbf}}{\text{in}^2}} + \frac{849.5\ \frac{\text{lbf}}{\text{in}^2}}{1564.8\ \frac{\text{lbf}}{\text{in}^2}} \\ &= 0.94 < 1.0 \quad \text{[OK]} \end{aligned}$$

$$f_b = f_{bx} = 849.5\ \text{lbf/in}^2 > f_t = 464.6\ \text{lbf/in}^2$$

Therefore, check NDS Eq. 3.9-2.

$$\frac{f_b - f_t}{F_b^{**}} \le 1.0$$

NDS Tbl 3.3.3

Find the beam stability factor, C_L.

The unbraced length is

$$\ell_u = (30\ \text{ft})\left(12\ \frac{\text{in}}{\text{ft}}\right) = 360\ \text{in}$$

$$\frac{\ell_u}{d} = \frac{360\ \text{in}}{10.5\ \text{in}} = 34.3 > 7.0$$

The effective length for uniformly distributed load is

$$\begin{aligned} \ell_e &= 1.63\ell_u + 3d \\ &= (1.63)(360\ \text{in}) + (3)(10.5\ \text{in}) \\ &= 618.3\ \text{in} \end{aligned}$$

NDS Eq 3.3-5

The slenderness ratio is

$$R_B = \sqrt{\frac{\ell_e d}{b^2}} = \sqrt{\frac{(618.3\ \text{in})(10.5\ \text{in})}{(5.125\ \text{in})^2}}$$

NDS Sec 3.3.3.7

$$= 15.72 < 50 \quad \text{[OK]}$$

NDS Tbl 5.3.1

The adjusted modulus of elasticity for beam stability is

$$\begin{aligned} E'_{\text{min}} &= E_{\text{min}} C_M C_t = \left(0.98 \times 10^6\ \frac{\text{lbf}}{\text{in}^2}\right)(0.833)(1.0) \\ &= 816{,}340\ \text{lbf/in}^2 \end{aligned}$$

NDS Sec 3.3.3.8

The critical buckling design value is

$$\begin{aligned} F_{bE} &= \frac{1.20 E'_{\text{min}}}{R_B^2} = \frac{(1.20)\left(816{,}340\ \frac{\text{lbf}}{\text{in}^2}\right)}{(15.72)^2} \\ &= 3964.1\ \text{lbf/in}^2 \end{aligned}$$

The reference bending design value without C_L or C_V is

$$\begin{aligned} F_{bx}^* &= F_{bx}C_DC_MC_t \\ &= \left(2000\ \frac{\text{lbf}}{\text{in}^2}\right)(1.0)(0.8)(1.0) \\ &= 1600\ \text{lbf/in}^2 \end{aligned}$$

$$\frac{F_{bE}}{F_{bx}^*} = \frac{3964.1\ \frac{\text{lbf}}{\text{in}^2}}{1600\ \frac{\text{lbf}}{\text{in}^2}} = 2.48$$

NDS Sec 3.3.3.8

The beam stability factor is

$$C_L = \frac{1+\frac{F_{bE}}{F_{bx}^*}}{1.9} - \sqrt{\left(\frac{1+\frac{F_{bE}}{F_{bx}^*}}{1.9}\right)^2 - \frac{\frac{F_{bE}}{F_{bx}^*}}{0.95}}$$

$$\begin{aligned} &= \frac{1+2.48}{1.9} - \sqrt{\left(\frac{1+2.48}{1.9}\right)^2 - \frac{2.48}{0.95}} \\ &= 0.97 \end{aligned}$$

NDS Sec 3.9.1

Calculate the reference bending design value multiplied by all applicable adjustment factors, except C_v.

$$\begin{aligned} F_{bx}^{**} &= F_{bx}^*C_L = \left(1600\ \frac{\text{lbf}}{\text{in}^2}\right)(0.97) \\ &= 1552.0\ \text{lbf/in}^2 \end{aligned}$$

NDS Eq 3.9-2

If the bending compressive stress, f_b, exceeds the axial tensile stress, f_t,

$$\begin{aligned} \frac{f_{bx}-f_t}{F_{bx}^{**}} &= \frac{849.5\ \frac{\text{lbf}}{\text{in}^2} - 464.6\ \frac{\text{lbf}}{\text{in}^2}}{1552.0\ \frac{\text{lbf}}{\text{in}^2}} \\ &= 0.25 < 1.0 \quad \text{[OK]} \end{aligned}$$

Therefore, a 30 ft long glulam tension member, 3DF, 5¹/₈ in × 10¹/₂ in, is adequate.

---

3. Compression Members (Columns)

[NDS Secs 3.6, 3.7]

A column is a relatively heavy, vertical compression member. Other types of compression members include *struts, studs,* and *truss compression members.* Struts are relatively smaller compression members, not necessarily in a vertical position (e.g., trench shores). Studs are light compression members in wood framing (e.g., wall studs). Truss compression members are those members of a truss that resist compressive forces.

Although only solid columns are treated in this book, columns are classified as follows.

Simple solid wood columns are made of a single piece of wood or properly glued pieces of wood to form a single member. (See NDS Secs. 3.6 and 3.7.)

Spaced columns are formed from two or more individual members that are spaced apart but act as a composite member. (See NDS Secs. 3.6.2.2 and 15.2.)

Built-up columns are individual laminations that are joined mechanically with no spaces between laminations. (See NDS Secs. 3.6.2.3 and 15.3.)

A *long column* is a column that will buckle before it reaches the crushing capacity of the wood (maximum compressive stress parallel to grain). A *short column*, however, is a column that will not buckle, and its strength is limited to the crushing capacity of the wood.

The check on the load capacity of a column is given as follows.

f_c	actual compression stress parallel to grain	lbf/in^2
F'_c	adjusted compression design value parallel to grain (See NDS Table 4.3.1 for sawn lumber and Table 5.3.1 for glued laminated timber.)	lbf/in^2
F_c	reference compression design value parallel to grain (See *NDS Supplement* Tables 4 and 5.)	lbf/in^2
C_P	column stability factor (For columns that are fully braced (laterally supported) for the entire column length, C_P is 1.0. See NDS Sec. 3.7.1.)	–

For sawn lumber,

$$f_c = \frac{P}{A} \le F'_c = F_C C_D C_M C_t C_F C_i C_P \quad \text{[NDS Table 4.3.1]}$$

For glulam,

$$f_c = \frac{P}{A} = F_C C_D C_M C_t C_P \quad \text{[NDS Table 5.3.1]}$$

Figure 6.1 Solid Column with Different Unbraced Lengths for Both Axes

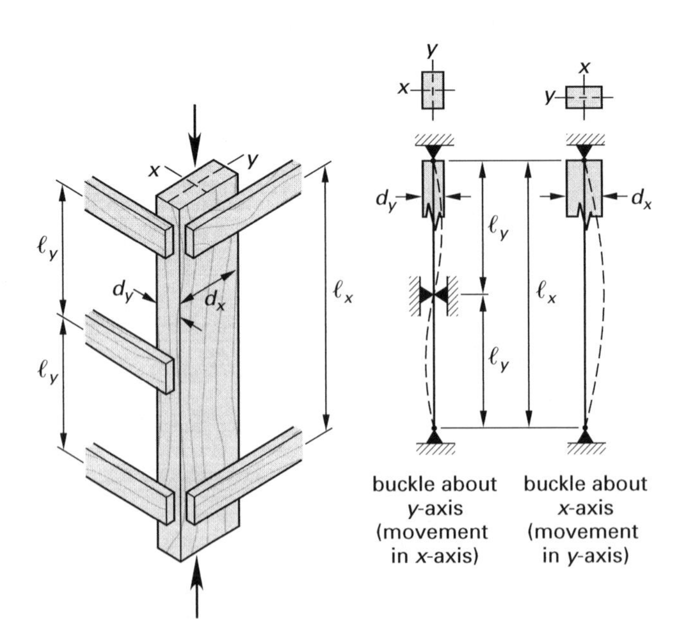

F_{cE}	Euler critical buckling stress for columns (See NDS Sec. 3.7.1.5.)	lbf/in^2
ℓ	unbraced column length	in
ℓ_e	effective column length	in
K_e	buckling length coefficient (See NDS App. G.)	–
F_c^*	allowable crushing strength (limiting compressive strength in column at zero slenderness ratio); tabulated compressive stress parallel to grain multiplied by all adjustment factors except C_P	lbf/in^2

E'_{min}	allowable modulus of elasticity for beam stability and column stability calculations	lbf/in^2
c	buckling and crushing interaction factor for columns (For round poles and piles, c is 0.85. For glulam columns, c is 0.9. For sawn lumber columns, c is 0.8.)	–
C_T	buckling stiffness factor for 2 × 4 or smaller truss compression chords subjected to combined flexure and compression with plywood nailed to narrow face of member (See NDS Sec. 4.4.2. For all other members, C_T is 1.0.)	
C_F	size factor for compression (Applies to visually graded dimension lumber, not glulam. See *NDS Supplement* Tables 4A, 4B, 4D, 4E, and 4F, NDS Sec. 4.3.6, NDS Table 4.3.1 footnote, and NDS Table 5.3.1.)	–
C_P	column stability factor (See NDS Eq. 3.7-1 and NDS App. H.)	–

The Euler critical buckling stress for columns is

$$F_{cE} = \frac{0.822E'_{\text{min}}}{\left(\frac{\ell_e}{d}\right)^2} \quad \text{[NDS Sec. 3.7.1.5]}$$

The effective column length is

$$\ell_e = K_e\ell$$

The allowable crushing strength, or the reference compressive stress parallel to the grain multiplied by all adjustment factors except C_P is

$$\begin{aligned} F_c^* &= F_cC_DC_MC_tC_FC_i \quad \text{[for sawn lumber; NDS Table 5.3.1]} \\ &= F_cC_DC_MC_t \quad \text{[for glued laminated timber]} \end{aligned}$$

The modulus of elasticity associated with the axis of column buckling is

$$\begin{aligned} E'_{\text{min}} &= E_{\text{min}}C_MC_tC_iC_T \quad \text{[for sawn lumber; NDS Table 4.3.1]} \\ &= E_{\text{min}}C_MC_t \quad \text{[for glued laminated timber; NDS Table 5.3.1]} \end{aligned}$$

For sawn lumber, $E_{\text{min}} = E_{x,\text{min}} = E_{y,\text{min}}$. For glulam, $E_{x,\text{min}}$ and $E_{y,\text{min}}$ are different. $E_{x,\text{min}}$ and $E_{y,\text{min}}$ are used for stability analysis. E_{axial} is for axial deformation calculations. E_x is for member deflection calculations.

The column stability factor is

$$C_P = \frac{1 + \frac{F_{cE}}{F_c^*}}{2c} - \sqrt{\left(\frac{1 + \frac{F_{cE}}{F_c^*}}{2c}\right)^2 - \frac{\frac{F_{cE}}{F_c^*}}{c}} \quad \text{[NDS Eq. 3.7-1]}$$

Example 6.5
Axially Loaded Column: Sawn Lumber

No. 3 spruce-pine-fir 2 × 4 bearing wall stud supports are shown in the illustration that follows. The total load, P, is caused by a combination dead load, w_D, and snow load, w_S, and is 1600 lbf. Assume dry service ($C_M = 1.0$) and normal temperature conditions ($C_t = 1.0$).

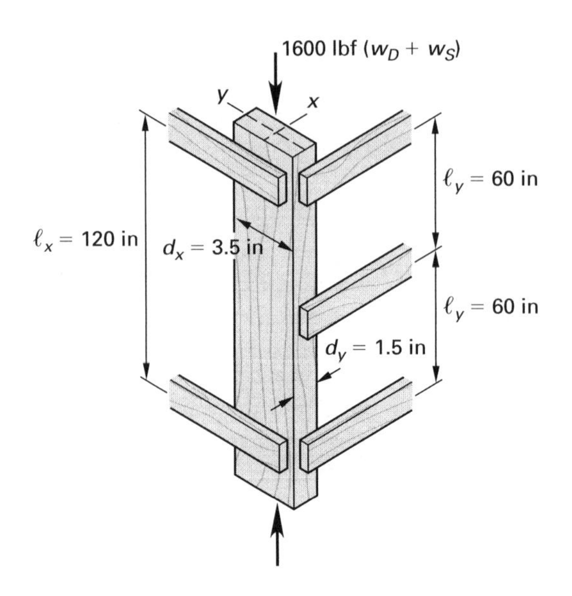

F_c	reference compression design value parallel to grain (see *NDS Supplement* Table 4A and adjustment factors)	650 lbf/in^2
C_F	size factor	1.15
E_{min}	modulus of elasticity associated with beam and column stability calculations	440,000 lbf/in^2

Check the stud capacity.

REFERENCE

Solution (ASD Method)

Calculate the actual compression stress parallel to grain.

$$f_c = \frac{P}{A} = \frac{1600 \text{ lbf}}{(1.5 \text{ in})(3.5 \text{ in})} = 304.8 \text{ lbf/in}^2$$

NDS Tbl 4.3.1

Use the following to calculate the adjusted compression design value parallel to the grain.

$$F'_c = F_c C_D C_M C_t C_F C_P = F^*_c C_P$$

NDS Tbl 2.3.2

For snow, the load duration factor, C_D, is 1.15.

NDS Supp Tbl 4A Adj Fac

The reference compression design value parallel to the grain, F_c, is 650 lbf/in^2. The size factor, C_F, is 1.15. The allowable crushing strength is

$$\begin{aligned} F^*_c &= F_c C_D C_M C_t C_F \\ &= \left(650 \ \frac{\text{lbf}}{\text{in}^2}\right)(1.15)(1.0)(1.0)(1.15) \\ &= 859.7 \text{ lbf/in}^2 \end{aligned}$$

The adjusted modulus of elasticity is

$$\begin{aligned} E'_{\text{min}} &= E_{\text{min}} C_M C_t \\ &= \left(440{,}000 \ \frac{\text{lbf}}{\text{in}^2}\right)(1.0)(1.0) \\ &= 440{,}000 \text{ lbf/in}^2 \end{aligned}$$

NDS Sec 3.7

Find the column stability factor, C_P.

$$\left(\frac{\ell_e}{d}\right)_{\max} = \begin{cases} \left(\frac{\ell_e}{d}\right)_x = \left(\frac{K_e\ell}{d}\right)_x = \frac{(1.0)(120 \text{ in})}{3.5 \text{ in}} = 34.3 \\ \left(\frac{\ell_e}{d}\right)_y = \left(\frac{K_e\ell}{d}\right)_y = \frac{(1.0)(60 \text{ in})}{1.5 \text{ in}} = 40 \quad \text{[controls]} \end{cases}$$

NDS Sec 3.7.1.5

The Euler critical buckling stress for columns is

$$F_{cE} = \frac{0.822E'_{\min}}{\left(\frac{l_e}{d}\right)^2} = \frac{(0.822)\left(440{,}000 \ \frac{\text{lbf}}{\text{in}^2}\right)}{(40)^2}$$
$$= 226 \text{ lbf/in}^2$$

For sawn lumber, the buckling and crushing iteration factor for columns, c, is 0.8.

NDS Eq 3.7-1

The stability factor is

$$C_P = \frac{1+\frac{F_{cE}}{F_c^*}}{2c} - \sqrt{\left(\frac{1+\frac{F_{cE}}{F_c^*}}{2c}\right)^2 - \frac{\frac{F_{cE}}{F_c^*}}{c}}$$

$$= \frac{1+\frac{226.0 \ \frac{\text{lbf}}{\text{in}^2}}{859.7 \ \frac{\text{lbf}}{\text{in}^2}}}{(2)(0.8)} - \sqrt{\left(\frac{1+\frac{226.0 \ \frac{\text{lbf}}{\text{in}^2}}{859.7 \ \frac{\text{lbf}}{\text{in}^2}}}{(2)(0.8)}\right)^2 - \frac{\frac{226.0 \ \frac{\text{lbf}}{\text{in}^2}}{859.7 \ \frac{\text{lbf}}{\text{in}^2}}}{0.8}}$$

$$= 0.246$$

The adjusted compression design value parrallel to the grain is

$$F'_c = F_c^* C_P = \left(859.7 \ \frac{\text{lbf}}{\text{in}^2}\right)(0.246)$$
$$= 211.5 \text{ lbf/in}^2 < f_c = 304.8 \text{ lbf/in}^2 \quad \text{[no good]}$$

REFERENCE

Solution (LRFD Method)

The total load is 1600 lbf, which is 100 lbf for the dead load and 1500 lbf for the snow load. The factored compression force is

$$P_u = 1.2w_D + 1.6w_S = (1.2)(100 \text{ lbf}) + (1.6)(1500 \text{ lbf})$$
$$= 2520 \text{ lbf}$$

NDS App Tbls N1, N2, N3

For F_c,

$$K_F = \frac{2.16}{\phi_c} = \frac{2.16}{0.9}$$
$$= 2.4$$

For the load combination $1.2w_D + 1.6w_S$, the time effect factor, λ, is 0.8.

For E_{min},

$$K_F = \frac{1.5}{\phi_s} = \frac{1.5}{0.85} = 1.765$$

NDS Supp Tbl 4A
Adj Fac

For no. 3 SPF 2 × 4, the reference compression design value, F_c, is 650 lbf/in^2. The modulus of elasticity, E_{min}, is 440,000 lbf/in^2.

NDS Tbl 4.3.1

Use the following to calculate the LRFD adjusted compression design value parallel to the grain.

$$F'_{c,\text{LRFD}} = F_c(K_F\phi_c\lambda)C_M C_t C_F C_P = F^*_{c,\text{LRFD}}C_P$$

NDS Tbl 4.3.1

The LRFD allowable compression stress multiplied by all applicable adjustment factors except C_P is

$$\begin{aligned} F^*_{c,\text{LRFD}} &= F_c(K_F\phi_c\lambda)C_M C_t C_F \\ &= \left(650\ \frac{\text{lbf}}{\text{in}^2}\right)\big((2.4)(0.9)(0.8)\big)(1.0)(1.0)(1.15)\left(\frac{1\text{ kip}}{1000\text{ lbf}}\right) \\ &= 1.291\text{ kips/in}^2 \end{aligned}$$

NDS Tbl 4.3.1

The LRFD modulus of elasticity for column stability is

$$\begin{aligned} E'_{\text{min,LRFD}} &= E_{\text{min}}(K_F\phi_s)C_M C_t \\ &= \left(440{,}000\ \frac{\text{lbf}}{\text{in}^2}\right)\big((1.765)(0.85)\big)(1.0)(1.0)\left(\frac{1\text{ kip}}{1000\text{ lbf}}\right) \\ &= 660.11\text{ kips/in}^2 \end{aligned}$$

From the ASD method,

$$\left(\frac{\ell_e}{d}\right)_{\text{max}} = 40$$

The LRFD Euler critical buckling stress for columns is

$$\begin{aligned} F_{cE,\text{LRFD}} &= \frac{0.822E'_{\text{min,LRFD}}}{\left(\frac{\ell_e}{d}\right)^2} = \frac{(0.822)\left(660.11\ \frac{\text{kips}}{\text{in}^2}\right)}{(40)^2} \\ &= 0.339\text{ kips/in}^2 \end{aligned}$$

The stability factor is

$$C_P = \frac{1 + \dfrac{F_{cE,\text{LRFD}}}{F^*_{c,\text{LRFD}}}}{2c} - \sqrt{\left(\frac{1 + \dfrac{F_{cE,\text{LRFD}}}{F^*_{c,\text{LRFD}}}}{2c}\right)^2 - \frac{\dfrac{F_{cE,\text{LRFD}}}{F^*_{c,\text{LRFD}}}}{c}}$$

$$= \frac{1 + \dfrac{0.339\ \dfrac{\text{kips}}{\text{in}^2}}{1.291\ \dfrac{\text{kips}}{\text{in}^2}}}{(2)(0.8)} - \sqrt{\left(\frac{1 + \dfrac{0.339\ \dfrac{\text{kips}}{\text{in}^2}}{1.291\ \dfrac{\text{kips}}{\text{in}^2}}}{(2)(0.8)}\right)^2 - \frac{\dfrac{0.339\ \dfrac{\text{kips}}{\text{in}^2}}{1.291\ \dfrac{\text{kips}}{\text{in}^2}}}{0.8}}$$

$$= 0.247$$

The LRFD adjusted compression design value parallel to the grain is

$$F'_{c,\text{LRFD}} = F^*_{c,\text{LRFD}} C_P = \left(1.291\ \frac{\text{kips}}{\text{in}^2}\right)(0.247)$$
$$= 0.319\ \text{kips/in}^2$$

The LRFD compressive force is

$$P_{\text{LRFD}} = F'_{c,\text{LRFD}} A = \left(0.319\ \frac{\text{kips}}{\text{in}^2}\right)(1.5\ \text{in})(3.5\ \text{in})$$
$$= 1.675\ \text{kips} < P_u = 2.52\ \text{kips} \quad \text{[no good]}$$

Alternatively, the LRFD actual compression stress parallel to the grain is

$$f_{c,\text{LRFD}} = \frac{P_u}{A} = \frac{2.520\ \text{kips}}{(1.5\ \text{in})(3.5\ \text{in})}$$
$$= 0.48\ \text{kips/in}^2 > F'_{c,\text{LRFD}} = 0.319\ \text{kips/in}^2 \quad \text{[no good]}$$

Example 6.6
Axially Loaded Column: Glulam

Assume a bending combination 24E-1.7E softwood 5 × 11 column. Assume that the moisture content, MC, is greater than 16% and that the temperature condition is normal.

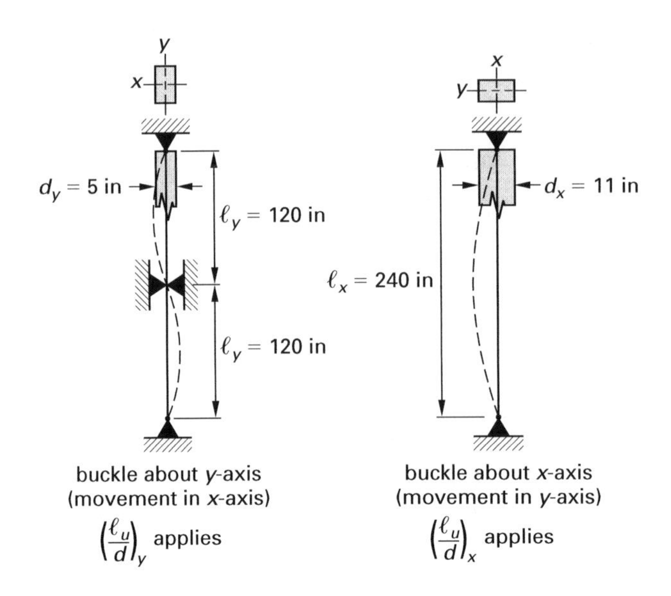

Find the allowable compressive stress, F_c', for combination dead load, w_D, and snow load, w_S.

Note that C_F is not applicable for glulam.

Using *NDS Supplement* Table 1D for a 5 × 11 beam,

$$A = 55 \text{ in}^2$$
$$I_x = 554.6 \text{ in}^4$$
$$I_y = 114.6 \text{ in}^4$$

REFERENCE	*Solution (ASD Method)*		
NDS Supp Tbl 5A	E_x	reference modulus of elasticity for x-axis	1.7×10^6 lbf/in^2
Adj Fac	$E_{x,\text{min}}$	reference modulus of elasticity for column buckling about the x-axis	0.88×10^6 lbf/in^2
	E_y	reference modulus of elasticity for y-axis	1.3×10^6 lbf/in^2
	$E_{y,\text{min}}$	reference modulus of elasticity for column buckling about the y-axis	0.67×10^6 lbf/in^2
	F_c	reference compression design value parallel to the grain	1000 lbf/in^2
	E_{axial}	reference modulus of elasticity axial deformation calculations	1.4×10^6 lbf/in^2

Since MC is assumed to be greater than 16%, the wet service factor, C_M, for the reference bending design value, F_b, is 0.8; for the reference compression design value, F_c is 0.73; and for the moduli of elasticity, E and E_{min}, is 0.833.

NDS Tbl 2.3.2

For the snow load, the load duration factor, C_D, is 1.15.

NDS App G

Assume that the buckling length coefficient for compression members, K_e, is 1.0.

Find the column stability factor, C_P, for x-axis buckling.

$$\left(\frac{\ell_e}{d}\right)_x = \frac{K_e \ell_x}{d_x} = \frac{(1.0)(240 \text{ in})}{11 \text{ in}} = 21.8$$

The adjusted modulus of elasticity for the x-axis is

$$\begin{aligned} E'_{x,\min} &= E_{x,\min} C_M C_t \\ &= \left(0.88 \times 10^6 \ \frac{\text{lbf}}{\text{in}^2}\right)(0.833)(1.0) \\ &= 0.733 \times 10^6 \text{ lbf/in}^2 \end{aligned}$$

The Euler critical buckling stress for columns is

$$\begin{aligned} F_{cE} &= \frac{0.822 E'_{x,\min}}{\left(\frac{\ell_e}{d}\right)_x^2} = \frac{(0.822)\left(0.733 \times 10^6 \ \frac{\text{lbf}}{\text{in}^2}\right)}{(21.8)^2} \\ &= 1267.8 \text{ lbf/in}^2 \end{aligned}$$

The reference design value multiplied by all applicable adjustment factors except C_P is

$$\begin{aligned} F_c^* &= F_c C_D C_M C_t = \left(1000 \ \frac{\text{lbf}}{\text{in}^2}\right)(1.15)(0.73) \\ &= 839.5 \text{ lbf/in}^2 \end{aligned}$$

For glulam, c is 0.9.

NDS Sec 3.7.1

The column stability factor is

$$\begin{aligned} C_P &= \frac{1 + \frac{F_{cE}}{F_c^*}}{2c} - \sqrt{\left(\frac{1 + \frac{F_{cE}}{F_c^*}}{2c}\right)^2 - \frac{\frac{F_{cE}}{F_c^*}}{c}} \\ &= \frac{1 + \frac{1267.8 \ \frac{\text{lbf}}{\text{in}^2}}{839.5 \ \frac{\text{lbf}}{\text{in}^2}}}{(2)(0.9)} - \sqrt{\left(\frac{1 + \frac{1267.8 \ \frac{\text{lbf}}{\text{in}^2}}{839.5 \ \frac{\text{lbf}}{\text{in}^2}}}{(2)(0.9)}\right)^2 - \frac{\frac{1267.8 \ \frac{\text{lbf}}{\text{in}^2}}{839.5 \ \frac{\text{lbf}}{\text{in}^2}}}{0.9}} \\ &= 0.88 \end{aligned}$$

NDS Tbl 5.3.1

The adjusted compression design value parallel to the grain is

$$\begin{aligned} F'_c &= F_c^* C_P = \left(839.5 \ \frac{\text{lbf}}{\text{in}^2}\right)(0.88) \\ &= 738.76 \text{ lbf/in}^2 \text{ for } x\text{-axis buckling} \end{aligned}$$

Find the column stability factor, C_P, for y-axis buckling.

$$\left(\frac{\ell_e}{d}\right)_y = \frac{K_e \ell_y}{d_y} = \frac{(1.0)(120 \text{ in})}{5 \text{ in}} = 24.0$$

The adjusted modulus of elasticity for the y-axis is

$$E'_{y,\text{min}} = E_{y,\text{min}} C_M C_t = \left(0.67 \times 10^6 \ \frac{\text{lbf}}{\text{in}^2}\right)(0.833)(1.0) = 0.558 \times 10^6 \text{ lbf/in}^2$$

The Euler critical buckling stress for the column is

$$F_{cE} = \frac{0.822 E'_{y,\text{min}}}{\left(\frac{\ell_e}{d}\right)_y^2} = \frac{(0.822)\left(0.558 \times 10^6 \ \frac{\text{lbf}}{\text{in}^2}\right)}{(24.0)^2} = 796.3 \text{ lbf/in}^2$$

NDS Sec 3.7.1

$$C_P = \frac{1 + \frac{F_{cE}}{F_c^*}}{2c} - \sqrt{\left(\frac{1 + \frac{F_{cE}}{F_c^*}}{2c}\right)^2 - \frac{\frac{F_{cE}}{F_c^*}}{c}}$$

$$= \frac{1 + \frac{796.3 \ \frac{\text{lbf}}{\text{in}^2}}{839.5 \ \frac{\text{lbf}}{\text{in}^2}}}{(2)(0.9)} - \sqrt{\left(\frac{1 + \frac{796.3 \ \frac{\text{lbf}}{\text{in}^2}}{839.5 \ \frac{\text{lbf}}{\text{in}^2}}}{(2)(0.9)}\right)^2 - \frac{\frac{796.3 \ \frac{\text{lbf}}{\text{in}^2}}{839.5 \ \frac{\text{lbf}}{\text{in}^2}}}{0.9}}$$

$$= 0.739$$

The adjusted compression design value parallel to the grain is

$$F'_c = F_c^* C_P = \left(839.5 \ \frac{\text{lbf}}{\text{in}^2}\right)(0.739) = 620.39 \text{ lbf/in}^2 \text{ for } y\text{-axis buckling}$$

Therefore, the y-axis buckling is critical, and F'_c for y-axis buckling, 620.39 lbf/in^2, controls.

REFERENCE

Solution (LRFD Method)

NDS App Tbls N1, N2, N3

The load combination is $1.2w_D + 1.6w_S$.

The time effect factor, λ, is 0.8 for this load combination.

For F_c,

$$K_F = \frac{2.16}{\phi_c} = \frac{2.16}{0.9} = 2.4$$

For $E_{\min}$,

$$K_F = \frac{1.5}{\phi_s} = \frac{1.5}{0.85} = 1.765$$

NDS Supp Tbl 5A Adj Fac

The adjusted modulus of elasticity for column buckling about the x-axis is

NDS Tbl 5.3.1

$$\begin{aligned} E'_{x,\text{min,LRFD}} &= E_{x,\min}(K_F\phi_s)C_MC_t \\ &= \left(0.88\times10^6\ \frac{\text{lbf}}{\text{in}^2}\right)\big((1.765)(0.85)\big)(0.833)(1.0) \\ &= 1099.7\times10^3\ \text{lbf/in}^2 \end{aligned}$$

The adjusted modulus of elasticity for column buckling about the y-axis is

$$\begin{aligned} E'_{y,\text{min,LRFD}} &= E_{y,\min}(K_F\phi_s)C_MC_t \\ &= \left(0.67\times10^6\ \frac{\text{lbf}}{\text{in}^2}\right)\big((1.765)(0.85)\big)(0.833)(1.0) \\ &= 0.837\times10^6\ \text{lbf/in}^2 \end{aligned}$$

Find the column stability factor, C_P, for x-axis buckling.

$$\left(\frac{\ell_e}{d}\right)_x = \frac{K_e\ell_x}{d_x} = \frac{(1.0)(240\ \text{in})}{11\ \text{in}} = 21.8$$

For glulam, c is 0.9. The Euler critical buckling stress for columns is

$$F_{cE,\text{LRFD}} = \frac{0.822E'_{x,\text{min,LRFD}}}{\left(\frac{\ell_e}{d}\right)_x^2} = \frac{(0.822)\left(1099.7\times10^3\ \frac{\text{lbf}}{\text{in}^2}\right)}{(21.8)^2} = 1902.1\ \text{lbf/in}^2$$

NDS Tbl 5.3.1

$$\begin{aligned} F^*_{c,\text{LRFD}} &= F_c(K_F\phi_c\lambda)C_MC_t \\ &= \left(1000\ \frac{\text{lbf}}{\text{in}^2}\right)\big((2.4)(0.9)(0.8)\big)(0.73)(1.0) \\ &= 1261\ \text{lbf/in}^2 \end{aligned}$$

The column stability factor for x-axis buckling is

$$C_P = \frac{1+\frac{F_{cE,\text{LRFD}}}{F^*_{c,\text{LRFD}}}}{2c} - \sqrt{\left(\frac{1+\frac{F_{cE,\text{LRFD}}}{F^*_{c,\text{LRFD}}}}{2c}\right)^2 - \frac{\frac{F_{cE,\text{LRFD}}}{F^*_{c,\text{LRFD}}}}{c}}$$

$$= \frac{1+\frac{1.9021\ \frac{\text{kips}}{\text{in}^2}}{1.261\ \frac{\text{kips}}{\text{in}^2}}}{(2)(0.9)} - \sqrt{\left(\frac{1+\frac{1.9021\ \frac{\text{kips}}{\text{in}^2}}{1.261\ \frac{\text{kips}}{\text{in}^2}}}{(2)(0.9)}\right)^2 - \frac{\frac{1.9021\ \frac{\text{kips}}{\text{in}^2}}{1.261\ \frac{\text{kips}}{\text{in}^2}}}{0.9}}$$

$$= 0.878$$

Find the column stability factor, C_P, for y-axis buckling.

$$\left(\frac{\ell_e}{d}\right)_y = \frac{K_e \ell_y}{d_y} = \frac{(1.0)(120 \text{ in})}{5 \text{ in}} = 24$$

$$F_{cE,\text{LRFD}} = \frac{0.822 E'_{y,\min,\text{LRFD}}}{\left(\frac{\ell_e}{d}\right)_y^2} = \frac{(0.822)\left(0.837 \times 10^6 \ \frac{\text{lbf}}{\text{in}^2}\right)}{(24)^2} = 1194.5 \text{ lbf/in}^2$$

$$C_P = \frac{1 + \frac{F_{cE,\text{LRFD}}}{F^*_{c,\text{LRFD}}}}{2c} - \sqrt{\left(\frac{1 + \frac{F_{cE,\text{LRFD}}}{F^*_{c,\text{LRFD}}}}{2c}\right)^2 - \frac{\frac{F_{cE,\text{LRFD}}}{F^*_{c,\text{LRFD}}}}{c}}$$

$$= \frac{1 + \frac{1194.5 \ \frac{\text{kips}}{\text{in}^2}}{1261 \ \frac{\text{kips}}{\text{in}^2}}}{(2)(0.9)} - \sqrt{\left(\frac{1 + \frac{1194.5 \ \frac{\text{kips}}{\text{in}^2}}{1261 \ \frac{\text{kips}}{\text{in}^2}}}{(2)(0.9)}\right)^2 - \frac{\frac{1194.5 \ \frac{\text{kips}}{\text{in}^2}}{1261 \ \frac{\text{kips}}{\text{in}^2}}}{0.9}}$$

$$= 0.739 \quad \text{[controls]}$$

The adjusted compression design value parallel to the grain is

$$F'_{c,\text{LRFD}} = F^*_{c,\text{LRFD}} C_P = \left(1261 \ \frac{\text{lbf}}{\text{in}^2}\right)(0.739) = 931.9 \text{ lbf/in}^2$$

4. Combined Compression and Bending

[NDS Sec 3.9.2; NDS Eq 3.9-3]

Figure 6.2 Combined Compression and Bending About x- and y-Axes

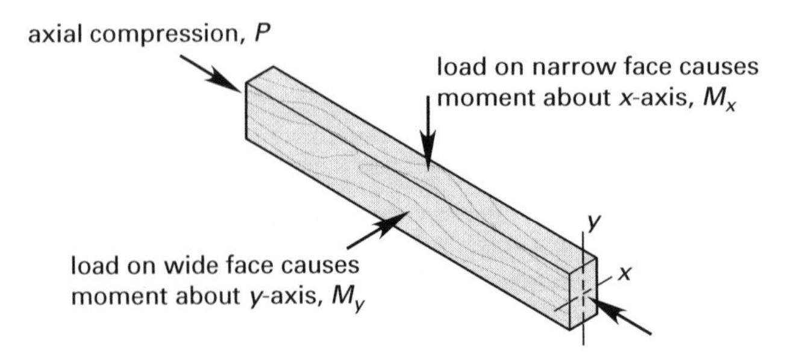

Some columns, in addition to carrying axial loads, support lateral loads and/or resist moments between the column ends. Such columns are called *beam-columns*. These columns are checked by the following.

NDS Eq 3.9-3

$$\left(\frac{f_c}{F'_c}\right)^2 + \frac{f_{bx}}{F'_{bx}\left(1 - \frac{f_c}{F_{cEx}}\right)} + \frac{f_{by}}{F'_{by}\left(1 - \frac{f_c}{F_{cEy}} - \left(\frac{f_{bx}}{F_{bE}}\right)^2\right)} \leq 1.0$$

The allowable compressive stress, F'_c, is based on the critical value of

$$\left(\frac{\ell_e}{d}\right)_x \text{ or } \left(\frac{\ell_e}{d}\right)_y$$

NDS Tbl 4.3.1

For sawn lumber,

$$F'_c = F_c C_D C_M C_t C_F C_i C_P$$

NDS Tbl 5.3.1

For glulam,

$$F'_c = F_c C_D C_M C_t C_P$$

The allowable flatwise bending design stress about the y-axis for all sizes of sawn lumber except beams and stringers category, F'_{by}, for sawn lumber is

NDS Tbl 4.3.1

$$F'_{by} = F_{by} C_D C_M C_t C_F C_{fu} C_i$$

For glulam beam,

NDS Tbl 5.3.1

$$F'_{by} = F_{by} C_D C_M C_t C_{fu}$$

NDS Tbls 4.3.1, 5.3.1

The allowable edgewise bending design stress about the x-axis, F'_{bx}, for sawn lumber is

$$F'_{bx} = F_{bx} C_D C_M C_t C_L C_F C_i C_r$$

For glulam beam,

$$F'_{bx} = F_{bx} C_D C_M C_t (C_L \text{ or } C_V)$$

For the value of C_r, see *NDS Supplement* Tables 4A, 4B, 4C, 4E, and 4F.

f_{bx}	edgewise bending stress (load applied to narrow face; i.e., strong axis bending)	lbf/in^2
f_{by}	flatwise bending stress (load applied to wide face; i.e., weak axis bending)	lbf/in^2

NDS Sec 3.9.2

The compressive stress parallel to grain, f_c, must be less than the critical buckling value for compression members, F_{cE}.

$$f_c < F_{cE_x} = \frac{0.822E'_{x,\min}}{\left(\frac{\ell_e}{d}\right)_x^2} \quad \left[\begin{array}{l}\text{for either uniaxial edgewise} \\ \text{bending or biaxial bending}\end{array}\right]$$

The edgewise bending stress, f_{bx}, must be less than the critical buckling value for bending member, F_{bE}.

NDS Secs 3.3.3.8, 3.9.2

$$f_{bx} < F_{bE} = \frac{1.20E'_{y,\min}}{(R_B)^2} \quad \text{[for biaxial bending]}$$

NDS Sec 3.7.1.4

The slenderness ratio for solid columns is $\ell_e/d \leq 50$.

The slenderness ratio for bending members must be less than or equal to 50.

NDS Secs 3.3.3.6, 3.3.3.7

$$R_B = \sqrt{\frac{\ell_e d}{b^2}} \leq 50$$

NDS Sec 3.3.3.8

The critical buckling design value for bending members is

$$F_{bE} = \frac{1.20E'_{\min}}{R_B^2}$$

For axial combination glulam, $E_{\min} = E_{x,\min} = E_{y,\min}$ and $E = E_{\text{axial}} = E_x$.

Example 6.7
Combined Compression and Bending: Sawn Lumber

[NDS Sec 3.9.2]

The roof rafter is shown to support a total roof load of 44.1 lbf/ft, which is the combined dead and snow load. Maximum forces and moments are as follows.

moment about x-axis, M_x	1940 ft-lbf
shear, V	391 lbf
axial compression, P	196 lbf

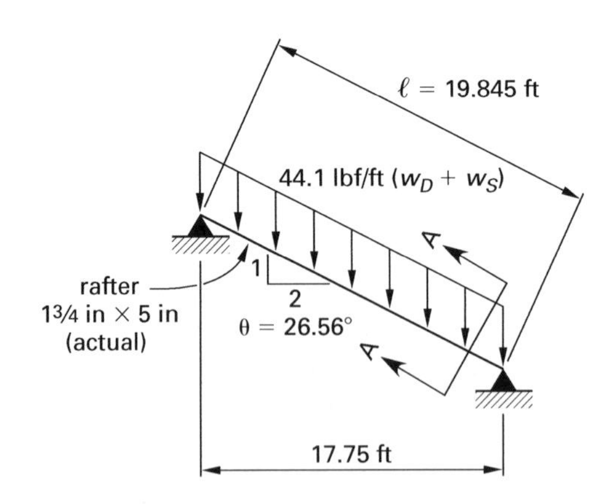

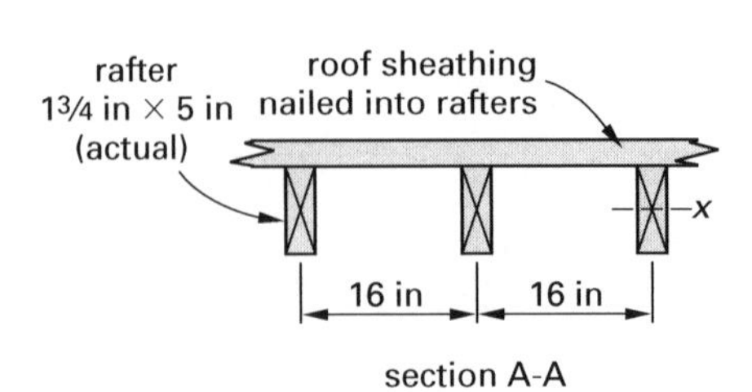

The rafter has dressed (actual) dimensions of 1¾ in × 5 in. Assume no. 2 SPF and a wet service condition. Assume roof sheathing provides continuous lateral support to rafter.

Determine if the rafter is adequately sized.

REFERENCE

Solution (ASD Method)

For a 1¾ in × 5 in section,

$$A = (1.75\ \text{in})(5\ \text{in}) = 8.75\ \text{in}^2$$

$$I_x = \frac{(1.75\ \text{in})(5\ \text{in})^3}{12} = 18.2\ \text{in}^4$$

$$S_x = \frac{(1.75\ \text{in})(5\ \text{in})^2}{6} = 7.29\ \text{in}^3$$

NDS Supp Tbl 4A Adj Fac

	C_F	C_r	C_M
$F_b = 875\ \text{lbf/in}^2$	1.4	1.15	0.85
$F_v = 135\ \text{lbf/in}^2$	–	–	0.97
$F_c = 1150\ \text{lbf/in}^2$	1.1	–	0.80
$E_{\text{min}} = 510{,}000\ \text{lbf/in}^2$	–	–	0.90

NDS Tbl 2.3.2

For snow load, C_D is 1.15.

1. Axial compression

The compression stress parallel to the grain is

$$f_c = \frac{P}{A} = \frac{196\ \text{lbf}}{(1.75\ \text{in})(5\ \text{in})} = 22.4\ \text{lbf/in}^2$$

NDS Tbl 4.3.1

$$F'_c = F_c C_D C_M C_t C_F C_i C_P = F_c^* C_P$$

The compression design value with all adjustments except C_P is

$$F_c^* = F_c C_D C_M C_t C_F C_i = \left(1150\ \frac{\text{lbf}}{\text{in}^2}\right)(1.15)(0.80)(1.0)(1.10)(1.0) = 1163.8\ \text{lbf/in}^2$$

NDS Sec 3.7.1

Find the column stability factor, C_P.

Because of continuous lateral support provided by roof sheathing, assume movement in the direction of the x-axis is prevented. Accordingly,

$$\left(\frac{\ell_e}{d}\right)_{\text{max}} = \begin{cases} \left(\frac{\ell_e}{d}\right)_y = 0 \\ \left(\frac{\ell_e}{d}\right)_x = \left(\frac{K_e \ell_u}{d}\right)_x = \dfrac{(1.0)(19.845\ \text{ft})\left(12\ \frac{\text{in}}{\text{ft}}\right)}{5\ \text{in}} \\ = 47.63 \quad \text{[controls]} \end{cases}$$

The adjusted modulus of elasticity for beam and column stability is

NDS Tbl 4.3.1

$$\begin{aligned} E'_{\text{min}} &= E_{\text{min}} C_M C_t C_i \\ &= \left(510{,}000\ \frac{\text{lbf}}{\text{in}^2}\right)(0.90)(1.0)(1.0) \\ &= 459{,}000\ \text{lbf/in}^2 \end{aligned}$$

NDS Sec 3.7.1.5

For sawn lumber, c is 0.8.

The Euler critical buckling stress for columns is

$$\begin{aligned} F_{cE} &= \frac{0.822 E'_{\text{min}}}{\left(\dfrac{\ell_e}{d}\right)^2_{\text{max}}} = \frac{(0.822)\left(459{,}000\ \dfrac{\text{lbf}}{\text{in}^2}\right)}{(47.63)^2} \\ &= 166.3\ \text{lbf/in}^2 \end{aligned}$$

NDS Eq 3.7-1

The column stability factor is

$$\begin{aligned} C_P &= \frac{1 + \dfrac{F_{cE}}{F_c^*}}{2c} - \sqrt{\left(\frac{1 + \dfrac{F_{cE}}{F_c^*}}{2c}\right)^2 - \frac{\dfrac{F_{cE}}{F_c^*}}{c}} \\ &= \frac{1 + \dfrac{166.3\ \dfrac{\text{lbf}}{\text{in}^2}}{1163.8\ \dfrac{\text{lbf}}{\text{in}^2}}}{(2)(0.8)} - \sqrt{\left(\frac{1 + \dfrac{166.3\ \dfrac{\text{lbf}}{\text{in}^2}}{1163.8\ \dfrac{\text{lbf}}{\text{in}^2}}}{(2)(0.8)}\right)^2 - \frac{\dfrac{166.3\ \dfrac{\text{lbf}}{\text{in}^2}}{1163.8\ \dfrac{\text{lbf}}{\text{in}^2}}}{0.8}} \\ &= 0.139 \end{aligned}$$

The adjusted compression design value is

$$\begin{aligned} F'_c &= F_c^* C_P = \left(1163.8\ \frac{\text{lbf}}{\text{in}^2}\right)(0.139) \\ &= 161.8\ \text{lbf/in}^2 > f_c = 22.4\ \text{lbf/in}^2 \quad \text{[OK]} \end{aligned}$$

2. Bending (there is no M_y)

The edgewise bending stress is

$$\begin{aligned} f_{bx} &= \frac{M_x}{S_x} = \frac{(1940\ \text{ft-lbf})\left(12\ \dfrac{\text{in}}{\text{ft}}\right)}{7.29\ \text{in}^3} \\ &= 3193.4\ \text{lbf/in}^2 \end{aligned}$$

NDS Tbl 4.3.1

The flatwise bending stress, f_{by}, is 0 lbf/in^2.

The rafter has full lateral support. Therefore, ℓ_u is 0 (so ℓ_e is 0) and C_L is 1.0.

The allowable bending stress about the x-axis is

$$F'_{bx} = F_b C_D C_M C_t C_F C_L C_r = \left(875\ \frac{\text{lbf}}{\text{in}^2}\right)(1.15)(0.85)(1.0)(1.4)(1.0)(1.15)$$

$$= 1377\ \text{lbf/in}^2 < f_{bx} = 3193.4\ \text{lbf/in} \quad \text{[no good]}$$

3. Combined compression and bending

The Euler critical buckling value for compression members is

NDS Sec 3.9.2

$$F_{cEx} = \frac{0.822E'_{\text{min}}}{\left(\frac{\ell_e}{d}\right)_x^2} = \frac{(0.822)\left(459{,}000\ \frac{\text{lbf}}{\text{in}^2}\right)}{(47.63)^2}$$

$$= 166.3\ \text{lbf/in}^2 > f_c = 22.4\ \frac{\text{lbf}}{\text{in}^2} \quad \text{[OK]}$$

NDS Sec 3.9.2; NDS Eq 3.9-3

The rafter shall be proportioned so that

$$\left(\frac{f_c}{F'_c}\right)^2 + \frac{f_{bx}}{F'_{bx}\left(1 - \frac{f_c}{F_{cEx}}\right)} \le 1.0$$

$$\left(\frac{22.4\ \frac{\text{lbf}}{\text{in}^2}}{161.8\ \frac{\text{lbf}}{\text{in}^2}}\right)^2 + \frac{3193.4\ \frac{\text{lbf}}{\text{in}^2}}{\left(1377\ \frac{\text{lbf}}{\text{in}^2}\right)\left(1 - \frac{22.4\ \frac{\text{lbf}}{\text{in}^2}}{166.3\ \frac{\text{lbf}}{\text{in}^2}}\right)}$$

$$= 2.69 > 1.0 \quad \text{[no good]}$$

Therefore, the rafter is not adquately sized.

Example 6.8
Combined Compression and Bending: Glulam

A glulam bending combination 20F-V2 (SP/SP) of 6³/₄ in × 12³/₈ in is laterally supported at its ends and at its midpoint. Assume $C_D = C_t = 1.0$ and MC > 16%.

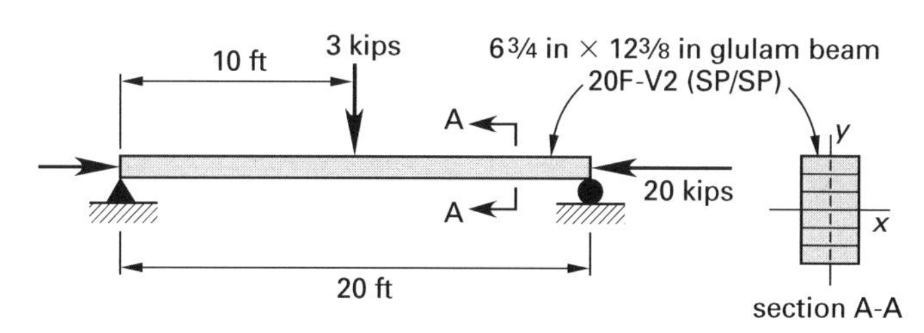

Check the adequacy of the member.

REFERENCE *Solution (ASD Method)*

NDS Supp Tbl 1D

For a 6³/₄ in × 12³/₈ in member and nine lams,

$$A = 83.53\ \text{in}^2$$

$$S_x = 172.3\ \text{in}^3$$

NDS Supp Tbl 5A Adj Fac

	C_M
$F_{bx}^{+} = F_{bx} = 2000\ \text{lbf/in}^2$	0.80
$F_{by} = 1450\ \text{lbf/in}^2$	0.80
$E_{x,\text{min}} = 780{,}000\ \text{lbf/in}^2$	0.833
$E_{y,\text{min}} = 730{,}000\ \text{lbf/in}^2$	0.833
$F_{vx} = 300\ \text{lbf/in}^2$	0.875
$F_{vy} = 260\ \text{lbf/in}^2$	0.875
$F_c = 1350\ \text{lbf/in}^2$	0.73

1. Axial compression

The actual compression stress parallel to the grain is

$$f_c = \frac{P}{A} = \frac{20{,}000\ \text{lbf}}{83.53\ \text{in}^2} = 239.4\ \text{lbf/in}^2$$

NDS Tbl 5.3.1

$$F_c' = F_c^* C_P$$

$$F_c^* = F_c C_D C_M C_t = \left(1350\ \frac{\text{lbf}}{\text{in}^2}\right)(1.0)(0.73)(1.0) = 985.5\ \text{lbf/in}^2$$

NDS Sec 3.7.1

Find the column stability factor, C_P.

NDS Eq 3.7-1

$$\left(\frac{\ell_e}{d}\right)_{\text{max}} = \begin{cases} \left(\frac{\ell_e}{d}\right)_x = \left(\frac{K_e \ell_u}{d}\right)_x = \dfrac{(1.0)(20\ \text{ft})\left(12\ \frac{\text{in}}{\text{ft}}\right)}{12.375\ \text{in}} \\ = 19.39 \quad \text{[controls]} \\ \left(\frac{\ell_e}{d}\right)_y = \left(\frac{K_e \ell_u}{d}\right)_y = \dfrac{(1.0)(10\ \text{ft})\left(12\ \frac{\text{in}}{\text{ft}}\right)}{6.75\ \text{in}} \\ = 17.8 \end{cases}$$

Therefore, the x-axis (the strong axis) of the beam-column is critical and $(\ell_e/d)_x$ is used to determine F_c'.

The adjusted modulus of elasticity for buckling about the x-axis is

$$E_{x,\text{min}}' = E_{x,\text{min}} C_M C_t = \left(780{,}000\ \frac{\text{lbf}}{\text{in}^2}\right)(0.833)(1.0) = 649{,}740\ \text{lbf/in}^2$$

NDS Sec 3.7.1.5

For glulam, $c = 0.9$.

The Euler critical buckling sstress for the x-axis is

$$F_{cEx} = \frac{0.822 E_{x,\text{min}}'}{\left(\frac{\ell_e}{d}\right)_x^2} = \frac{(0.822)\left(649{,}740\ \frac{\text{lbf}}{\text{in}^2}\right)}{(19.39)^2} = 1420.55\ \text{lbf/in}^2$$

The column stability factor is

$$C_P = \frac{1+\dfrac{F_{cEx}}{F_c^*}}{2c} - \sqrt{\left(\frac{1+\dfrac{F_{cEx}}{F_c^*}}{2c}\right)^2 - \frac{\dfrac{F_{cEx}}{F_c^*}}{c}}$$

$$= \frac{1+\dfrac{1420.55\ \frac{\text{lbf}}{\text{in}^2}}{985.5\ \frac{\text{lbf}}{\text{in}^2}}}{(2)(0.9)} - \sqrt{\left(\frac{1+\dfrac{1420.55\ \frac{\text{lbf}}{\text{in}^2}}{985.5\ \frac{\text{lbf}}{\text{in}^2}}}{(2)(0.9)}\right)^2 - \frac{\dfrac{1420.55\ \frac{\text{lbf}}{\text{in}^2}}{985.5\ \frac{\text{lbf}}{\text{in}^2}}}{0.9}}$$

$$= 0.869$$

The adjusted compression design value parallel to the grain is

$$F_c' = F_c^* C_P = \left(985.5\ \frac{\text{lbf}}{\text{in}^2}\right)(0.869)$$
$$= 856.40\ \text{lbf/in}^2 > f_c = 239.4\ \text{lbf/in}^2 \quad \text{[OK]}$$

2. Bending (about x-axis only)

The maximum bending moment for the x-axis is

$$M_x = \frac{P\ell_x}{4} = \frac{(3000\ \text{lbf})(240\ \text{in})}{4}$$
$$= 180{,}000\ \text{in-lbf}$$

The bending stress for the x-axis is

$$f_{bx} = \frac{M_x}{S_x} = \frac{180{,}000\ \text{in-lbf}}{172.3\ \text{in}^3}$$
$$= 1044.7\ \text{lbf/in}^2$$

NDS Tbl 5.3.1 Ftn 1

For glulam timber bending, the smaller of the two values, C_L and C_V, controls the following.

$$F_{bx}' = \begin{cases} F_{bx}C_D C_M C_t C_L = F_{bx}^* C_L \\ F_{bx}C_D C_M C_t C_V = F_{bx}^* C_V \end{cases}$$

NDS Sec 3.3.3

Find the beam stability factor, C_L.

NDS Eq 3.3-6

Lateral-torsional buckling will occur in the plane of the y-axis.

NDS Tbl 3.3.3

For a concentrated load at center, the effective length is

$$\ell_e = 1.11\ell_u = (1.11)(10\ \text{ft})\left(12\ \frac{\text{in}}{\text{ft}}\right)$$
$$= 133.2\ \text{in}$$

NDS Eq 3.3-5

The slenderness ratio is

$$R_B = \sqrt{\frac{\ell_e d}{b^2}} = \sqrt{\frac{(133.2 \text{ in})(12.375 \text{ in})}{(6.75 \text{ in})^2}}$$
$$= 6.02$$

The adjusted modulus of elasticity for beam stability for the y-axis is

$$E'_{y,\min} = E_{y,\min} C_M C_t = \left(730{,}000 \ \frac{\text{lbf}}{\text{in}^2}\right)(0.833)(1.0)$$
$$= 608{,}090 \text{ lbf/in}^2$$

The critical buckling design value for bending is

NDS Sec 3.3.3.8

$$F_{bE} = \frac{1.20 E'_{y,\min}}{R_B^2} = \frac{(1.20)\left(608{,}090 \ \frac{\text{lbf}}{\text{in}^2}\right)}{(6.02)^2}$$
$$= 20{,}135.2 \text{ lbf/in}^2$$

The bending design stress with all adjustments except C_V and C_L is

$$F_{bx}^* = F_{bx} C_D C_M C_t = \left(2000 \ \frac{\text{lbf}}{\text{in}^2}\right)(1.0)(0.80)(1.0)$$
$$= 1600 \text{ lbf/in}^2$$

NDS Eq 3.3-6

The beam stability factor is

$$C_L = \frac{1 + \frac{F_{bE}}{F_{bx}^*}}{1.9} - \sqrt{\left(\frac{1 + \frac{F_{bE}}{F_{bx}^*}}{1.9}\right)^2 - \frac{\frac{F_{bE}}{F_{bx}^*}}{0.95}}$$

$$= \frac{1 + \frac{20{,}135.2 \ \frac{\text{lbf}}{\text{in}^2}}{1600 \ \frac{\text{lbf}}{\text{in}^2}}}{1.9} - \sqrt{\left(\frac{1 + \frac{20{,}135.2 \ \frac{\text{lbf}}{\text{in}^2}}{1600 \ \frac{\text{lbf}}{\text{in}^2}}}{1.9}\right)^2 - \frac{\frac{20{,}135.2 \ \frac{\text{lbf}}{\text{in}^2}}{1600 \ \frac{\text{lbf}}{\text{in}^2}}}{0.95}}$$

$$= 0.99 \le 1.0$$

NDS Sec 5.3.6;
NDS Supp Tbl 5A

Find the volume factor of glulam, C_V.

For southern pine, $x = 20$.

NDS Eq 5.3-1

$$C_V = \left(\frac{21}{L}\right)^{1/x} \left(\frac{12}{d}\right)^{1/x} \left(\frac{5.125}{b}\right)^{1/x}$$
$$= \left(\frac{21}{20 \text{ ft}}\right)^{1/20} \left(\frac{12}{12.375 \text{ in}}\right)^{1/20} \left(\frac{5.125}{6.75 \text{ in}}\right)^{1/20}$$
$$= 0.987 < C_L = 0.99$$

Therefore, C_V governs lateral stability.

Find the bending design stress with all applicable adjustments,

$$F'_{bx} = F^*_{bx}C_V = \left(1600\ \frac{\text{lbf}}{\text{in}^2}\right)(0.987)$$
$$= 1579.2\ \text{lbf/in}^2$$

3. Combined compression and bending

The glulam beam must be proportioned that

NDS Eq 3.9-3

$$\left(\frac{f_c}{F'_c}\right)^2 + \frac{f_{bx}}{F'_{bx}\left(1 - \dfrac{f_c}{F_{cEx}}\right)} \leq 1.0$$

$$\left(\frac{239.4\ \dfrac{\text{lbf}}{\text{in}^2}}{985.5\ \dfrac{\text{lbf}}{\text{in}^2}}\right)^2 + \frac{1044.7\ \dfrac{\text{lbf}}{\text{in}^2}}{\left(1579.2\ \dfrac{\text{lbf}}{\text{in}^2}\right)\left(1 - \dfrac{239.4\ \dfrac{\text{lbf}}{\text{in}^2}}{1420.55\ \dfrac{\text{lbf}}{\text{in}^2}}\right)}$$
$$= 0.855 < 1.0 \quad [\text{OK}]$$

7

Mechanical Connections

NDS Secs. 10–13, cover the design of mechanical connections, bolts, lag screws/bolts, wood screws, nails and spikes, and other connectors. Reference design values for laterally loaded bolts, lag screws, wood screws, and nails and spikes are based on yield limit equations, which model the various yield modes. (See NDS App. I.) These design values in a given species apply to all grades unless otherwise noted.

1. Typical Lateral (Shear) Connections

Figure 7.1 Mechanical Connectors: Shear

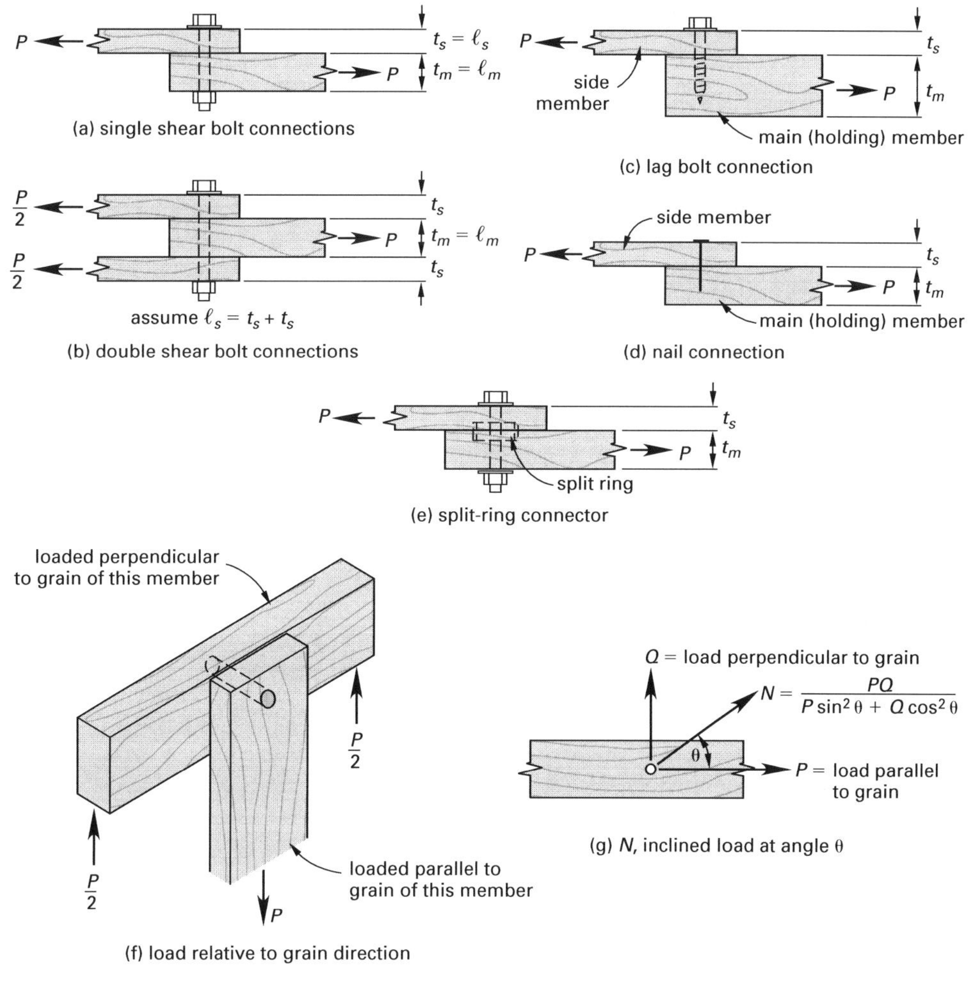

(a) single shear bolt connections

(b) double shear bolt connections

(c) lag bolt connection

(d) nail connection

(e) split-ring connector

(f) load relative to grain direction

(g) N, inclined load at angle θ

2. Typical Tension (Withdrawal) Connections

Withdrawal refers to the pulling out of a connector from the wood as caused by tensile forces, as shown in Fig. 7.2.

Figure 7.2 Mechanical Connectors: Withdrawal

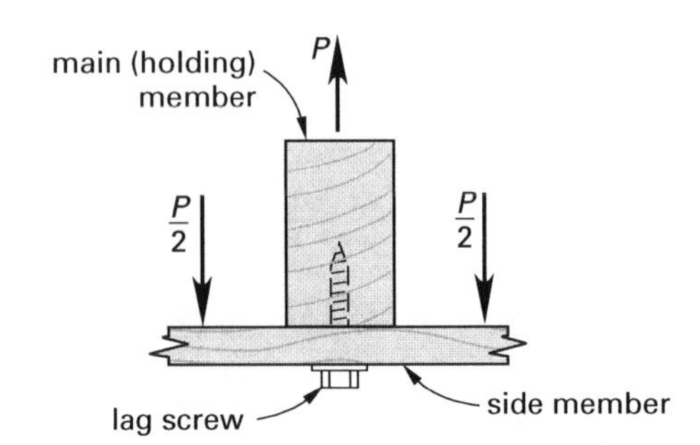

(a) side grain loads (perpendicular to grain)

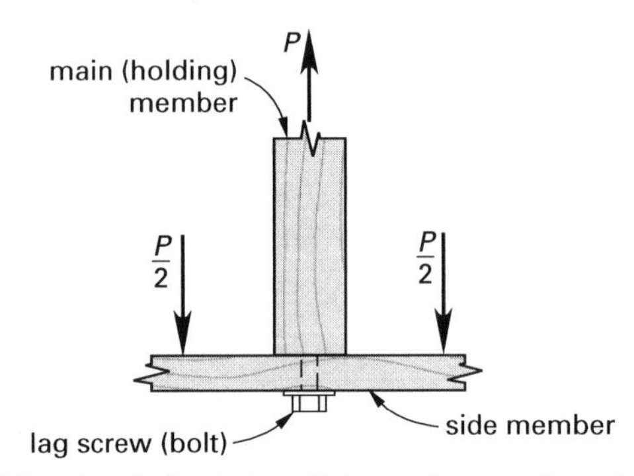

(b) end grain loads (parallel to grain, not allowed for nails and wood screws)

3. Adjustment Factors

[NDS Sec 10.3; NDS Tbls 10.3.1, 10.3.3]

Reference design values, Z and W, are multiplied by all applicable adjustment factors (NDS Sec. 10.3) to determine adjusted design values, Z' and W' (see Tables 7.1 and 7.2).

Z, Z' reference and adjusted lateral design value for a single fastener connection

W, W' reference and adjusted withdrawal design value for a fastener per inch of penetration

A. Load Duration Factor, C_D (ASD only)

[NDS Tbl 2.3.2; NDS App B; NDS Sec 10.3.2]

C_D values used previously for beams and columns are also applicable to the wood parts of the connections except when the connector capacity is controlled by the metal parts of the connections (plates, hangers, fasteners, and so on). Furthermore, the impact load duration factor is not applicable to connections.

B. Wet Service Factor, C_M

[NDS Tbl 10.3.3]

Reference design values are for connections in wood with a moisture content of 19% or less, at the time of fabrication and in service, and are multiplied by the wet service factors, C_M, when different conditions are encountered (see Table 7.2).

Table 7.1 Applicability of Adjustment Factors for Connections (NDS Table 10.3.1)

		ASD only	ASD and LRFD									LRFD only		
		load duration factor[a]	wet service factor[b]	temperature factor	group action factor[c]	geometry factor	penetration depth factor[c]	end grain factor[c]	metal side plate factor[c]	diaphragm factor[c]	toe-nail factor[c]	format conversion factor	resistance factor	time effect factor
		lateral loads												
dowel-type fasteners	$Z' = Z \times$	C_D	C_M	C_t	C_g	C_Δ	–	C_{eg}	–	C_{di}	C_{tn}	K_F	ϕ_z	λ
split ring and shear plate connectors	$P' = P \times$	C_D	C_M	C_t	C_g	C_Δ	C_d	–	C_{st}	–	–	K_F	ϕ_z	λ
	$Q' = Q \times$	C_D	C_M	C_t	C_g	C_Δ	C_d	–	–	–	–	K_F	ϕ_z	λ
timber rivets	$P' = P \times$	$C_D{}^d$	C_M	C_t	–	–	–	–	$C_{st}{}^e$	–	–	K_F	ϕ_z	λ
	$Q' = Q \times$	$C_D{}^d$	C_M	C_t	–	$C_\Delta{}^f$	–	–	$C_{st}{}^e$	–	–	K_F	ϕ_z	λ
metal plate connectors	$Z' = Z \times$	C_D	C_M	C_t	–	–	–	–	–	–	–	K_F	ϕ_z	λ
spike grids	$Z' = Z \times$	C_D	C_M	C_t	–	C_Δ	–	–	–	–	–	K_F	ϕ_z	λ
		withdrawal loads												
nails, spikes, lag screws, wood screws, and drift pins	$W' = W \times$	C_D	C_M	C_t	–	–	–	C_{eg}	–	–	C_{tn}	K_F	ϕ_z	λ

[a]The load duration factor, C_D, shall not exceed 1.6 for connections (see NDS Sec. 10.3.2).

[b]The wet service factor, C_M, shall not apply to toe-nails loaded in withdrawal (see NDS Sec. 11.5.4.1).

[c]Specific information concerning geometry factors, C_Δ, penetration depth factors, C_d, end grain factors, C_{eg}, metal side plate factors, C_{st}, diaphragm factors, C_{di}, and toe-nail factors, C_{tn}, is provided in NDS Chs. 11, 12, and 13.

[d]The load duration factor, C_D, is only applied when wood capacity (P_w, Q_w) controls (see NDS Ch. 13).

[e]The metal side plate factor, C_{st}, is only applied when rivet capacity (P_r, Q_r) controls (see NDS Ch. 13).

[f]The geometry factor, C_Δ, is only applied when wood capacity, Q_w, controls (see NDS Ch. 13).

Table 7.2 Wet Service Factors, C_M, for Connections (NDS Table 10.3.3)

	moisture content		
fastener type	at time of fabrication	in-service	C_M
	lateral loads		
shear plates and split rings[a]	$\leq$ 19%	$\leq$ 19%	1.0
	> 19%	$\leq$ 19%	0.8
	any	> 19%	0.7
metal connector plates	$\leq$ 19%	$\leq$ 19%	1.0
	> 19%	$\leq$ 19%	0.8
	any	> 19%	0.7
dowel-type fasteners	$\leq$ 19%	$\leq$ 19%	1.0
	> 19%	$\leq$ 19%	0.4[b]
	any	> 19%	0.7
timber rivets	$\leq$ 19%	$\leq$ 19%	1.0
	$\leq$ 19%	> 19%	0.8
	withdrawal loads		
lag screws and wood screws	any	$\leq$ 19%	1.0
	any	> 19%	0.7
nails and spikes	$\leq$ 19%	$\leq$ 19%	1.0
	> 19%	$\leq$ 19%	0.25
	$\leq$ 19%	> 19%	0.25
	> 19%	> 19%	1.0
threaded hardened nails	any	any	1.0

[a]For split ring or shear plate connectors, moisture content limitations apply to a depth of 3/4 in below the surface of the wood.
[b]$C_M = 0.7$ for dowel type fasteners with diameter, D, less than 1/4 in. $C_M = 1.0$ for dowel type fastener connections with:
1) one fastener only, or
2) two or more fasteners placed in a single row parallel to grain, or
3) fasteners placed in two or more rows parallel to grain with separate splice plates for each row

C. Group Action Factor, C_g

[NDS Sec 10.3.6; Eq 10.3-1; NDS Tbls 10.3.6A–10.3.6D]

Reference design values are multiplied by group action factors, C_g, specified in NDS Sec. 10.3.6. NDS Eq. 10.3-1 and NDS Tables 10.3.6A through 10.3.6D provide C_g values for various connection geometries.

D. Other Adjustment Factors

[NDS Sec 11.5]

Adjustment factors, such as connection geometry, C_Δ, end grain factor, C_{eg}, penetration factor, C_d, and so on, will be covered as they are needed.

4. Laterally Loaded Fasteners

[NDS Sec 11.3 and NDS App I]

Reference design values for dowel-type fasteners (bolts, lag screws, wood screws, nails, and spikes) are based on yield limit models, using engineering mechanics approach and yield limit theory. NDS App. I provides more detail on the failure mechanisms considered for development of the yield limit equations for connections. In addition to these equations for laterally loaded dowel-type fasteners, the NDS also provides a limited set of load tables for common, limited conditions (as noted in the footnotes to each table). While these tables are based on the yield limit equations, they have been simplified by grouping parameters together. (See NDS Tables 11A through 11R.)

5. Fasteners Loaded in Withdrawal and Tension

[NDS Sec 11.2]

The reference design values in NDS Tables 11.2A, 11.2B, and 11.2C are for lag screw (lag bolt), wood screws, and nails and spikes inserted in side grain of the main member (that receives the pointed end of the fastener). The fastener axis is perpendicular to the wood fibers.

8

Nails and Spikes

The four basic types of nails are box nails, common wire nails, common wire spikes, and threaded hardened-steel nails (see NDS Sec. 11.1.5). For a given penny size, all four have the same length. The first three nails have the same form but different diameters. Box nails have the smallest diameter and common wire spikes have the largest diameter. The threaded hardened-steel nails are made from high-carbon steel and have an annularly, or helically, threaded shank that provides better withdrawal capacity than other nails.

The sizes of these nails are given in NDS App. Tables L4 and NDS Tables 11N, 11P, 11Q, and 11R (which also list lateral design values). Frequently used nail diameters vary from 3/32 in to 1/4 in, and their lengths can be from 2 in to 6 in depending upon their size classifications. A summary of commonly used nail and spike sizes is given in Table 8.1.

Table 8.1 Nail and Spike Sizes

		wire diameter (in)			
penny size	length (in)	box nails	common wire nails	threaded hardened-steel nails	common wire spikes
6d	2	0.099	0.113	0.120	–
8d	2 1/2	0.113	0.131	0.120	–
10d	3	0.128	0.148	0.135	0.192
12d	3 1/4	0.128	0.148	0.135	0.192
16d	3 1/2	0.135	0.162	0.148	0.207
20d	4	0.148	0.192	0.177	0.225
30d	4 1/2	0.148	0.207	0.177	0.244
40d	5	0.162	0.225	0.177	0.263
50d	5 1/2	–	0.244	0.177	0.283
60d	6	–	0.263	0.177	0.283
70d	7	–	–	0.207	–
80d	8	–	–	0.207	–
90d	9	–	–	0.207	–
5/16 in	7	–	–	–	0.312
3/8 in	8 1/2	–	–	–	0.375

Common wire nails and spikes are essentially the same, except that common wire spikes have a larger diameter. For example, a 12d (12-penny) common wire nail has a diameter of 0.148 in and a 12d common wire spike has a diameter of 0.192 in.

1. Withdrawal Design Values

[NDS Sec 11.2.3]

A. Reference Withdrawal Design Value, W

[NDS Tbl 11.2C]

For a single nail or spike driven in the side grain of the main member, the reference withdrawal design values, W, depend upon the specific gravities of the wood species (NDS Table 11.3.2A) and are given in pounds per inch of penetration in NDS Table 11.2C or by NDS Eq. 11.2-3, as follows.

$$W = 1380G^{5/2}D$$

It is not permitted for nails and spikes to be loaded in withdrawal from the end grain of the wood.

B. Minimum Penetration Depth, p_{min}

[NDS Sec 11.1.5.5]

The minimum penetration depth of a nail or spike into the main member for a single shear connection or the side member for a double shear connector is $p_{\text{min}} = 6D$.

C. Toe-Nail Factor, C_{tn}

[NDS Secs 11.1.5.4, 11.5.4]

The toe-nail factor, C_{tn}, to be multiplied to the reference withdrawal design value, W, is 0.67 (see NDS Sec. 11.5.4.1). The wet service factor, C_M, does not apply to toe-nails loaded in withdrawal (see NDS Sec. 11.5.4.1). When toe-nailed connections are used, the reference lateral design value, Z, shall be multiplied by the toe-nail factor, 0.83 (see Table 7.1, NDS Table 10.3.1, and NDS Sec. 11.5.4.2).

D. Adjusted Withdrawal Design Value, W'

[NDS Sec 10.3.1]

The allowable design value, W', in lbf per inch of penetration into the main member depends on the reference design value, W, obtained from the table. The value, p, is the depth of the fastener into the holding (main) member. p is given in inches.

$$W' = WpC_DC_MC_tC_{eg}C_{tn} \quad \text{[ASD; NDS Table 10.3.1]}$$

For LRFD, C_D, is replaced with $(K_F\phi_z\lambda)$.

Example 8.1
Withdrawal Capacity of Nails

Two 16d (16-penny) common wire nails connect 2×6 to 4×8 hem-fir lumbers as shown. The connected lumbers will be subjected to normal temperatures ($C_t = 1.0$). Assume the combined loads consist of 10 lbf dead load and 20 lbf snow load.

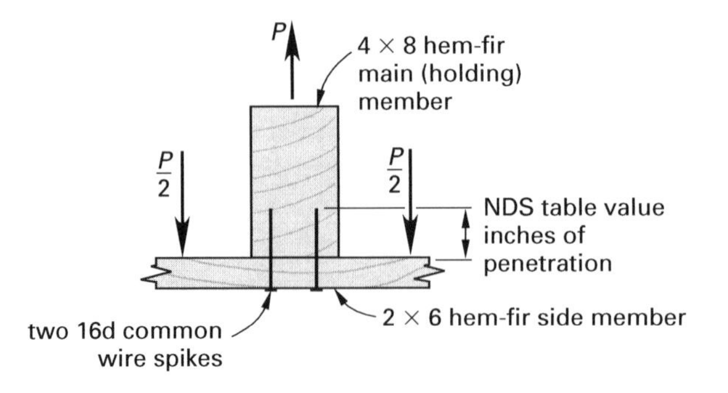

Determine the adjusted withdrawal value of two 16d common wire nails for combined snow and dead loads. Assume wet service conditions in service, and dry service conditions at time of fabrication.

REFERENCE — *Solution (ASD Method)*

$$P = 10\text{ lbf} + 20\text{ lbf} = 30\text{ lbf}$$

NDS Tbl 2.3.2 — For snow load, C_D is 1.15.

NDS Tbl 10.3.3 — C_M is 0.25.

NDS Tbl 11.3.2A — For hem-fir, the specific gravity, G, is 0.43.

NDS App Tbl L4 — The diameter, D, of a 16d common wire nail is 0.162 in. Its length, L, is 3.5 in.

NDS Tbl 11.2C — For the given specifc gravity, G, and nail diameter, the reference withdrawal design value, W, is 27 lbf per inch of penetration into main member.

The actual penetration, p, into a 4 × 8 holding member is

$$\begin{aligned} p &= 3.5\text{ in} - 1.5\text{ in} \\ &= 2.0\text{ in} \quad \begin{bmatrix}\text{dressed depth for a} \\ 2 \times 6 \text{ board is } 1.5\text{ in}\end{bmatrix} \end{aligned}$$

The adjusted withdrawal design value is

$$\begin{aligned} W' = WpC_DC_MC_t &= \left(27\ \frac{\text{lbf}}{\text{in}}\right)(2.0\text{ in})(1.15)(0.25)(1.0) \\ &= 15.5\text{ lbf per nail} \\ &= \left(15.5\ \frac{\text{lbf}}{\text{nail}}\right)(2\text{ nails}) \\ &= 31\text{ lbf} > P = 30\text{ lbf} \quad [\text{OK}] \end{aligned}$$

REFERENCE — *Solution (LRFD Method)*

Assume the combined loads consist of 10 lbf dead load and 20 lbf snow load.

NDS App Tbls N1, N2, N3

$$\begin{aligned} P_u &= 1.2w_D + 1.6w_S = (1.2)(10\text{ lbf}) + (1.6)(20\text{ lbf}) \\ &= 44\text{ lbf} \end{aligned}$$

For all connections, $K_F = 2.16/\phi_z$, $\phi_z = 0.65$. The time effect factor, λ, for the load combination is 0.80.

Calculate the adjusted withdrawal design value for LRFD.

The number of nails, N, is 2. The reference withdrawal design value, W, is 27 lbf per inch of penetration.

$$W'_{\text{LRFD}} = Wp(K_F\phi_z\lambda)C_MC_tN$$

NDS Tbls 11.2C, 10.3.1

$$\begin{aligned} W'_{\text{LRFD}} &= \left(27\ \frac{\text{lbf}}{\text{in}}\right)(2.0\text{ in})\left(\left(\frac{2.16}{\phi_z}\right)\phi_z(0.8)\right)(0.25)(1.0)(2\text{ nails}) \\ &= 46.7\text{ lbf} > P_u = 44\text{ lbf} \quad [\text{OK}] \end{aligned}$$

2. Lateral Design Values

[NDS Sec 11.3; Eqs 11.3-1–11.3-10; Tbls 11.3.1A, 11.3.1B; 11N, 11P, 11Q, and 11R; App. I]

Reference Lateral Design Value, Z

The reference lateral design value, Z, is the smallest value calculated by NDS yield limit Eqs. 11.3-1 through 11.3-10. For usual connections, Z is given in NDS Tables 11N through 11R, in accordance with yield mode equations. Z is used for the minimum nail penetration, $p = 10D$, into the main member. If the actual nail length, p, into the main member is insufficient to provide $10D$ penetration but is greater than $6D$, reference lateral design values, Z, shall be adjusted by $p/10D$. Z is given for nails inserted in side grain with nail axis perpendicular to wood fibers.

NDS Tables 11N–11R footnotes indicate that the following must be true.

$$6D \leq p < 10D$$

Table 8.2 Yield Limit Equations (Single Shear) (NDS Tables 11.3.1A and 11.3.1B and App. I)

yield mode	Z	
I_m	$\dfrac{D\ell_m F_{em}}{R_d}$	[NDS Eq. 11.3-1]
I_s	$\dfrac{D\ell_s F_{es}}{R_d}$	[NDS Eq. 11.3-2]
II	$\dfrac{k_1 D\ell_s F_{es}}{R_d}$	[NDS Eq. 11.3-3]
III_m	$\dfrac{k_2 D\ell_m F_{em}}{(1+2R_e)R_d}$	[NDS Eq. 11.3-4]
III_s	$\dfrac{k_3 D\ell_s F_{em}}{(2+R_e)R_d}$	[NDS Eq. 11.3-5]
IV	$\dfrac{D^2}{R_d}\sqrt{\dfrac{2F_{em}F_{yb}}{3(1+R_e)}}$	[NDS Eq. 11.3-6]

The following notes for NDS Table 11.3.1A apply to these equations.

$$k_1 = \frac{\sqrt{R_e + 2R_e^2(1+R_t+R_t^2) + R_t^2R_e^3} - R_e(1+R_t)}{1+R_e}$$

$$k_2 = -1 + \sqrt{2(1+R_e) + \frac{2F_{yb}(1+2R_e)D^2}{3F_{em}\ell_m^2}}$$

$$k_3 = -1 + \sqrt{\frac{2(1+R_e)}{R_e} + \frac{2F_{yb}(1+2R_e)D^2}{3F_{em}\ell_s^2}}$$

$$R_e = \frac{F_{em}}{F_{es}}$$

R_d	reduction terms based on dowel member diameter (see NDS Table 11.3.1B)	–
F_{em}	main member dowel bearing strength (see NDS Tables 11.3.2A and 11.3.2)	lbf/in^2
F_{es}	side member dowel bearing strength (see NDS Tables 11.3.2A and 11.3.2)	lbf/in^2

ℓ_m	length of dowel bearing in wood main member	in
ℓ_s	length of dowel bearing in wood side member	in
F_{yb}	dowel (fastener) bending yield strength (see footnote of NDS Tables 11A–11I)	lbf/in^2
D	diameter (see NDS Sec. 11.3.6 and NDS Table L4)	in

A. For Wood-to-Wood Connections

For NDS Eqs. 11.3-1 through 11.3-6, the dowel bearing strengths, F_e, from NDS Table 11.3.2 can be used. The nail bending strengths, F_{yb}, vary from 70,000 lbf/in^2 to 100,000 lbf/in^2, depending on the nail diameter, D (see NDS Table 11N footnotes).

Alternatively, NDS Table 11N contains reference lateral design values, Z, for usual connections. For those cases not contained in the table, the yield limit equations must be used. The reference lateral design values are for the minimum nail penetration, $p = 10D$, into the main member. If the nail length into the main member, p, is insufficient to provide $10D$ penetration, but is greater than $6D$, Z shall be adjusted by $p/10D$. The reference lateral design values are for nails inserted in the side grain with nail axis perpendicular to the wood fibers.

B. For Wood-to-Metal Connections

[NDS Tbls 11.3.1A, 11.3.1B, 11N, 11.3.2]

For NDS Eqs. 11.3-1 through 11.3-6, the dowel bearing strength of the steel side plate, F_{es} is 61,850 lbf/in^2 for ASTM A653 Grade 33. The nail bending yield strengths, F_{yb}, vary from 70,000 lbf/in^2 to 100,000 lbf/in^2, depending upon the nail diameter, D (see NDS Table 11P footnotes).

Alternatively, NDS Table 11P contains reference lateral design values, Z, for usual connections. For those cases not contained in this table, the yield limit equations must be used. The reference lateral design values are for the minimum nail penetration, $p = 10D$, into the main member. If the nail length into the main member is insufficient to provide $10D$ penetration, but is greater than $6D$, Z, shall be adjusted by $p/10D$. The reference lateral design value, Z is given for nails inserted in side grain with nail axis perpendicular to wood fibers.

C. Adjusted Lateral Design Value, Z', and Adjustment Factors

[NDS Tbls 10.3.1, 10.3.3; NDS Sec 11.5]

The adjusted lateral design values, Z', are determined by multiplying all applicable adjustment factors as given in Tables 7.1 and 7.2 (NDS Tables 10.3.1 and 10.3.3) to the reference lateral design values.

$$Z' = ZC_DC_MC_tC_gC_\Delta C_{eg}C_{di}C_{tn}$$

C_g	group action factor (NDS Sec. 10.3.6)	–
C_Δ	geometry factor (NDS Sec. 11.5.1)	–
C_{eg}	end grain factor (NDS Sec. 11.5.2)	0.67
C_{di}	diaphragm factor (NDS Sec. 11.5.3)	1.1
C_{tn}	toe-nail factor for connections (NDS Sec. 11.5.4)	0.83
p	actual penetration into main member	–

When $6D < p < 10D$, the reference lateral design values, Z, shall be multiplied by $p/10D$.

Example 8.2
Lateral Design Capacity by NDS Table 11N: Tension Splice

Six 40d common wire nails hold two members, 2×6 and 4×8, together as shown. Both members are douglas fir-larch under wet service conditions. The load is a combination dead load and snow load. Assume $C_t = 1.0$.

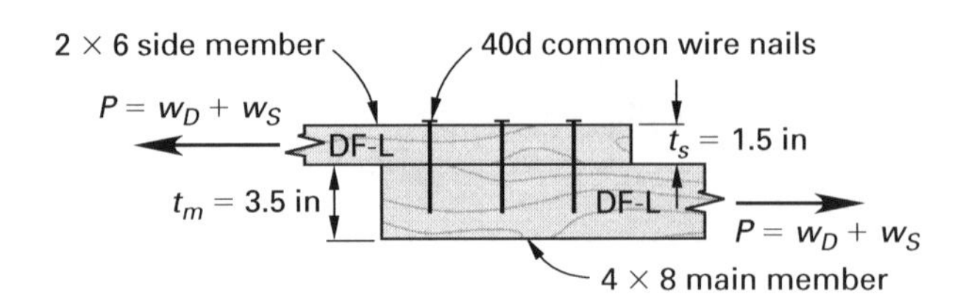

Determine the adjusted lateral design load for the six 40d common wire nails using NDS Table 11N.

REFERENCE *Solution (ASD Method)*

For douglas fir-larch, the side member thickness, t_s, is 1.5 in.

NDS App Tbl L4 — The diameter, D, of a 40d common nail is 0.225 in. The length, L, is 5 in.

NDS Tbl 11N — The nominal design value, Z, is 205 lbf for douglas fir-larch.

NDS Tbl 2.3.2 — For snow load, C_D is 1.15.

NDS Tbl 10.3.3 — For wet service connection, C_M is 0.7.

$$p = 10D = 2.25 \text{ in}$$

$$p_{\text{actual}} = 5 \text{ in} - 1.5 \text{ in} = 3.5 \text{ in} > p = 2.25 \text{ in} \quad \text{[OK]}$$

NDS Tbl 7.3.1 — The adjusted lateral design value is

$$Z' = ZC_DC_MC_tN = (205 \text{ lbf})(1.15)(0.7)(1.0)(6 \text{ nails}) = 990.15 \text{ lbf}$$

Example 8.3
Lateral Design Capacity by NDS Equations 11.3-1–11.3-6: Tension Splice

For the configuration shown in Ex. 8.2, determine the reference lateral design value, Z, using NDS Eqs. 11.3-1 through 11.3-6.

REFERENCE *Solution (ASD Method)*

NDS App Tbl L4 — The diameter, D, of 40d common wire nails is 0.225 in. The length, L, is 5 in.

NDS Tbl 11N, Ftn

$$p = 5 \text{ in} - 1.5 \text{ in} = 3.5 \text{ in}$$

$$6D = (6)(0.225 \text{ in}) = 1.35 \text{ in}$$

$$10D = (10)(0.225 \text{ in}) = 2.25 \text{ in}$$

$$p > 10D = 2.25 \text{ in}$$

According to NDS Table 11N, the reference lateral design value, Z, is 205 lbf.

Calculate the reference lateral design value, Z, using the yield mode equations.

NDS Tbl 11.3.1B

Since $D < 0.25$ in,

$$R_d = K_D = 10D + 0.5 = (10)(0.225 \text{ in}) + 0.5 = 2.75$$

NDS Tbls 11.3.2A, 11.3.2

For douglas fir-larch, the dowel bearing strength of the side member, F_{es}, is the same as that of the main member, F_{em}. With $D < 1/4$ in, the dowel bearing strength is 4650 lbf/in^2 for $G = 0.50$.

NDS Tbl 11N Ftn; NDS App Tbl I1

For a 40d nail, the bending yield strength, F_{yb} is 80,000 lbf/in^2.

NDS Eq 11.3-1

The reference lateral design value for yield mode I_m is

$$\begin{aligned} Z &= \frac{D\ell_m F_{em}}{R_d} \\ &= \frac{(0.225 \text{ in})(3.5 \text{ in})\left(4650 \ \frac{\text{lbf}}{\text{in}^2}\right)}{2.75} \\ &= 1331.6 \text{ lbf} \end{aligned}$$

NDS Eq 11.3-2

The reference lateral design value for yield mode I_s is

$$\begin{aligned} Z &= \frac{D\ell_s F_{es}}{R_d} \\ &= \frac{(0.225 \text{ in})(1.5 \text{ in})\left(4650 \ \frac{\text{lbf}}{\text{in}^2}\right)}{2.75} \\ &= 570.68 \text{ lbf} \end{aligned}$$

Calculate the reference lateral design value for yield mode II.

$$R_e = \frac{F_{em}}{F_{es}} = \frac{4650 \ \frac{\text{lbf}}{\text{in}^2}}{4650 \ \frac{\text{lbf}}{\text{in}^2}} = 1$$

$$R_t = \frac{\ell_m}{\ell_s} = \frac{3.5}{1.5} = 2.33$$

$$\begin{aligned} k_1 &= \frac{\sqrt{R_e + 2R_e^2(1 + R_t + R_t^2) + R_t^2 R_e^3} - R_e(1 + R_t)}{1 + R_e} \\ &= \frac{\sqrt{1 + (2)(1)^2\left(1 + 2.33 + (2.33)^2\right) + (2.33)^2(1)^3} - (1)(1 + 2.33)}{1 + 1} \\ &= 0.785 \end{aligned}$$

NDS Eq 11.3-3

$$\begin{aligned} Z &= \frac{k_1 D\ell_s F_{es}}{R_d} \\ &= \frac{(0.785)(0.225 \text{ in})(1.5 \text{ in})\left(4650 \ \frac{\text{lbf}}{\text{in}^2}\right)}{2.75} \\ &= 447.98 \text{ lbf} \end{aligned}$$

Calculate the reference lateral design value for yield mode III_m.

NDS Eq 11.3-4

$$k_2 = -1 + \sqrt{(2)(1+R_e) + \frac{2F_{yb}(1+2R_e)D^2}{3F_{em}\ell_m^2}}$$
$$= -1 + \sqrt{(2)(1+1) + \frac{(2)\left(80{,}000\ \frac{\text{lbf}}{\text{in}^2}\right)(1+(2)(1))(0.225\ \text{in})^2}{(3)\left(4650\ \frac{\text{lbf}}{\text{in}^2}\right)(3.5\ \text{in})^2}}$$
$$= 1.035$$

$$Z = \frac{k_2 D\ell_m F_{em}}{(1+2R_e)R_d}$$
$$= \frac{(1.03)(0.225\ \text{in})(3.5\ \text{in})\left(4650\ \frac{\text{lbf}}{\text{in}^2}\right)}{(1+(2)(1))(2.33)}$$
$$= 539.59\ \text{lbf}$$

NDS Eq 11.3-5

Calculate the reference lateral design value for yield mode III_s.

$$k_3 = -1 + \sqrt{\frac{2(1+R_e)}{R_e} + \frac{2F_{yb}(1+2R_e)D^2}{3F_{em}\ell_s^2}}$$
$$= -1 + \sqrt{\frac{(2)(1+1)}{1} + \frac{(2)\left(80{,}000\ \frac{\text{lbf}}{\text{in}^2}\right)(2+(1))(0.225\ \text{in})^2}{(3)\left(4650\ \frac{\text{lbf}}{\text{in}^2}\right)(1.5\ \text{in})^2}}$$
$$= 1.185$$

$$Z = \frac{k_3 D\ell_s F_{em}}{(2+R_e)R_d}$$
$$= \frac{(1.185)(0.225\ \text{in})(1.5\ \text{in})\left(4650\ \frac{\text{lbf}}{\text{in}^2}\right)}{(2+1)(2.75)}$$
$$= 225.42\ \text{lbf}$$

NDS Eq 11.3-6

The reference lateral design value for yield mode IV is

$$Z = \frac{D^2}{R_d}\sqrt{\frac{2F_{em}F_{yb}}{3(1+R_e)}}$$
$$= \frac{(0.225\ \text{in})^2}{2.75}\sqrt{\frac{(2)\left(4650\ \frac{\text{lbf}}{\text{in}^2}\right)\left(80{,}000\ \frac{\text{lbf}}{\text{in}^2}\right)}{(3)(1+1)}}$$
$$= 204.99\ \text{lbf} \quad [\text{controls}]$$

Z is 204.99 lbf (compared to $Z = 205$ lbf from NDS Table 11N).

9

Bolts

[NDS Secs 11.3 and 11.5; Tbls 11A–I]

Bolts, which are dowel-type connections, are used for laterally loaded joints. Connections can be single-shear (two-member), double-shear (three-member), or multiple-shear (several-member) joints. They can also be used for wood-to-wood or wood-to-metal connections.

Figure 9.1 Bolt Connections

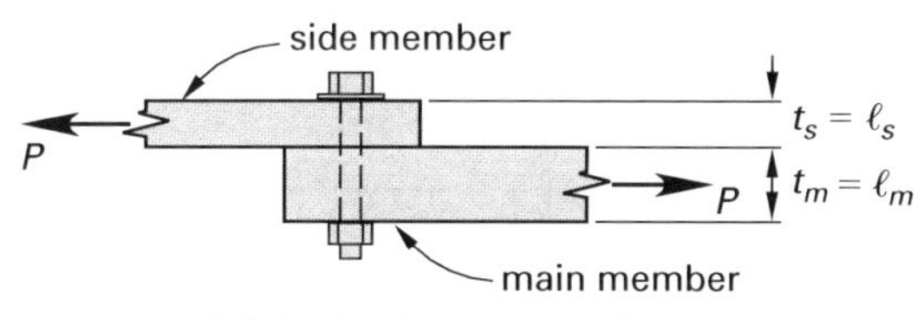

(a) single-shear connection

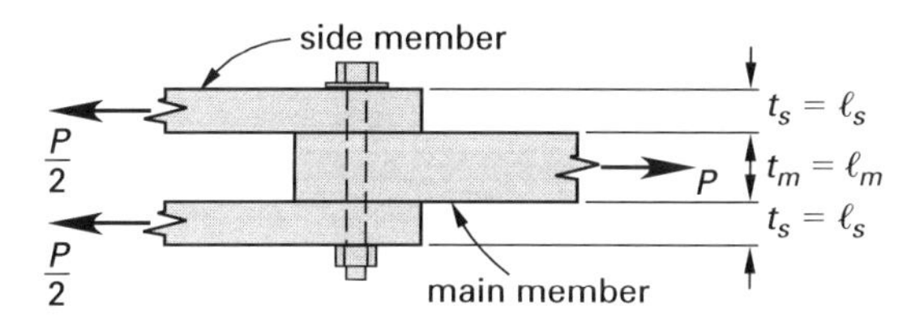

(b) double-shear connection

The yield limit equations for the lateral capacity of bolts in single shear are the same as that of all dowel-type fasteners and can be found in NDS Tables 11.3.1A and 11.3.1B. Numerical values are also listed in NDS Tables 11A–11I, for usual connections.

1. Reference Design Values for Single-Shear Connections

[NDS Tbls 11.3.1A–E]

A. Reference Lateral Design Values, *Z*, for Wood-to-Wood Connections

The reference lateral design value, Z, for one bolt in a single-shear connection is the smallest load capacity obtained from the following six equations. These are the same equations listed in Ch. 8 Sec. 2 (for nails) and will be used for all dowel-type fasteners in single shear.

B. Reference Lateral Design Values, *Z*, for Wood-to-Metal Connections

[NDS Tbls 11B and 11D]

For NDS Eqs. 11.3–11.3.6 the dowel bearing strength of the steel side plate, F_{es}, is 87,000 lbf/in^2 for ASTM A36 steel and a bolt bending strength, F_{yb}, of 45,000 lbf/in^2 (see NDS Table 11B footnotes). As an alternative to these equations, NDS Tables 11B and 11D contain reference lateral design values, Z, for usual connections. However, for those cases not contained in this table, the yield limit equations in NDS Tables 11.3.1A and 11.3.1B must be used.

C. Dowel Bearing Design Values at an Angle to the Grain

[NDS Sec 11.3.3; NDS App. J]

The dowel bearing strength at an angle of load to grain, θ, is given by the Hankinson formula in NDS App. J.

$$F_{e\theta} = \frac{F_{e\parallel}F_{e\perp}}{F_{e\parallel}\sin^2\theta + F_{e\perp}\cos^2\theta} \quad \text{[NDS Eq. 11.3-11]}$$

Different reduction terms, R_d, are applicable to different angles of load to grain. NDS Table 11.3.1B lists the various reduction terms and respective yield modes.

D. Allowable Lateral Design Values for Single-Shear Connections and Adjustment Factors

[Tbl 7.1 (NDS Tbl 10.3.1); NDS Tbls 11A–11E; NDS Sec 11.5]

The allowable lateral design value, Z, is the reference lateral design value multiplied by the product of adjustment factors (see NDS Table 10.3.1).

$$Z' = ZC_DC_MC_tC_gC_\Delta C_{eg}$$

C_g is the group action factor for connections (see NDS Sec. 10.3.6), and C_Δ is the geometry factor for base dimensions of bolts (see NDS Sec. 11.5.1). If the bolt is inserted in the endgrain of the main member with bolt axis parallel to the wood fibers, the end grain factor, C_{eg}, is included as 0.67 (see NDS Sec. 11.5.2).

Reference lateral design values determined using the equations or tables are for bolts with edge and end distances and spacings equal to or greater than the minimum required for full design values (see NDS Tables 11.5.1A–11.5.1D). When these edge and end distances and spacings are less than the minimum required, but greater than the minimum for reduced lateral design values, the reference lateral design values are multiplied by the smallest geometry factor obtained as follows.

$$C_\Delta = \begin{cases} \dfrac{\text{actual end distance}}{\text{minimum end distance required for full design value}} \\ \dfrac{\text{actual bolt spacing}}{\text{minimum spacing required for full design value}} \end{cases}$$

Figure 9.2 Edge, End, and Spacing Requirements for Bolts (NDS Tables 11.5.1A–11.5.1D and Figure 11G)

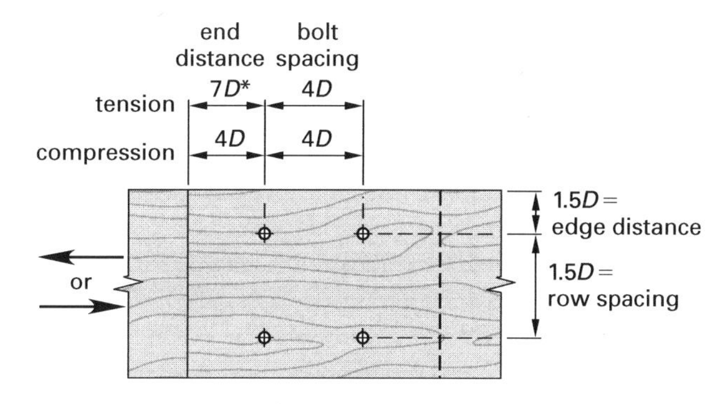

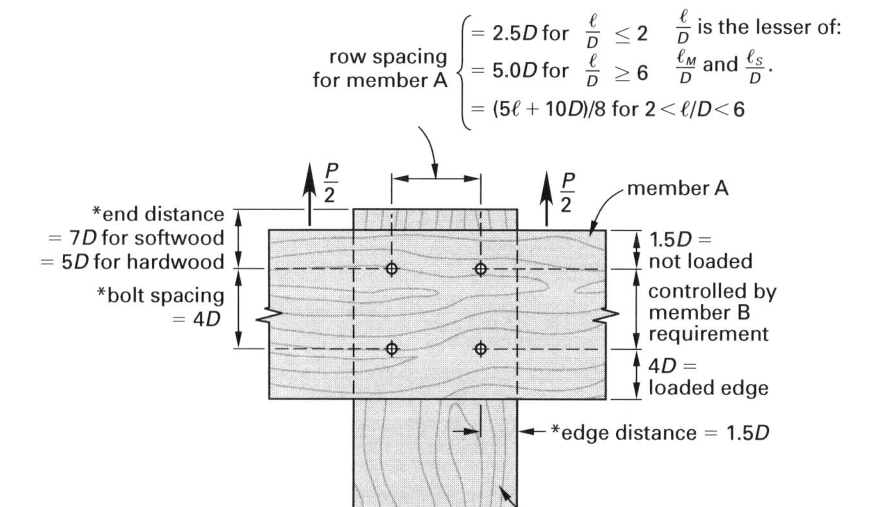

Adapted from *National Design Specification for Wood Construction*, 2005 Edition, courtesy, American Forest & Paper Association, Washington, D.C.

Example 9.1
Wood-to-Wood Single-Shear Connection by NDS Tables for Sawn Lumber

A two-member (single-shear) joint is made with 2×6 and 3×6 douglas fir-larch members, select structural grade lumber, and four $^3/_4$ in bolts in $^1/_8$ in oversized holes as shown. The tension load is caused by a combination dead load, w_D, and wind load, w_W. Assume normal temperature ($C_t = 1.0$) and wet service conditions.

Determine the allowable tension load, P_{allow}, using NDS Table 11 values.

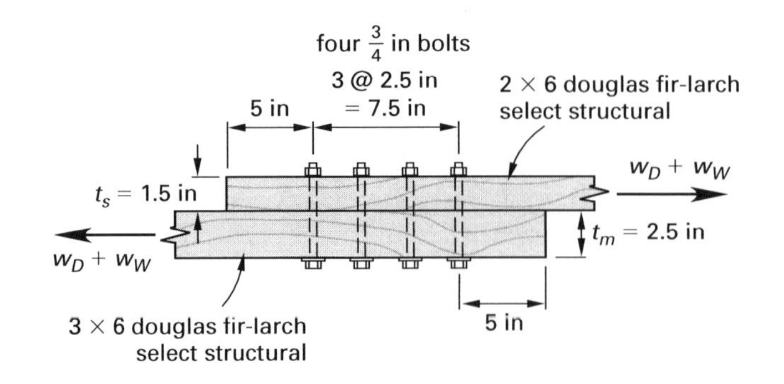

Assume that the total load, $w_D + w_W$, is

$$1000 \text{ lbf} + 2000 \text{ lbf} = 3000 \text{ lbf}$$

REFERENCE

Solution (ASD Method)

1. Find the tension load as determined by bolt capacity.

Find the adjusted bolt design value using the following equation.

NDS Tbl 10.3.1

$$Z' = ZC_DC_MC_tC_gC_\Delta C_{eg}$$

NDS Tbl 11A

For $t_m = 2^1/_2$ in, $t_s = 1^1/_2$ in, $D = {}^3/_4$ in, and douglas fir-larch, the lateral reference design value is
$Z = Z_{\parallel} = 1020$ lbf

NDS Tbl 2.3.2

For wind load, C_D is 1.6.

NDS Tbl 10.3.3

For wet service, C_M is 0.7.

NDS Tbl 10.3.6A

The group action factor, C_g, for four ${}^3/_4$ in bolts in a row (by using NDS Table 10.3.6A, conservatively) is calculated as follows.

The cross-sectional area of the main wood member is

$$\begin{aligned} A_m &= (2.5 \text{ in})(5.5 \text{ in}) \\ &= 13.75 \text{ in}^2 \end{aligned}$$

The sum of the gross cross-sectional areas of side members is

$$\begin{aligned} A_s &= (1.5 \text{ in})(5.5 \text{ in}) \\ &= 8.25 \text{ in}^2 \end{aligned}$$

$$\begin{aligned} \frac{A_s}{A_m} &= \frac{8.25 \text{ in}^2}{13.75 \text{ in}^2} \\ &= 0.6 \end{aligned}$$

NDS Tbl 10.3.6A, Ftn 2

After interpolation, $C_g = 0.88$.

NDS Sec 11.5.1

The geometry factor is C_Δ.

Find the end distance factor.

The actual end distance provided is 5 in.

NDS Tbl 11.5.1B

For full design value, minimum end distance is

$$7D = (7)(0.75\text{ in}) = 5.25\text{ in}$$

For the reduced design value, minimum end distance is

$$3.5D = (3.5)(0.75\text{ in}) = 2.625\text{ in}$$

$$C_\Delta = \frac{\text{actual end distance}}{\text{minimum end distance required for full design value}} = \frac{5\text{ in}}{5.25\text{ in}} = 0.952$$

Find the bolt spacing factor.

The actual bolt spacing is 2.5 in.

NDS Tbl 11.5.1C

For full design value, minimum spacing is

$$4D = (4)(0.75\text{ in}) = 3\text{ in}$$

For reduced design value, minimum spacing is

$$3D = (3)(0.75\text{ in}) = 2.25\text{ in} < \text{actual bolt spacing 2.5 in} \quad [\text{OK}]$$

The lesser of the two results controls.

NDS 11.5.1.2

$$C_\Delta = \frac{\text{actual spacing}}{\begin{array}{c}\text{minimum spacing required}\\ \text{for full design value}\end{array}} = \frac{2.5\text{ in}}{4D} = \frac{2.5\text{ in}}{(4)(0.75\text{ in})} = 0.833 \quad [\text{controls}]$$

NDS Tbl 11.5.1A

Determine the minimum required edge distance.

ℓ_m	length of dowel bearing in wood main member ($= t_m$)	2.5 in
ℓ_s	length of dowel bearing in wood side member ($= t_s$)	1.5 in

$$\frac{\ell_m}{D} = \frac{2.5\text{ in}}{0.75\text{ in}} = 3.33$$

$$\frac{\ell_s}{D} = \frac{1.5\text{ in}}{0.75\text{ in}} = 2 \quad [\text{controls for } \ell/D]$$

The minimum edge distance required for $\ell/D = 2$ is $1.5D$.

$$1.5D = (1.5)(0.75\text{ in}) = 1.13$$

The actual edge distance is

$$\frac{5.5 \text{ in}}{2} = 2.75 \text{ in} > 1.13 \text{ in} \quad \text{[OK]}$$

NDS Tbl 11.5.1D

Row spacing is not applicable.
The adjusted lateral bolt design value is

NDS Tbl 10.3.1

$$\begin{aligned} Z' &= ZC_DC_MC_tC_gC_\Delta C_{eg} \\ &= (1020 \text{ lbf})(1.6)(0.7)(1.0)(0.88)(0.833)(1.0) \\ &= 837.4 \text{ lbf} \end{aligned}$$

The allowable tension load as determined by the bolts is

$$\begin{aligned} P_{\text{allow}} &= Z'(\text{no. of bolts}) = \left(837.4 \ \frac{\text{lbf}}{\text{bolt}}\right)(4 \text{ bolts}) \\ &= 3349.6 \text{ lbf} \end{aligned}$$

2. Calculate the tension load as determined by wood tension capacity (for 2 × 6 member).

NDS Supp Tbl 4A Adj Fac

Calculate the adjusted tension design value as determined by the wood tension design capacity, F_t'.

The reference tension design value, F_t, is 1000 lbf/in^2. The wet service factor, C_M, is 1.0. For 2 × 6 or 3 × 6, the size factor for sawn lumber, C_F, is 1.3.

NDS Tbl 4.3.1

$$\begin{aligned} F_t' &= F_tC_DC_MC_tC_FC_i \\ &= \left(1000 \ \frac{\text{lbf}}{\text{in}^2}\right)(1.6)(1.0)(1.0)(1.3)(1.0) \\ &= 2080 \text{ lbf/in}^2 \end{aligned}$$

The allowable tension load as determined by wood tension capacity is

$$\begin{aligned} P_{\text{allow}} &= F_t'A_n \\ &= \left(2080 \ \frac{\text{lbf}}{\text{in}^2}\right)\big(5.5 \text{ in} - (0.75 \text{ in} + 0.125 \text{ in})\big)(1.5 \text{ in}) \\ &= 14{,}430 \text{ lbf} \end{aligned}$$

The allowable tension load by the bolt capacity is 3349.6 lbf.

$$P_{\text{applied}} = w_D + w_W = 3000 \text{ lbf} < P_{\text{allow}} = 3349.6 \text{ lbf} \quad \text{[OK]}$$

REFERENCE

Solution (LRFD Method)

NDS App Tbl N3

For the load combination of dead load and wind load,

$$P_u = 1.2w_D + 1.6w_W = (1.2)(1000 \text{ lbf}) + (1.6)(2000 \text{ lbf}) = 4400 \text{ lbf}$$

The time effect factor, λ, is 1.0 for this load combination.

NDS App Tbls N1, N2, N3

$$K_F = \frac{2.16}{\phi_Z}; \quad \phi_Z = 0.65 \text{ for connections}$$

1. Find the tension load as determined by bolt capacity.

NDS Tbl IIA

$Z = Z_{\parallel} = 1020 \text{ lbf}$

NDS Tbl 10.3.1

The wet service factor, C_M, the group action factor, C_g, and the geometry factor, C_Δ, are the same as in the ASD method.

$$C_M = 0.7,\ C_g = 0.88,\ C_\Delta = 0.833$$

$$\begin{aligned} Z'_{\text{LRFD}} &= Z(K_F\phi_z\lambda)C_MC_tC_gC_\Delta \\ &= (1020 \text{ lbf})\left(\left(\frac{2.16}{0.65}\right)(0.65)(1.0)\right)(0.7)(1.0)(0.88)(0.833) \\ &= 1130.5 \text{ lbf} \end{aligned}$$

The allowable tension load as determined by the bolts is

$$\begin{aligned} P_{\text{allow,LRFD}} &= Z'_{\text{LRFD}}N = (1130.5 \text{ lbf})(4 \text{ bolts}) \\ &= 4522 \text{ lbf} \quad \text{[controls]} \end{aligned}$$

2. Calculate the tension load as determined by wood tension capacity (for 2 × 6 member).

The wet service factor, C_M, the temperature factor, C_t, and the size factor, C_F, are the same as in the ASD method.

$$F_t = 1000 \ \frac{\text{lbf}}{\text{in}^2},\ C_M = C_t = 1.0,\ C_F = 1.3$$

NDS Tbl 4.3.1; NDS App N

$$\begin{aligned} F'_{t,\text{LRFD}} &= F_t(K_F\phi_t\lambda)C_MC_tC_F \\ &= \left(1000 \ \frac{\text{lbf}}{\text{in}^2}\right)\left(\left(\frac{2.16}{\phi_t}\right)\phi_t(1.0)\right)(1.0)(1.0)(1.3) \\ &= 2808 \text{ lbf/in}^2 \end{aligned}$$

The allowable tension load as determined by the wood tension capacity is

$$\begin{aligned} P_{\text{allow,LRFD}} &= F'_{t,\text{LRFD}}A_{\text{net}} \\ &= \left(2808 \ \frac{\text{lbf}}{\text{in}^2}\right)(6.9375 \text{ in}^2) \\ &= 19{,}481 \text{ lbf} \end{aligned}$$

$$P_{\text{applied}} = P_u = 4400 \text{ lbf} < P_{\text{allow,LRFD}} = 4522 \text{ lbf} \quad \text{[OK]}$$

Example 9.2

Wood-to-Wood Single-Shear Connection by NDS Tables for Glulam

Given the same configuration as Ex. 9.1 except that the main member is a $2^1/_2$ in × 6 in glulam tension member made of combination 28 douglas fir-larch, determine the allowable tension load, P_{allow}, by using NDS table values (not by using NDS equations).

REFERENCE

Solution (ASD Method)

1. Bolt capacity

NDS Tbl 11C

The main member thickness, t_m, the side member thickness, t_s, and the bolt diameter, D, are the same as in Ex. 9.1.

$$t_m = 2\tfrac{1}{2}\text{ in (glulam)}$$
$$t_s = 1\tfrac{1}{2}\text{ in } (2 \times 6)$$
$$D = \tfrac{3}{4}\text{ in}$$

For douglas fir-larch, the lateral design value is

$$Z = Z_{\parallel} = 1020\text{ lbf}$$

$$\begin{aligned} Z' &= ZC_DC_MC_tC_gC_\Delta \quad \begin{bmatrix}\text{refer to Ex. 9.1 for}\\ \text{adjustment factors}\end{bmatrix} \\ &= (1020\text{ lbf})(1.6)(0.7)(1.0)(0.88)(0.833) \\ &= 837.4\text{ lbf per bolt} \end{aligned}$$

The allowable tension load as determined by the bolts is

$$\begin{aligned} P_{\text{allow}} &= Z'(\text{no. of bolts}) = \left(837.4\ \frac{\text{lbf}}{\text{bolt}}\right)(4\text{ bolts}) \\ &= 3349.6\text{ lbf} \end{aligned}$$

2. Side member (2 × 6) wood capacity

The adjusted tension design value is

$$\begin{aligned} F_t' &= F_tC_DC_MC_tC_FC_i \\ &= \left(1000\ \frac{\text{lbf}}{\text{in}^2}\right)(1.6)(1.0)(1.0)(1.3)(1.0) \quad \text{[from Ex. 9.1]} \\ &= 2080\text{ lbf/in}^2 \end{aligned}$$

The allowable tension load as determined by the wood tension is

$$\begin{aligned} P_{\text{allow}} &= F_t'A_n \\ &= 14{,}430\text{ lbf} \quad \text{[from Ex. 9.1]} \end{aligned}$$

The allowable tension load by the bolt capacity is

$$P_{\text{allow}} = 3349.6\text{ lbf} \quad \text{[controls]}$$

Example 9.3

Wood-to-Wood Single-Shear Connection by NDS Yield Limit Equations for Sawn Lumber

A joint is made as shown.

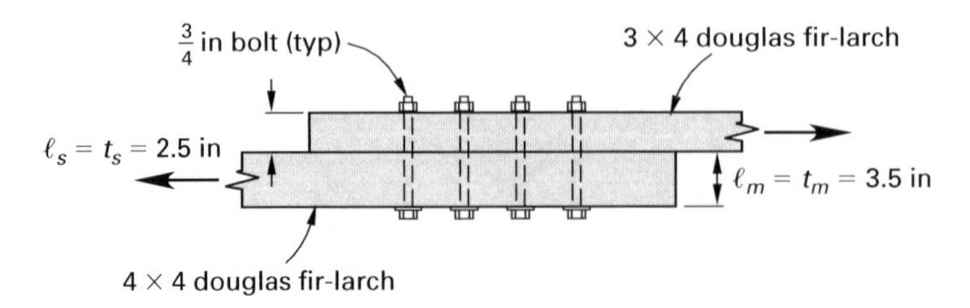

Determine the reference design value for a single bolt in shear using NDS yield limit equations and also using NDS Tables. Assume all spacing and edge distance requirements are met for full design values.

REFERENCE

Solution (ASD Method)

NDS Tbl 11.3.2A

For douglas fir-larch, $D = 3/4$ in and $G = 0.5$.

NDS Tbl 11.3.2, Ftn 2

θ, the angle between the direction of load and the direction of grain, is 0°.

The dowel bearing strength of the main member is

$$F_{em} = F_{es} = F_{es\parallel} = F_{em\parallel} = \left(11{,}200\ \frac{\text{lbf}}{\text{in}^2}\right) G = \left(11{,}200\ \frac{\text{lbf}}{\text{in}^2}\right)(0.5) = 5600\ \text{lbf/in}^2 \quad [\text{for } D \geq 1/4\ \text{in}]$$

NDS Tbl 11.3.1B

The angle to grain coefficient is

$$K_\theta = 1 + \frac{\theta}{360°} = 1 + 0 = 1$$

NDS Tbl 11.3.1A, Notes

$$R_e = \frac{F_{em}}{F_{es}} = \frac{5600\ \frac{\text{lbf}}{\text{in}^2}}{5600\ \frac{\text{lbf}}{\text{in}^2}} = 1$$

$$R_t = \frac{\ell_m}{\ell_s} = \frac{t_m}{t_s} = \frac{3.5\ \text{in}}{2.5\ \text{in}} = 1.4$$

NDS Tbl 11A, Ftn 2

The bending yield strength of the fastener, F_{yb}, is 45,000 lbf/in^2.

NDS Tbl 11.3.1B

$0.25\ \text{in} \leq D \leq 1.0\ \text{in}$

Therefore, the reduction term, R_d, varies for different yield modes accordingly with NDS Table 11.3.1B.

Substitution of the previously given values results in the following.

NDS Eq 11.3-1

The reference lateral design value for yield mode, I_m, is

$$Z = \frac{D\ell_m F_{em}}{R_d} = \frac{D\ell F_{em}}{4K_\theta} = \frac{(0.75\ \text{in})(3.5\ \text{in})\left(5600\ \frac{\text{lbf}}{\text{in}^2}\right)}{(4)(1)} = 3675\ \text{lbf}$$

NDS Eq 11.3-2

The reference lateral design value for yield mode, I_s, is

$$Z = \frac{D\ell_s F_{es}}{R_d} = \frac{D\ell_s F_{es}}{4K_\theta} = \frac{(0.75\ \text{in})(2.5\ \text{in})\left(5600\ \frac{\text{lbf}}{\text{in}^2}\right)}{(4)(1)} = 2625\ \text{lbf}$$

NDS Eq 11.3-3

Calculate the reference lateral design value for yield mode II.

$$k_1 = \frac{\sqrt{R_e + 2R_e^2(1+R_t+R_t^2) + R_t^2R_e^3} - R_e(1+R_t)}{1+R_e}$$

$$= \frac{\sqrt{1.0 + (2)(1.0)^2(1+1.4+(1.4)^2) + (1.4)^2(1.0)^3} - (1.0)(1+1.4)}{1+1}$$

$$= 0.509$$

$$Z = \frac{k_1 D \ell_s F_{es}}{R_d} = \frac{k_1 D \ell F_{es}}{3.6K_\theta}$$

$$= \frac{(0.509)(0.75\text{ in})(2.5\text{ in})\left(5600\ \frac{\text{lbf}}{\text{in}^2}\right)}{(3.6)(1)}$$

$$= 1484.58\text{ lbf}$$

NDS Eq 11.3-4

Calculate the reference lateral design value for yield mode III_m.

$$k_2 = -1 + \sqrt{2(1+R_e) + \frac{2F_{yb}(1+2R_e)D^2}{3F_{em}\ell_m^2}}$$

$$= -1 + \sqrt{(2)(1+1.0) + \frac{(2)\left(45{,}000\ \frac{\text{lbf}}{\text{in}^2}\right)(1+(2)(1.0))(0.75\text{ in})^2}{(3)\left(5600\ \frac{\text{lbf}}{\text{in}^2}\right)(3.5\text{ in})^2}}$$

$$= 1.17$$

$$Z = \frac{k_2 D \ell_m F_{em}}{(1+2R_e)R_d} = \frac{k_2 D \ell_m F_{em}}{(1+2R_e)(3.2K_\theta)}$$

$$= \frac{(1.177)(0.75\text{ in})(3.5\text{ in})\left(5600\ \frac{\text{lbf}}{\text{in}^2}\right)}{(1+(2)(1))(3.2)(1)}$$

$$= 1802.28\text{ lbf}$$

NDS Eq 11.3-5

Calculate the reference lateral design value for yield mode III_s.

$$k_3 = -1 + \sqrt{\frac{2(1+R_e)}{R_e} + \frac{2F_{yb}(1+2R_e)D^2}{3F_{em}t_s^2}}$$

$$= -1 + \sqrt{\frac{(2)(1+1)}{1} + \frac{(2)\left(45{,}000\ \frac{\text{lbf}}{\text{in}^2}\right)(2+(2)(1))(0.75\text{ in})^2}{(3)\left(5600\ \frac{\text{lbf}}{\text{in}^2}\right)(2.5\text{ in})^2}}$$

$$= 1.334$$

$$Z = \frac{k_3 D \ell_s F_{em}}{(2+R_e)R_d} = \frac{k_3 D \ell_s F_{em}}{(2+R_e)(3.2K_\theta)}$$

$$= \frac{(1.334)(0.75\text{ in})(2.5\text{ in})\left(5600\ \frac{\text{lbf}}{\text{in}^2}\right)}{(2+1)(3.2)(1)}$$

$$= 1459\text{ lbf}$$

NDS Eq 11.3-6

The lateral design value for yield mode IV is

$$Z = \frac{D^2}{R_d}\sqrt{\frac{2F_{em}F_{yb}}{3(1+R_e)}} = \frac{D^2}{3.2K_\theta}\sqrt{\frac{2F_{em}F_{yb}}{3(1+R_e)}}$$

$$= \frac{(0.75\ \text{in})^2}{(3.2)(1)}\sqrt{\frac{(2)\left(5600\ \frac{\text{lbf}}{\text{in}^2}\right)\left(45{,}000\ \frac{\text{lbf}}{\text{in}^2}\right)}{(3)(1+1)}}$$

$$= 1611\ \text{lbf}$$

Case III$_s$ controls. The reference lateral design value is 1459 lbf.

NDS Tbl 11A

NDS Table 11A for douglas fir-larch gives

t_m (in)	t_s (in)	$Z_{\parallel}$ (lbf)
3.5	1.5	1200
3.5	3.5	1610

By interpolation for $t_s = 2.5$ in for a 3/4 in bolt,

$$Z_{\parallel} = \frac{1200\ \text{lbf} + 1610\ \text{lbf}}{2}$$

$$= 1405\ \text{lbf} \quad \left[\begin{array}{c}\text{as compared with the 1459 lbf}\\ \text{determined by NDS yield limit equations}\end{array}\right]$$

Example 9.4
Wood-to-Wood Single-Shear Connection at an Angle

An assembly consisting of a 4×6 douglas fir-larch at an angle to grain of 60° with a 6×8 southern pine horizontal member is joined by a 1 in bolt as shown.

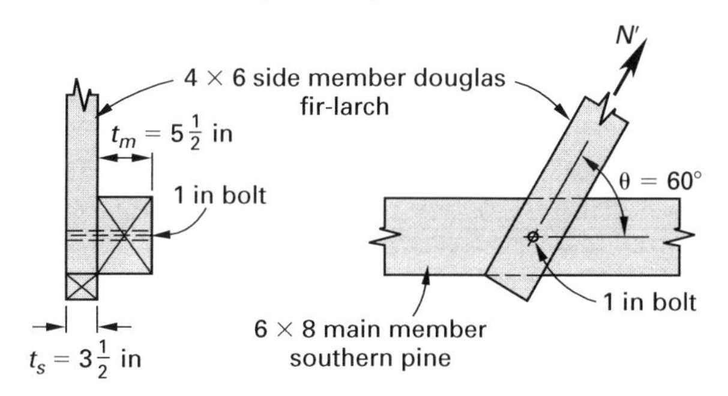

Using NDS tables, determine the allowable design value, N', for the joint. Assume that all adjustment factors are 1.0.

REFERENCE

Solution (ASD Method)

1. Bolt capacity in the 6 × 8 main southern pine member

For $t_m = 5^1/_2$ in and $t_s = 3^1/_2$ in,

NDS Tbl 11A

$P = Z_{\parallel} = 2870$ lbf members loaded $\parallel$ to grain (0°)

$Q = Z_{m\perp} = 1550$ lbf main member loaded $\perp$ to grain (90°)

The Hankinson formula (NDS App. J, Eq. J-4) gives the reference lateral design value at an angle to grain as follows.

NDS App J, Eq J-4; Tbl 10.3.1

The adjusted lateral design value at an angle to grain is

$$N' = N(C_D C_M C_t C_g C_\Delta C_{eq}) = \left(\frac{PQ}{P\sin^2\theta + Q\cos^2\theta}\right) \times (C_D C_M C_t C_g C_\Delta C_{eg})$$

$$= \left(\frac{(2870\text{ lbf})(1550\text{ lbf})}{(2870\text{ lbf})(\sin^2 60°) + (1550\text{ lbf})(\cos^2 60°)}\right) \times (1.0)(1.0)(1.0)(1.0)(1.0)(1.0)$$

$$= 1751.4\text{ lbf}$$

2. Bolt capacity in the 4 × 6 side douglas fir-larch member

NDS Tbl 11A

For a 4 × 6 side douglas fir-larch member, the adjusted lateral design value at an angle to grain is

$$N' = N(\text{adjustment factors}) = Z_{\|}(\text{adjustment factors})$$

$$= (2660\text{ lbf})(1.0)(1.0)(1.0)(1.0)(1.0)(1.0)$$

$$= 2660\text{ lbf}$$

$Z_{s\perp}$ is not applicable.

The allowable bolt capacity is 1751.4 lbf because the smallest allowable bolt capacity controls.

Example 9.5
Wood-to-Metal Single-Shear Connection at an Angle

A wood-to-metal connection is shown.

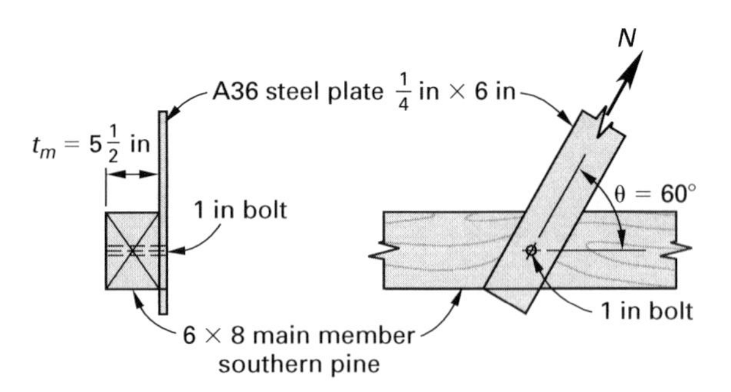

Determine the reference design value, N, for the connection by using NDS yield limit equations and NDS tables.

Note that the NDS tables can only be used for approximations because of simplifying assumptions for certain cases. However, use of these table values is usually sufficient for ordinary designs.

REFERENCE

Solution (ASD Method)

For a 6 × 8 southern pine main member,

$$t_m = 5.5\text{ in}$$

NDS Tbl 11B

$$P = Z_{\|} = 2980 \text{ lbf}$$
$$Q = Z_{\perp} = 1320 \text{ lbf}$$
$$\theta = 60°$$

Use the Hankinson formula to calculate the reference lateral design value at an angle to the grain.

NDS App Eq J-4

$$N = \frac{PQ}{P\sin^2\theta + Q\cos^2\theta} = \frac{(2980 \text{ lbf})(1320 \text{ lbf})}{(2980 \text{ lbf})(\sin^2 60°) + (1320 \text{ lbf})(\cos^2 60°)}$$
$$= 1533.57 \text{ lbf} \quad \text{[by NDS table values]}$$

NDS App Figs I1 and Mode I_s

Recall that only NDS Eq. 11.3-2 does not apply to wood-to-metal connections and does not have to be checked.

By checking NDS App. Eqs. 11.3-1, 11.3-3, 11.3-4, 11.3-5, and 11.3-6, Eq. 11.3-3 controls.

NDS Tbl 11.3.2A

$G = 0.55$ for southern pine, and $D = 1$ in.

NDS Tbl 11.3.2

$$F_{em\|} = 6150 \text{ lbf/in}^2 \text{ for } D \geq 1/4 \text{ in}$$
$$F_{em\perp} = 2550 \text{ lbf/in}^2 \text{ for } D = 1 \text{ in}$$

NDS Tbl 11B; Ftn 2

$$t_s = 0.25 \text{ in}$$

For ASTM A36 steel, dowel bearing strength, F_{es}, is 87,000 lbf/in^2.

From the Hankinson formula,

NDS App J or Eq 11.3-11

$$F_{e\theta} = \frac{F_{em\|}F_{em\perp}}{F_{em\|}\sin^2\theta + F_{em\perp}\cos^2\theta}$$
$$= \frac{\left(6150 \frac{\text{lbf}}{\text{in}^2}\right)\left(2550 \frac{\text{lbf}}{\text{in}^2}\right)}{\left(6150 \frac{\text{lbf}}{\text{in}^2}\right)(\sin^2 60°) + \left(2550 \frac{\text{lbf}}{\text{in}^2}\right)(\cos^2 60°)}$$
$$= 2987.1 \text{ lbf/in}^2$$

NDS Sec 11.3.1

$$R_e = \frac{F_{em}}{F_{es}} = \frac{F_{e\theta}}{F_{es}} = \frac{2987.1 \frac{\text{lbf}}{\text{in}^2}}{87{,}000 \frac{\text{lbf}}{\text{in}^2}} = 0.0343$$

$$R_t = \frac{\ell_m}{\ell_s} = \frac{5.5 \text{ in}}{0.25 \text{ in}} = 22$$

NDS Tbl 11.3.1B

$$K_\theta = 1 + (0.25)\left(\frac{60°}{90°}\right) = 1.1667$$

NDS Eq 11.3-3

Calculate the reference lateral design value for yield mode II.

$$k_1 = \frac{\sqrt{R_e + 2R_e^2(1 + R_t + R_t^2) + R_t^2R_e^3} - R_e(1 + R_t)}{1 + R_e}$$
$$= \frac{\sqrt{0.0343 + (2)(0.0343)^2\left(1 + 22 + (22)\right)^2 + (22)^2(0.0343)^3} - (0.0343)(1 + 22)}{1 + 0.0343}$$
$$= 0.3171$$

NDS Eq 11.3-3, Mode II

$$Z = \frac{k_1 D \ell_s F_{es}}{R_D} = \frac{k_1 D t_s F_{es}}{3.6 K_\theta} \quad [t_s = \ell_s]$$
$$= \frac{(0.3171)(1 \text{ in})(0.25 \text{ in})\left(87{,}000 \ \frac{\text{lbf}}{\text{in}^2}\right)}{(3.6)(1.1667)}$$
$$= 1642 \text{ lbf}$$

$$N = Z$$
$$= 1642 \text{ lbf} \quad \left[\begin{array}{c}\text{as compared with 1533.57 lbf} \\ \text{by the NDS table values}\end{array}\right]$$

NDS Eqs 11.3-1, 11.3-2, 11.3-4, 11.3-5, 11.3-6

Calculate the lateral reference design value for yield modes I_m, I_s, III_m, III_s, and IV.

$$k_2 = -1 + \sqrt{2(1+R_e) + \frac{F_{yb}(1+2R_e)D^2}{3F_{em}\ell_m^2}}$$
$$= -1 + \sqrt{(2)(1+0.03433) + \frac{\left(45{,}000 \ \frac{\text{lbf}}{\text{in}^2}\right)(1+(2)(0.03433))(1 \text{ in})^2}{(3)\left(2987 \ \frac{\text{lbf}}{\text{in}^2}\right)(5.5 \text{ in})^2}}$$
$$= 0.4987$$

$$k_3 = -1 + \sqrt{\frac{2(1+R_e)}{R_e} + \frac{2F_{yb}(2+R_e)D^2}{3F_{em}\ell_s^2}}$$
$$= -1 + \sqrt{\frac{(2)(1+0.03433)}{0.03433} + \frac{(2)\left(45{,}000 \ \frac{\text{lbf}}{\text{in}^2}\right)(2+0.03433)(1 \text{ in})^2}{(3)\left(2987 \ \frac{\text{lbf}}{\text{in}^2}\right)(0.25 \text{ in})^2}}$$
$$= 18.68$$

For yield mode I_m,

$$Z = \frac{D\ell_m F_{em}}{4K_\theta} = \frac{(1 \text{ in})(5.5 \text{ in})\left(2987 \ \frac{\text{lbf}}{\text{in}^2}\right)}{(4)(1.1667)}$$
$$= 3520 \text{ lbf}$$

For yield mode I_s,

$$Z = \frac{D\ell_s F_{es}}{4K_\theta} = \frac{(1 \text{ in})(0.25 \text{ in})\left(87{,}000 \ \frac{\text{lbf}}{\text{in}^2}\right)}{(4)(1.1667)}$$
$$= 4661 \text{ lbf}$$

For yield mode III_m,

$$Z = \frac{k_2 D\ell_m F_{em}}{(1+2R_e)3.2K_\theta} = \frac{(0.4987)(1 \text{ in})(5.5 \text{ in})\left(2987 \ \frac{\text{lbf}}{\text{in}^2}\right)}{(1+(2)(0.03433))(3.2)(1.1667)}$$
$$= 2053 \text{ lbf}$$

For yield mode III_s,

$$Z = \frac{k_3 D\ell_s F_{em}}{(2+R_e)3.2K_\theta} = \frac{(18.68)(1\text{ in})(0.25\text{ in})\left(2987\ \frac{\text{lbf}}{\text{in}^2}\right)}{(2+0.03433)(3.2)(1.1667)}$$
$$= 1837\text{ lbf}$$

For yield mode IV,

$$Z = \frac{D^2}{3.2K_\theta}\sqrt{\frac{2F_{em}F_{yb}}{3(1+R_e)}} = \frac{(1\text{ in})^2}{(3.2)(1.1667)}\sqrt{\frac{(2)\left(2987\ \frac{\text{lbf}}{\text{in}^2}\right)\left(45{,}000\ \frac{\text{lbf}}{\text{in}^2}\right)}{(3)(1+0.03433)}}$$
$$= 2493\text{ lbf}$$

Therefore, Z for yield mode II by the yield limit equations, 1642 lbf, controls. The value from NDS Table 11B and App. Eq. J-4 is 1533.57 lbf.

Note that for a thorough analysis, the load capacity of the A36 steel plate must also be checked.

2. Reference Design Values for Double-Shear Connections

[NDS 11.3; Eqs 11.3-7–11.3-10; Tbls 11F–11I]

The yield limit equations for bolt connections in double shear are different from those for bolts in single shear since these two cases have different yield modes (see NDS Fig. 11C and App. I).

Figure 9.3 Double-Shear Connection

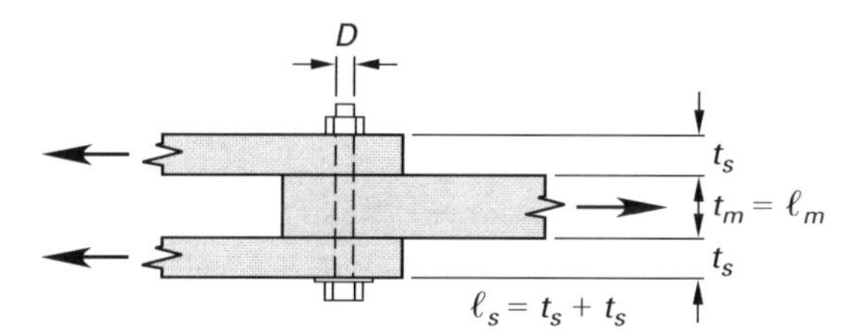

A. Wood-to-Wood Connections

The reference design value for one bolt in a double-shear connection between three wood members is the smallest load capacity obtained from the following four equations. The thicknesses of the side members are assumed to be equal. The reference design value for double shear connections is the smallest of the following equations from NDS Table 11.3.1A.

NDS Eq 11.3-7 $$Z = \frac{D\ell_m F_{em}}{R_d} \quad \text{[yield mode I}_m\text{]}$$

NDS Eq 11.3-8 $$Z = \frac{2D\ell_s F_{es}}{R_d} \quad \text{[yield mode I}_s\text{]}$$

NDS Eq 11.3-9 $$Z = \frac{2k_3 D\ell_s F_{em}}{(2+R_e)R_d} \quad \text{[yield mode III}_s\text{]}$$

NDS Eq 11.3-10

$$Z = \frac{2D^2}{R_d}\sqrt{\frac{2F_{em}F_{yb}}{3(1+R_e)}} \quad \text{[yield mode IV]}$$

The following notes from NDS Table 11.3.1A apply to these equations.

$$k_3 = -1 + \sqrt{\frac{2(1+R_e)}{R_e} + \frac{2F_{yb}(2+R_e)D^2}{3F_{em}l_s^2}}$$

$$R_e = \frac{F_{em}}{F_{es}}$$

$$K_\theta = (1+0.25)\left(\frac{\theta}{90^\circ}\right)$$

NDS App J or Eq 11.3-11

$$F_{e\theta} = \frac{F_{e\parallel}F_{e\perp}}{F_{e\parallel}\sin^2\theta + F_{e\perp}\cos^2\theta}$$

F_{em}	dowel bearing strength of main (center) member (See NDS Table 11.3.2.)	lbf/in^2
F_{es}	side member dowel bearing strength (see NDS Table 11.3.2.)	lbf/in^2
F_{yb}	bending yield strength of bolt (see NDS Table 11G Footnote 2)	45,000 lbf/in^2
θ	maximum angle of load to grain ($0 \leq \theta \leq 90^\circ$) for any member in connection	degree

$F_{em} = F_{e\parallel}$ for load parallel to grain

$= F_{e\perp}$ for load perpendicular to grain

$= F_{e\theta}$ for load at angle to grain θ (see the Hankinson formula)

$F_{es} = F_{e\parallel}$ for load parallel to grain

$= F_{e\perp}$ for load perpendicular to grain

$= F_{e\theta}$ for load at angle to grain θ (see the Hankinson formula)

For a steel side member, F_e = 87,000 lbf/in^2 for ASTM A36 steel (see NDS Table 11G Footnote 2).

The dowel bearing strength at an angle of load to grain θ is given by the Hankinson formula.

B. Wood-to-Metal Connections

For one bolt in a double-shear connection between a wood main member and two steel side plates, the reference design value is taken as the smallest load capacity obtained from the three equations previously given for a wood-to-wood connection (NDS Eqs. 11.3-7, 11.3-9, and 11.3-10). NDS Eq. 11.3-8 is not used to evaluate this type of wood-to-metal connection. In most cases, NDS Eq. 11.3-7 is applicable (see mode I_m in NDS App. Fig. I1). The designer is responsible for making sure that the bearing capacity of the steel side members is not exceeded as required by prevailing steel design practices.

A less frequently encountered double-shear connection involves a steel main (center) member and two wood side members. In this situation, NDS Eqs. 11.3-8, 11.3-9, and 11.3-10 are applicable. NDS Eq. 11.3-7 is not used to evaluate this type of wood-to-metal connection. The designer is responsible for making sure that the bearing capacity of the steel main member is not exceeded as required by prevailing steel design practices.

C. NDS Tables

As an alternative to NDS yield limit Eqs. 11.3-7–11.3-10, NDS Tables 11F–11I provide reference design values for double-shear connections for most common cases. The cases not included in these tables must be determined using the NDS yield limit equations.

3. Reference Design Values for Multiple-Shear Connections

[NDS Sec 11.3.8]

The reference design value for a multiple-shear connection is the smallest value obtained for a single-shear plane multiplied by the number of shear planes.

Figure 9.4 Multiple-Shear Connection

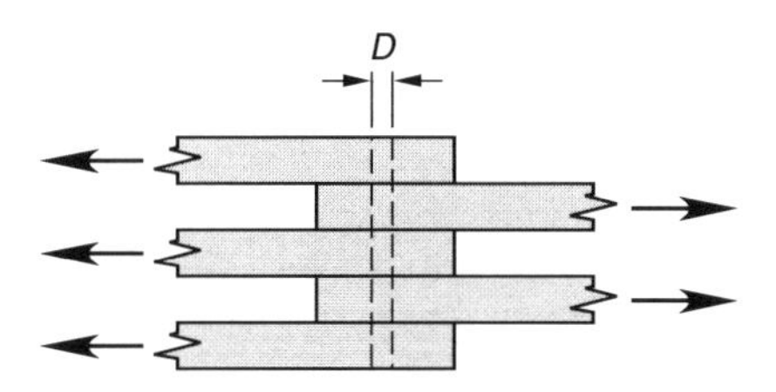

Example 9.6
Wood-to-Wood Double-Shear Connection for Sawn Lumber

A three-member joint is made of douglas fir-larch, select structural grade lumbers, 2 × 6 for side members and 3 × 6 for the main member, with two 3/4 in bolts as shown. A wet service condition exists. Assume all other adjustment factors are 1.0.

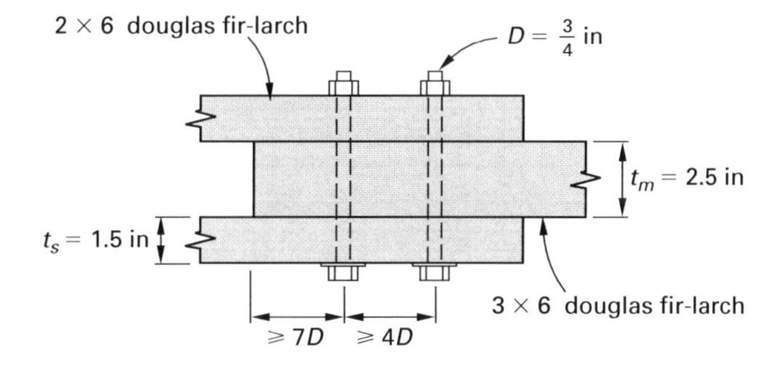

Determine the allowable tension load.

REFERENCE

Solution (ASD Method)

1. Bolt capacity

Find the adjusted lateral bolt design value, Z.

NDS Tbl 10.3.3 — $C_M = 0.7$

NDS Tbl 11F — For $t_m = 2^1/_2$ in, $t_s = 1^1/_2$ in, $^3/_4$ in bolt, and douglas fir-larch, the reference lateral bolt design value, $Z = Z_{\|}$, is 2400 lbf.

NDS Tbls 11.5.1B, 11.5.1C

The geometry factor, C_Δ is 1.0.

NDS Sec 10.3.6
NDS Tbls 10.3.6A, Ftn 1, 10.3.6.2B, 10.3.6.3

$$A_m = (2.5 \text{ in})(5.5 \text{ in}) = 13.75 \text{ in}^2$$

$$A_s = (2)\big((1.5 \text{ in})(5.5 \text{ in})\big) = 16.5 \text{ in}^2$$

$$\frac{A_s}{A_m} = \frac{16.5 \text{ in}^2}{13.75 \text{ in}^2} = 1.2 > 1.0$$

Use A_m instead of A_s.

$$\frac{A_m}{A_s} = \frac{13.75 \text{ in}^2}{16.5 \text{ in}^2} = 0.833 < 1.0$$

By interpolation, $C_g = 1.0$.

NDS Tbl 10.3.1

$$\begin{aligned} Z' &= ZC_DC_MC_tC_gC_\Delta \\ &= (2400 \text{ lbf})(1.0)(0.7)(1.0)(1.0)(1.0) \\ &= 1680 \text{ lbf/bolt} \end{aligned}$$

The allowable tension load as determined by the bolts is

$$\begin{aligned} P_{\text{allow}} &= Z'(\text{no. of bolts}) = \left(1680 \ \frac{\text{lbf}}{\text{bolt}}\right)(2 \text{ bolts}) \\ &= 3360 \text{ lbf} \end{aligned}$$

2. Lumber capacity

Find the adjusted tension design value, F_t', as determined by the lumber capacity. (The 3 × 6 main member controls the capacity.)

NDS Supp Tbl 4A Adj Fac

The reference tension design value, F_t, is 1000 lbf/in^2.

$$C_M = 1.0$$

$$C_F = 1.3$$

$$\begin{aligned} F_t' &= F_tC_DC_MC_tC_F \\ &= \left(1000 \ \frac{\text{lbf}}{\text{in}^2}\right)(1.0)(1.0)(1.0)(1.3) \\ &= 1300 \text{ lbf/in}^2 \end{aligned}$$

The allowable tension load as determined by wood tension capacity is

$$\begin{aligned} P_{\text{allow}} &= F_t'A_n = \left(1300 \ \frac{\text{lbf}}{\text{in}^2}\right)(2.5 \text{ in})\big(5.5 \text{ in} - (0.75 \text{ in} + 0.125 \text{ in})\big) \\ &= 15{,}031 \text{ lbf} \end{aligned}$$

3. Allowable tension load

The allowable tension load is 3360 lbf.

Example 9.7
Wood-to-Metal Double-Shear Connection for Glulam

A three-member tension joint consists of a 2$^1/_2$ in × 6 in glulam main member, combination 28 douglas fir (DF), and two 3 in wide by $^1/_4$ in thick A36 steel side plates. Assume wet service conditions.

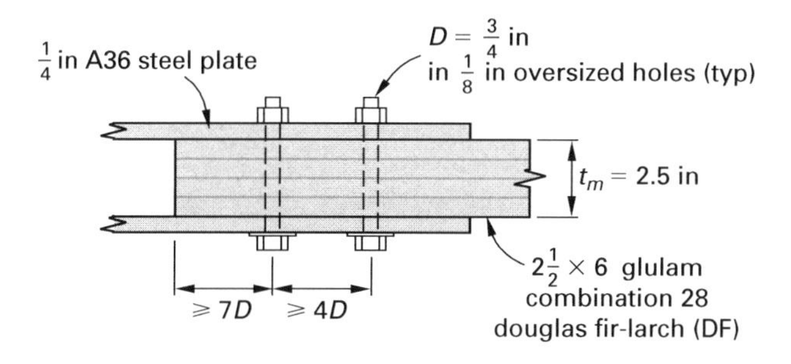

Determine the allowable tension load as based on bolt and glulam member capacities.

REFERENCE

Solution (ASD Method)

1. Bolt capacity

Find the adjusted bolt lateral design value, Z.

NDS Tbl 10.3.1

$$Z' = ZC_DC_MC_tC_gC_\Delta$$

NDS Tbl 10.3.3

$$C_M = 0.70$$

NDS Tbl 11I

For $t_m = 2^1/2$ in, $t_s = ^1/4$ in, douglas fir glulam combination, and $^3/4$ in bolt, the reference lateral bolt design value is

$$Z = Z_{\|} = 2630 \text{ lbf per bolt}$$

NDS Tbls 11.5.1C, 11.5.1B

The geometry factor, C_Δ, is 1.0.

NDS Eq 10.3-1
Table 10.3.6C

Calculate the group action factor C_g.

$$\begin{aligned}A_m &= (2.5 \text{ in})(6 \text{ in}) \\ &= 15 \text{ in}^2 \\ A_s &= (2)\big((3 \text{ in})(0.25 \text{ in})\big) \\ &= 1.5 \text{ in}^2 \\ \frac{A_m}{A_s} &= \frac{15 \text{ in}^2}{1.5 \text{ in}^2} \\ &= 10 < 12\end{aligned}$$

NDS Eq 10.3-1;
NDS Supp Tbl 5B

R_{EA} is the lesser of E_sA_s/E_mA_m and E_mA_m/E_sA_s.
The modulus of elasticity for the main member, E_m, is 2×10^6 lbf/in^2.
The modulus of elasticity for the side member, E_s, is 29×10^6 lbf/in^2.

$$R_{EA} = \begin{cases} \dfrac{E_sA_s}{E_mA_m} = \dfrac{\left(29 \times 10^6 \ \dfrac{\text{lbf}}{\text{in}^2}\right)(1.5 \text{ in}^2)}{\left(2 \times 10^6 \ \dfrac{\text{lbf}}{\text{in}^2}\right)(15 \text{ in}^2)} = 1.45 \\[2em] \dfrac{E_mA_m}{E_sA_s} = \dfrac{\left(2 \times 10^6 \ \dfrac{\text{lbf}}{\text{in}^2}\right)(15 \text{ in}^2)}{\left(29 \times 10^6 \ \dfrac{\text{lbf}}{\text{in}^2}\right)(1.5 \text{ in}^2)} = 0.689 \quad \text{[controls]} \end{cases}$$

The load/slip modulus is

$$\gamma = (270{,}000)D^{1.5} = (270{,}000)(0.75 \text{ in})^{1.5} = 175{,}370.1 \text{ in}$$

The center to center spacing between adjacent fasteners is

$$s = 4D = (4)(0.75 \text{ in}) = 3 \text{ in}$$

The number of fasteners in a row, n, is 2.

$$\mu = 1 + \gamma\frac{s}{2}\left(\frac{1}{E_mA_m} + \frac{1}{E_sA_s}\right) = 1.014$$

$$m = \mu - \sqrt{\mu^2 - 1} = 1.014 - \sqrt{(1.014)^2 - 1} = 0.846$$

NDS Eq 10.3-1

The group action factor is

$$C_g = \left(\frac{m(1-m^{2n})}{n(1+R_{EA}m^n)(1+m) - 1 + m^{2n}}\right)\left(\frac{1+R_{EA}}{1-m}\right)$$

$$= \left(\frac{(0.846)\left(1-(0.846)^{(2)(2)}\right)}{(2)\left(1+(0.689)(0.846)^2\right)(1+0.846) - 1 + (0.846)^{(2)(2)}}\right) \times \left(\frac{1+0.689}{1-0.846}\right)$$

$$= 1.13 \quad [\text{Use } C_g = 1.0]$$

The adjusted lateral bolt design value is

$$Z' = ZC_DC_mC_tC_gC_\Delta = (2630 \text{ lbf})(1.0)(0.7)(1.0)(1.0)(1.0) = 1841 \text{ lbf/bolt}$$

The allowable tension load as determined by the bolts is

$$P_{\text{allow}} = Z'(\text{no. of bolts}) = \left(1841 \ \frac{\text{lbf}}{\text{bolt}}\right)(2 \text{ bolts}) = 3682 \text{ lbf}$$

2. Tension capacity of glulam member

Find the adjusted tension design value, F_t', as determined by the glulam tension capacity.

NDS Supp Tbl 1C

$2^1/_2$ in × 6 in is made up of four laminations.

NDS Supp Tbl 5B Adj Fac

The reference tension design value, F_t, is 1100 lbf/in^2.
$C_M = 0.8$

NDS Tbl 5.3.1

$$F_t' = F_tC_DC_MC_t \quad [\text{for glulam members}]$$

$$= \left(1100 \ \frac{\text{lbf}}{\text{in}^2}\right)(1.0)(0.8)(1.0) = 880 \text{ lbf/in}^2$$

The allowable tension load as determined by glulam tension capacity is

$$P_{\text{allow}} = F_t' A_n = \left(800\ \frac{\text{lbf}}{\text{in}^2}\right)(2.5\ \text{in})\big(6\ \text{in} - (0.75\ \text{in} + 0.125\ \text{in})\big)$$
$$= 10{,}250\ \text{lbf}$$

3. Allowable tension load

The allowable tension load is 3682 lbf. Note that although this problem did not require that the tension capacities of the steel side members be checked, they should be checked in an actual design and/or analysis situation.

10

Lag Screws and Wood Screws

1. Lag Screws

[NDS Sec 11.1.3, App Tbls L2, 11.2.1, 11.4.1]

Lag screws, also called *lag bolts*, are more similar to bolts than they are to wood screws. Lag screws, which have a hex or square head, consist of nonsmooth and smooth shanks. Their abilities to resist withdrawal and lateral loads are superior to those of nails. Their diameter ranges from $^1/_4$ in to $1^1/_4$ in, and their length varies from 1 in to 12 in (see NDS Sec. 11.1.3 and NDS App. L). Wood screw diameters range from $^1/_8$ in to $^3/_8$ in (circular round or flat head) and $^1/_2$ in to 4 in long.

Lag screws are inserted into lead holes. However, lead holes are not required for $^3/_8$ in and smaller diameter lag screws loaded primarily in withdrawal with $G < 0.50$ (see NDS Sec. 11.1.3.3).

2. Withdrawal Design Values for Lag Screws

[NDS Secs 11.2.1, 11.4.1, 11.5.1, 11.5.2; NDS Tbls 10.3.1, 11.2A]

The reference design value for withdrawal specified in NDS Table 11.2A and NDS Eq. 11.2-1 is the withdrawal value per inch of thread penetration into the side grain of the main (holding) member.

The reference withdrawal design value is

$$W = 1800G^{3/2}D^{3/4} \quad \text{[NDS Eq. 11.2-1 or Table 11.2A]}$$

The allowable withdrawal design value is

$$\begin{aligned} W' &= \text{(reference design values)(product of adjustment factors)} \\ &= WpC_DC_MC_tC_{eg} \quad \text{[NDS Table 10.3.1]} \end{aligned}$$

W	reference withdrawal design value (NDS Table 11.2A)	lbf/in of penetration
p	effective threaded penetration	in
C_D	load duration factor (NDS Table 2.3.2; NDS Sec. 10.3.2)	–
C_M	wet service factor (NDS Table 10.3.3)	–
C_t	temperature factor (NDS Sec. 10.3.4)	–
C_{eg}	end grain factor if loaded in withdrawal from end grain (NDS Sec. 11.5.2 or Sec. 11.2.1.2)	0.75
G	specific gravity (NDS Table 11.3.2A)	–

Figure 10.1 Lag Screws: Effective Thread Penetration into Main Member, $p = T - E$ (based on NDS Appendix Table L2)

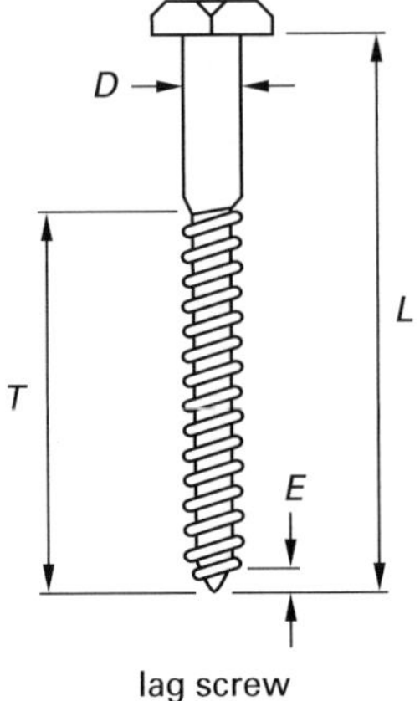

(Note that the minimum thread length, T, for lag screw length, L, is 6 in or 1/2 the lag screw length plus 0.5 in, whichever is smaller.)

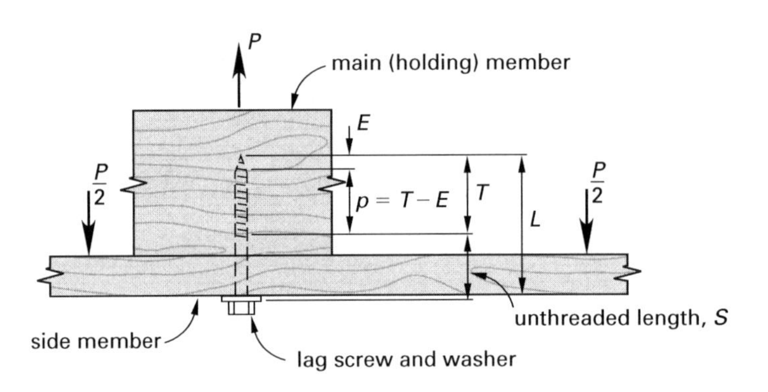

Example 10.1
Lag Screw Connection for Withdrawal

A 1/2 in × 6 in lag screw is used to attach a 4 × 16 southern pine side member to a 12 × 12 southern pine beam as shown. The lumber is exposed to weather in service. Assume that the load is permanent ($C_D = 0.9$ as per NDS Table 2.3.2) and the connection is under normal temperature conditions ($C_t = 1.0$).

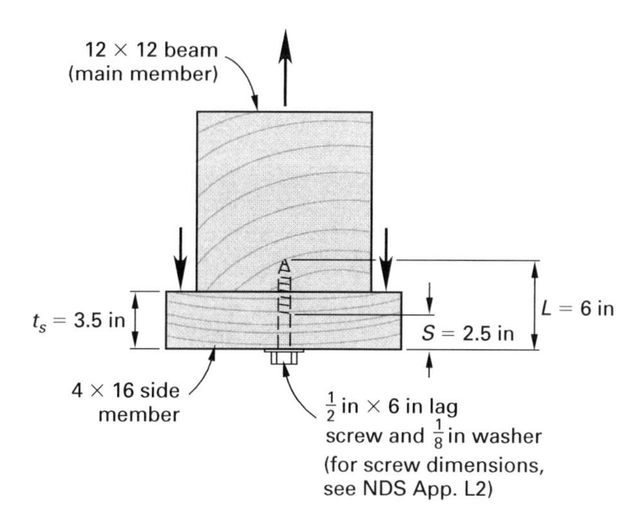

Determine the allowable withdrawal load.

REFERENCE *Solution (ASD Method)*

NDS App Tbl L2 For a 1/2 in × 6 in lag screw with an 1/8 in washer, the length of lag screw into beam (main member) is

$$\begin{aligned}\text{length of lag screw into main beam} &= 6\text{ in} - 0.125\text{ in} - 3.5\text{ in} \\ &= 2.375\text{ in}\end{aligned}$$

NDS App Tbl L2 The length of the tapered tip, E, is 5/16 in.

The effective thread penetration length into beam (main member) is

$$\begin{aligned}p &= 2.375\text{ in} - E \\ &= 2.375\text{ in} - \tfrac{5}{16}\text{ in} \\ &= 2.06\text{ in}\end{aligned}$$

NDS Tbl 2.3.2 For the permanent dead load, the load duration factor, C_D, is 0.9.

NDS Tbl 10.3.3 With exposure to weather, the wet service factor, C_M, is 0.7.

NDS Sec 11.5.2 With no end grain application, the end grain factor, C_{eg}, is 1.0.

NDS Tbl 11.3.2A For southern pine, the specific gravity, G, is 0.55.

NDS Tbl 11.2A The reference design load, W, is 437 lbf per inch effective penetration.

The allowable design load is

$$\begin{aligned}W' &= WpC_DC_MC_tC_{eg} \\ &= \left(437\ \frac{\text{lbf}}{\text{in}}\right)(2.06\text{ in})(0.9)(0.7)(1.0)(1.0) \\ &= 567.1\text{ lbf}\end{aligned}$$

3. Lateral Design Values

[NDS Sec 11.3; NDS Tbls 10.3.1, 10.3.3, 11J, 11K]

The yield limit equations for lag screws in single shear are the same as those for bolt connections (see NDS Sec. 11.3).

The minimum penetration excluding the length of tapered tip into the main member is $p_{\text{min}} = 4D$ (see NDS Sec. 11.1.3.6).

The reference lateral design values apply when the minimum lag screw penetration, p, into the main member is $8D$ (excluding the length of tapered tip). When $4D \leq p < 8D$, reference values must be multiplied by $p/8D$. (See NDS Tables 11J and 11K, Footnotes 2 and 3.)

A. Wood-to-Wood Connections

[NDS Tbl 11J]

For one lag screw in a single-shear connection between two wood members, the reference design value is taken as the smallest load capacity obtained from NDS Eq. 11.3-1 for yield mode I_m, NDS Eq. 11.3-4 for yield mode III_m, and NDS Eq. 11.3-6 for yield mode IV. Find these equations and related notes in Table 8.1 (NDS Tables 11.3.1A and 11.3.1B). The reference design values used in common applications are also given in NDS Table 11J.

NDS Eq 11.3-1

$$Z = \frac{D\ell_m F_{em}}{R_d} \quad \text{[yield mode I}_m\text{]}$$

NDS Eq 11.3-4

$$Z = \frac{k_2 D\ell_m F_{em}}{(1 + 2R_e)R_d} \quad \text{[yield mode III}_m\text{]}$$

NDS Eq 11.3-6

$$Z = \frac{D^2}{R_d}\sqrt{\frac{2F_{em}F_{yb}}{3(1 + R_e)}} \quad \text{[yield mode IV]}$$

The terms are defined as follows.

D	diameter (see NDS Sec. 11.3.6 and NDS App. Table L2)	in
ℓ_m	length of dowel bearing in wood main member	in
ℓ_s	length of dowel bearing in wood side member	in
R_d	reduction terms based on dowel member diameter (see NDS Table 11.3.1B)	–

The value for the dowel (fastener) bending yield strength, F_{yb}, varies as follows. (See footnote of NDS Table 11J.)

$$\begin{aligned} F_{yb} &= 70{,}000\ \text{lbf/in}^2 \text{ when } D = 1/4\ \text{in} \\ &= 60{,}000\ \text{lbf/in}^2 \text{ when } D = 5/16\ \text{in} \\ &= 45{,}000\ \text{lbf/in}^2 \text{ when } D \geq 3/8\ \text{in} \end{aligned}$$

The value for the main member dowel bearing strength, F_{em}, varies as follows. (See NDS Tables 11.3.2 and 11.3.2A.)

$$\begin{aligned} F_{em} &= F_{e\|} \text{ for parallel-to-grain loading} \\ &= F_{e\perp} \text{ for perpendicular-to-grain loading} \\ &= F_{e\theta} \text{ for load at angle to grain } \theta \text{ (see the Hankinson formula)} \end{aligned}$$

The side member dowel bearing strength, F_{es}, varies as follows. (See NDS Tables 11.3.2 and 11.3.2A.)

$$\begin{aligned} F_{es} &= F_{e\|} \text{ for parallel-to-grain loading} \\ &= F_{e\perp} \text{ for perpendicular-to-grain loading} \\ &= F_{e\theta} \text{ for load at angle to grain } \theta \text{ (see NDS Table 11.3.2)} \\ &= F_u \text{ for a steel member} \end{aligned}$$

The dowel bearing strength at an angle of load to grain, θ, is given by the Hankinson formula.

$$F_{e\theta} = \frac{F_{e\|}F_{e\perp}}{F_{e\|}\sin^2\theta + F_{e\perp}\cos^2\theta} \quad \text{[NDS Eq. 11.3-11]}$$

B. Wood-to-Metal Connections

[NDS Tbl 11K]

For one lag screw in a single-shear connection between a wood main member and a steel side plate, the reference design value, Z, is the smallest load capacity obtained from the NDS yield Eqs. 11.3-1, 11.3-4, and 11.3-6. NDS Eqs. 11.3-2, 11.3-3, and 11.3-5 are eliminated for wood-to-metal connections. In applying the yield limit equations, F_{es} may conservatively be taken as equal to the ultimate tensile strength, F_u, of the steel side plate. The designer is responsible for ensuring that the bearing capacity of the steel side member is not exceeded in accordance with recognized steel design practices.

Dowel bearing strength, F_e, of steel side plates is 61,850 lbf/in^2 for ASTM A653 Grade 33 steel, and 87,000 lbf/in^2 for ASTM A36 steel (NDS Table 11K, Footnote 2).

As an alternative to solving these yield limit equations, the nominal design values for lag screws used in common applications are given in NDS Tables 11J and 11K. These values are given for parallel- and perpendicular-to-grain loadings with the following definitions of the notations used.

For F_{es} values, NDS Table 11K footnotes are referenced as follows.

$Z_\|$	loading parallel to grain in both main and side members	lbf
$Z_{s\perp}$	loading perpendicular to grain in side member and parallel to grain in main member	lbf
$Z_{m\perp}$	loading perpendicular to grain in main member and parallel to grain in side member	lbf

The reference design values in NDS Tables 11J and 11K are for the base conditions: dry at fabrication and in service, normal temperature, penetration of lag screw into the main member with a minimum penetration length of eight times the size of the shank diameter, and other minimum parameters necessary for base (minimum) conditions.

See footnotes for NDS Tables 11J and 11K for information regarding the penetration depth factor, C_d.

C. Allowable Design Value, Z'

[NDS Tbls 10.3.1, 10.3.3]

The allowable design values are given by

$$\begin{aligned} Z' &= (\text{reference design values})(\text{product of adjustment factors}) \quad \text{[NDS Table 10.3.1]} \\ &= ZC_DC_MC_tC_gC_\Delta C_dC_{eg} \end{aligned}$$

C_D	load duration factor (NDS Sec. 2.3.2)	–
C_M	wet service factor (NDS Sec. 10.3.3)	–
C_t	temperature factor (NDS Sec. 10.3.3)	–
C_g	group action factor (NDS Sec. 10.3.6; NDS Tables 10.3.6A, 10.3.6C)	–
C_Δ	geometry factor (same as for bolted connections; NDS Sec. 11.5.1)	–
C_{eg}	end grain factor (NDS Sec. 11.5.2)	0.67
C_d	penetration factor (NDS erroneously excludes C_d for lag screws)	$\frac{p}{8D}$

Example 10.2
Lateral Loads on Lag Screws for Wood-to-Metal Connection

In the illustration shown, two 5/8 in × 5 in lag screws are used to attach a 1/4 in × 4 in steel strap to a 5 1/8 in × 6 in douglas fir-larch glulam beam. The beam is exposed to weather in service. Assume normal temperature ($C_t = 1.0$). The total load consists of dead load and wind load ($w_D + w_W$).

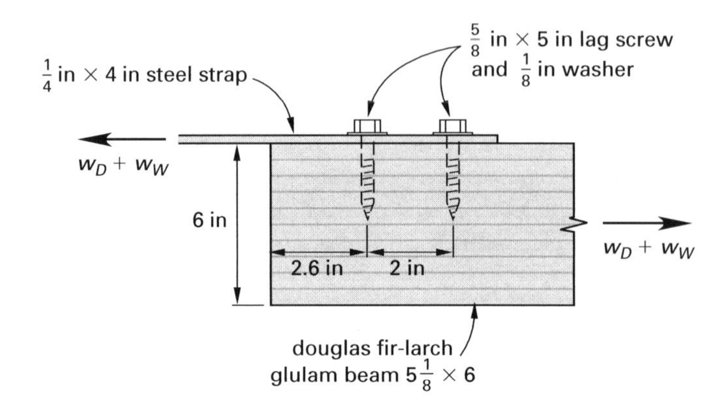

Determine the allowable lateral load that can be carried by the two lag screws.

REFERENCE

Solution (ASD Method)

Calculate the reference design value, Z, using NDS Eqs. 11.3-2, 11.3-3, and 11.3-4.

$\theta = 0°$

The diameter, D, is 0.625 in.

NDS Tbl 11K, Ftn 2 — For $D \geq$ 3/8 in, the bending yield strength is $F_{yb} = 45{,}000$ lbf/in^2.

NDS Tbl 11.3.2A — The specific gravity for the main member, G, is 0.5.

NDS Tbl 11.3.2 — For parallel-to-grain loading, the main member dowel strength is

$$F_{em} = F_{e\parallel} = 5600 \text{ lbf/in}^2 \text{ for } D \geq 1/4 \text{ in}$$

NDS Tbl 11K, Ftn 2 — The side member dowel strength, F_{es}, is 61,850 lbf/in^2 for ASTM A653 Grade 33 steel side plate.

The side member thickness, t_s, is 0.25 in.

The length of the dowel bearing in the wood main member is

$$\begin{aligned} \ell_m = p &= L - t_s - t_{\text{washer}} - E_{\text{tip}} \\ &= 5 \text{ in} - \tfrac{1}{4} \text{ in} - \tfrac{1}{8} \text{ in} - \tfrac{13}{32} \text{ in} \\ &= 4.22 \text{ in} \\ \ell_s &= 0.25 \text{ in} \end{aligned}$$

NDS Tbl 11K, Ftn 3 — $4D = 2.5 \text{ in} < p = 4.22 \text{ in} < 8D = 5 \text{ in}$

The penetration depth factor, C_d, is

$$\frac{p}{8D} = \frac{4.22 \text{ in}}{5 \text{ in}} = 0.844$$

NDS Tbl 11.3.1A, notes

$$R_e = \frac{F_{em}}{F_{es}} = \frac{5600 \ \frac{\text{lbf}}{\text{in}^2}}{61{,}850 \ \frac{\text{lbf}}{\text{in}^2}} = 0.09054$$

NDS Tbl 11.3.1B

$D \geq 0.25$ in and ≤ 1.0 in. Therefore, the reduction term, R_d, varies for different yield modes.

NDS Eq 11.3-1, Mode I_m

The reference design value for yield mode I_m is

$$Z = \frac{D\ell_m F_{em}}{R_d} = \frac{D\ell F_{em}}{4K_\theta} = \frac{(0.625 \text{ in})(4.22 \text{ in})\left(5600 \ \frac{\text{lbf}}{\text{in}^2}\right)}{(4)(1)} = 3692.5 \text{ lbf}$$

NDS Eq 11.3-2, Mode I_s

The reference design value for yield mode I_s is

$$Z = \frac{D\ell_s F_{es}}{R_d} = \frac{D\ell_s F_{es}}{4K_\theta} = \frac{(0.625 \text{ in})(0.25 \text{ in})\left(61{,}850 \ \frac{\text{lbf}}{\text{in}^2}\right)}{(4)(1)} = 2416 \text{ lbf}$$

Calculate the reference design value for yield mode II.

$$k_1 = \frac{\sqrt{R_e + 2R_e^2(1 + R_t + R_t^2) + R_t^2 R_e^3} - R_e(1 + R_t)}{1 + R_e}$$

$$= \frac{\sqrt{0.09054 + (2)(0.09054)^2\left(1 + 18.5 + (18.5)^2\right) + (18.5)^2(0.09054)^3} - (0.09054)(1 + 18.5)}{1 + 0.054} = 0.6782$$

NDS Eq 11.3-3, Mode II

$$Z = \frac{k_1 D\ell_s F_{es}}{R_d} = \frac{k_1 D\ell_s F_{es}}{3.6K_\theta} = \frac{(0.6782)(0.625 \text{ in})(0.25 \text{ in})\left(61{,}850 \ \frac{\text{lbf}}{\text{in}^2}\right)}{(3.6)(1)} = 1821 \text{ lbf}$$

$$k_2 = -1 + \sqrt{2(1+R_e) + \frac{2F_{yb}(1+2R_e)D^2}{3F_{em}l_m^2}}$$

$$= -1 + \sqrt{(2)(1+0.09054) + \frac{(2)\left(45{,}000\ \frac{\text{lbf}}{\text{in}^2}\right)(1+(2)(0.09054)) \times (0.625\ \text{in})^2}{(3)\left(5600\ \frac{\text{lbf}}{\text{in}^2}\right)(4.22)^2}}$$

$$= 0.568$$

NDS Eq 11.3-4, Mode III_m

$$Z = \frac{k_2 D\ell_m F_{em}}{(1+2R_e)R_d} = \frac{k_2 D\ell_m F_{em}}{(1+2R_e)K_\theta}$$

$$= \frac{(0.568)(0.625\ \text{in})(4.22\ \text{in})\left(5600\ \frac{\text{lbf}}{\text{in}^2}\right)}{(1+(2)(0.09054))(3.2)(1)}$$

$$= 2138\ \text{lbf}$$

NDS Eq 11.3-5, Mode III_s

Calculate the reference design value for yield mode III_s.

$$k_3 = -1 + \sqrt{\frac{2(1+R_e)}{R_e} + \frac{2F_{yb}(2+R_e)D^2}{3F_{em}\ell_s^2}}$$

$$= -1 + \sqrt{\frac{(2)(1+0.09054)}{1} + \frac{(2)\left(45{,}000\ \frac{\text{lbf}}{\text{in}^2}\right)(2+0.09054) \times (0.625\ \text{in})^2}{(3)\left(5600\ \frac{\text{lbf}}{\text{in}^2}\right)(0.25\ \text{in})^2}}$$

$$= 7.5$$

$$Z = \frac{k_3 D\ell_s F_{em}}{(2+R_e)R_d} = \frac{k_3 D\ell_s F_{em}}{(2+R_e)3.2K_\theta}$$

$$= \frac{(7.5)(0.625\ \text{in})(0.25\ \text{in})\left(5600\ \frac{\text{lbf}}{\text{in}^2}\right)}{(2+0.09054)(3.2)(1)}$$

$$= 981\ \text{lbf}$$

NDS Eq 11.3-6

Calculate the reference design value for yield mode IV.

$$Z = \frac{D^2}{R_d}\sqrt{\frac{2F_{em}F_{yb}}{3(1+R_e)}} = \frac{D_2}{3.2K_\theta}\sqrt{\frac{2F_{em}F_{es}}{3(1+R_e)}}$$

$$= \frac{(0.625\ \text{in})^2}{(3.2)(1)}\sqrt{\frac{(2)\left(5600\ \frac{\text{lbf}}{\text{in}^2}\right)\left(45{,}000\ \frac{\text{lbf}}{\text{in}^2}\right)}{(3)(1+0.09054)}}$$

$$= 1515.1\ \text{lbf}$$

Yield mode III_s controls. The reference design value is 981 lbf.

Compare this value to the reference design value of 750 lbf found using NDS Table 11K for douglas fir-larch with a specific gravity of 0.5 and a lag screw root diameter of $D_r = 0.471$ in ($D = 5/8$ in).

Note that NDS Table 11K values are for "reduced body diameter" lag screws, as indicated in footnote 2 of that table. Footnote 1 of

NDS App. Table L2 indicates that the "reduced body diameter" is approximately equal to the screw root diameter, D_r. For a 5/8 in lag ($D = 0.625$ in) screw, as in this example, the screw root diameter is 0.471 in.

Therefore, while the NDS yield limit equations used $D = 5/8$ in for 981 lbf, the NDS Table 11K values used $D_r = 0.471$ in for 750 lbf.

Calculate the adjustment factors.

NDS Sec 2.3.2

For wind loads, the load diameter factor, C_D, is 1.6.

NDS Tbl 10.3.3

For exposed-to-weather conditions, the wet service factor, C_M, is 0.7.

NDS Tbl 10.3.6A

Determine the group factor, C_g.

The spacing, s, is 2 in.

For a $5^1/8$ in × 6 in glulam beam,

$$\begin{aligned} A_m &= (5.125 \text{ in})(6 \text{ in}) \\ &= 30.75 \text{ in}^2 \end{aligned}$$

For a 1/4 in × 4 in steel strap,

$$\begin{aligned} A_s &= (0.25 \text{ in})(4 \text{ in}) \\ &= 1 \text{ in}^2 \end{aligned}$$

$$\begin{aligned} \frac{A_m}{A_s} &= \frac{30.75 \text{ in}^2}{1 \text{ in}^2} \\ &= 30.75 \end{aligned}$$

For two lag screws in a row, $C_g = 1.0$.

NDS Sec 11.5.1

Calculate the geometry factor, C_Δ.

NDS Sec 11.5.1A

Check the edge distance.

The minimum edge distance required is

$$\begin{aligned} 1.5D &= (1.5)(0.625 \text{ in}) \\ &= 0.94 \text{ in} \end{aligned}$$

The actual edge distance is

$$\begin{aligned} &\left(6 \text{ in} - \left(\tfrac{5}{8} \text{ in} + \tfrac{1}{8} \text{ in}\right)\right)\left(\tfrac{1}{2}\right) \\ &= 2.62 \text{ in} > 1.5D \quad \text{[OK]} \end{aligned}$$

NDS Sec 11.5.1B

Determine the end distance factor. The actual end distance is 2.6 in.

The minimum end distance required for full design value is

$$7D = (7)(0.625 \text{ in}) = 4.38 \text{ in}$$

The minimum end distance required for reduced design value is

$$\begin{aligned} 3.5D &= (3.5)(0.625 \text{ in}) \\ &= 2.19 \text{ in} < \text{actual end distance} = 2.6 \text{ in} \quad \text{[OK]} \end{aligned}$$

The geometry factor is

$$C_\Delta = \frac{\text{actual end distance}}{\text{minimum end distance required for full design values}} = \frac{2.6 \text{ in}}{4.38 \text{ in}} = 0.594 \quad \text{[controls]}$$

NDS Sec 11.5.1C

Find the spacing factor, C_Δ.

The minimum spacing required for full design value is

$$4D = (4)(0.625 \text{ in}) = 2.5 \text{ in}$$

$$8D = (8)(0.625 \text{ in}) = 5 \text{ in}$$

For reduced design value,

$$3D = (3)(0.625 \text{ in}) = 1.88 \text{ in} < \text{actual spacing} = 2 \text{ in} \quad \text{[OK]}$$

The spacing factor is

$$C_\Delta = \frac{\text{actual spacing}}{\text{minimum spacing required for full design values}} = \frac{2 \text{ in}}{2.5 \text{ in}} = 0.80$$

The smaller C_Δ controls. Therefore, use $C_\Delta = 0.594$.

NDS Eq 9.3-5

Find the penetration depth factor, C_d.

The actual penetration, p, into the beam is

NDS App Tbl L2

For a 5/8 in × 5 in lag screw,

$$p = L - t_s - t_{\text{washer}} - E_{\text{tip}} = 5 \text{ in} - \tfrac{1}{4} \text{ in} - \tfrac{1}{8} \text{ in} - \tfrac{13}{32} \text{ in} = 4.22 \text{ in}$$

NDS Tbl 11K, Ftns 2 and 3

Since $4D \le p \le 8D$, the value must be multiplied by $p/8D$ in accordance with NDS Table 11K footnotes.

The penetration depth factor is

$$C_d = \frac{p}{8D} = \frac{4.22 \text{ in}}{5 \text{ in}} = 0.844$$

NDS Sec 11.5.2

Find the end grain factor, C_{eg}.

Lag screws are inserted into the side grain of the main member. Therefore, $C_{eg} = 1.0$.

NDS Tbl 10.3.1

The allowable design value is

$$Z' = ZC_DC_MC_tC_gC_\Delta C_dC_{eg} = (981 \text{ lbf})(1.6)(0.7)(1.0)(1.0)(0.594)(0.844)(1.0) = 641 \text{ lbf per lag screw}$$

The allowable lateral load is

$$P_{\text{allow}} = \left(641\ \frac{\text{lbf}}{\text{screw}}\right)(2\ \text{lag screws})$$
$$= 1282\ \text{lbf}$$

Note that a thorough analysis also requires that the load capacity of the $^1/_4$ in × 4 in steel strap be checked.

Example 10.3
Lateral Loads on Lag Screws for Wood-to-Wood Connection

A $^1/_2$ in × 6 in lag screw connects a 2 × 6 × 14 shelf beam to a 2 × 4 wall stud as shown. The total load of 250 lbf consists of dead load and live load ($w_D + w_L$). Assume a dry condition and douglas fir-larch lumber, and that $C_D = C_M = C_t = C_g = 1.0$.

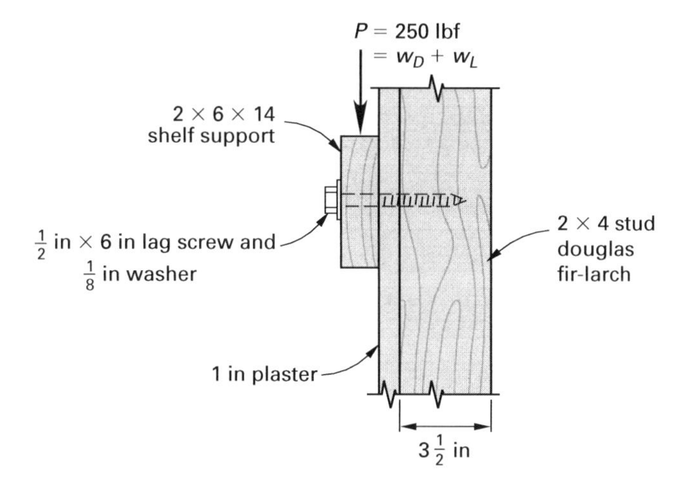

Determine whether the connection is strong enough to support the total load and calculate the allowable lag screw capacity.

REFERENCE *Solution (ASD Method)*

NDS Tbl 10.3.1 $Z' = ZC_DC_MC_tC_gC_\Delta C_dC_{eg}$

NDS Tbl 11J With a side member thickness of 1.5 in, $Z = Z_{s\perp} = 220$ lbf. The main member is loaded parallel to the grain, and the side member is loaded perpendicular to the grain.

Calculate the adjustment factors.

The wet service factor, the temperature factor, and the load duration factor are each 1.0.

1. Check the geometry factors for the main member (loaded parallel to grain).

NDS Sec 11.5.1 The geometry factor, C_Δ, is found as follows.

NDS Sec 11.5.1B C_Δ is determined from the end distance requirement.

The actual end distance provided is several inches.

The minimum end distance required for full design value is

$$\begin{aligned}4D &= (4)\left(\tfrac{1}{2}\text{ in}\right)\\ &= 2.0\text{ in}\end{aligned}$$

For reduced design value,

$$\begin{aligned}2D &= (2)\left(\tfrac{1}{2}\text{ in}\right)\\ &= 1.0\text{ in}\end{aligned}$$

The geometry factor is

$$C_\Delta = \frac{\text{actual end distance}}{2.0\text{ in}} > 1.0$$

Therefore, $C_\Delta = 1.0$.

NDS Sec 11.5.1C

The bolt spacing factor, C_Δ, is not applicable. Therefore, $C_\Delta = 1.0$.

NDS Sec 11.5.1A

Check the edge distance.

$$\text{actual edge distance} = \frac{1.5\text{ in}}{2} = 0.75\text{ in}$$

$$\begin{aligned}\text{required edge distance} &= 1.5D\\ &= (1.5)\left(\tfrac{1}{2}\text{ in}\right)\\ &= 0.75\text{ in}\end{aligned}$$

Therefore, the edge distance is acceptable.

NDS Sec 11.5.1A

2. Check the edge distance for the side member (loaded perpendicular to grain).

The required loaded edge distance is

$$\begin{aligned}4D &= (4)\left(\tfrac{1}{2}\text{ in}\right)\\ &= 2.0\text{ in}\end{aligned}$$

The actual loaded edge distance is

$$\frac{5.5\text{ in}}{2} = 2.75\text{ in} > 4D$$

Therefore, the loaded edge distance is acceptable.

NDS Tbl 10.3.6

Since there is one lag screw, the group action factor, C_g, is 1.0.

Find the penetration depth factor, C_d.

NDS App Tbl L2

For $\frac{1}{2}$ in × 6 in lag screw, $L = 6$ in; $E_{\text{tip}} = \frac{5}{16}$ in.

The actual penetration into the beam is

$$\begin{aligned}p &= L - (t_s + t_{\text{washer}}) - 1\text{ in plaster} - E_{\text{tip}}\\ &= 6\text{ in} - \left(1.5\text{ in} + \tfrac{1}{8}\text{ in}\right) - 1\text{ in} - \tfrac{5}{16}\text{ in}\\ &= 3.06\text{ in}\end{aligned}$$

NDS Tbl 11J Ftns 2 and 3

The minimum penetration required for full design value is

$$8D = (8)\left(\tfrac{1}{2}\text{ in}\right) = 4\text{ in}$$

For the reduced design value,

$$4D = (4)\left(\tfrac{1}{2}\text{ in}\right) = 2\text{ in}$$

$$2\text{ in} < 3.06\text{ in}$$

$$4D < p \quad [\text{OK}]$$

$$C_d = \frac{p}{8D} = \frac{3.06\text{ in}}{4\text{ in}} = 0.765$$

The allowable lag screw capacity is

NDS Tbl 10.3.1

$$\begin{aligned} Z' &= ZC_DC_MC_tC_gC_\Delta C_dC_{eg} \\ &= (220\text{ lbf})(1.0)(1.0)(1.0)(1.0)(1.0)(0.765)(1.0) \\ &= 168.3\text{ lbf} < \text{total load of }250\text{ lbf} \quad [\text{no good}] \end{aligned}$$

Example 10.4
Lag Screws Loaded Laterally at an Angle to Grain

A 1/2 in × 4 in lag screw connects a 1/4 in thick steel plate to a 6 3/4 × 11 southern pine glulam beam (20F-V2) as shown. Assume all adjustment factors, except C_d, are 1.0. Use NDS table values for an approximation.

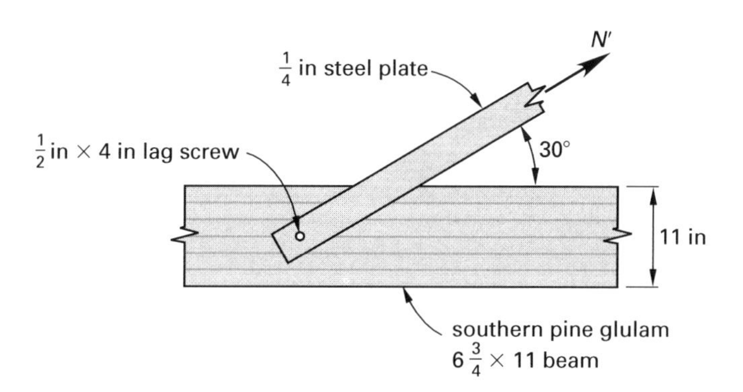

Determine the maximum allowable tensile design force, N', that can be applied to the steel plate as based on bolt and glulam member capacities.

REFERENCE

Solution (ASD Method)

Calculate the adjusted lateral design value, N'.

NDS Tbl 11K

$Z_{\|} = 540\text{ lbf};\ Z_{\perp} = 340\text{ lbf}$

NDS App J

Application of the Hankinson formula gives a reference tensile design force of

$$N = Z = \frac{Z_{\parallel} Z_{\perp}}{Z_{\parallel} \sin^2 \theta + Z_{\perp} \cos^2 \theta}$$

$$= \frac{(540 \text{ lbf})(340 \text{ lbf})}{(540 \text{ lbf})(\sin^2 30°) + (340 \text{ lbf})(\cos^2 30°)}$$

$$= 470.8 \text{ lbf}$$

Find the penetration depth factor, C_d.

Actual penetration, p, into the beam is

NDS App Tbl L2

$$p = L - (t_s - t_{\text{washer}}) - E_{\text{tip}}$$

$$= 4 \text{ in} - (0.25 \text{ in} - 0 \text{ in}) - \tfrac{5}{16} \text{ in}$$

$$= 3.44 \text{ in}$$

NDS Tbl 11K, Ftn 3

The minimum penetration required for the full design value is

$$8D = (8)\left(\tfrac{1}{2} \text{ in}\right)$$

$$= 4 \text{ in}$$

The minimum penetration required for the reduced design value is

$$4D = (4)\left(\tfrac{1}{2} \text{ in}\right)$$

$$= 2 \text{ in}$$

$$2 \text{ in} < 3.44 \text{ in}$$

$$4D < p \quad [\text{OK}]$$

The penetration depth factor is

$$C_d = \frac{p}{8D} = \frac{3.44 \text{ in}}{4 \text{ in}}$$

$$= 0.86$$

NDS Tbl 10.3.1

Find the allowable load, N'.

$$N = Z \text{ and } N' = Z'$$

$$N' = N C_D C_M C_t C_g C_\Delta C_d C_{eg}$$

$$= N(1.0)(1.0)(1.0)(1.0)(1.0)(0.86)(1.0)$$

$$= (470.8 \text{ lbf})(0.86)$$

$$= 404.9 \text{ lbf}$$

Note that although it is not required for this problem, the steel plate capacity should also be checked in an actual design and/or analysis situation.

---

4. Wood Screws

[NDS Secs 11.1.4, 11.2.2, 11.3, 11.4, 11.5; NDS App Tbl L3; NDS Tbls 11L, 11M]

Wood screws are similar to lag screws, but are smaller in diameter. Wood screws have either a circular flat head or a circular round head, while lag screws have either a hex head or a square head as described previously. The load capacities of wood screws are relatively small compared with those of lag screws.

Figure 10.2 Wood Screws (based on NDS Appendix L3)

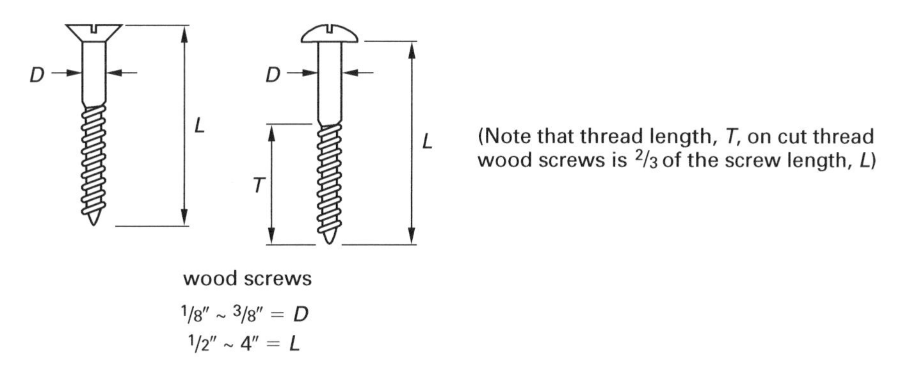

Wood screws are designated by gage, which is an indicator of the diameter of the shank (the shank is the smooth, unthreaded part of the screw length). For example, no. 10 and no. 20 gages have shank diameters of 0.190 in and 0.320 in, respectively. Common wood screws have a diameter ranging from 1/8 in to 3/8 in diameter and have a length of 1/2 in to 4 in. The thread length in cut thread wood screws is two-thirds of the screw length (see NDS App. L3).

5. Withdrawal Design Values for Wood Screws

[NDS Sec 11.2.2; NDS Tbls 10.3.1, 11.2B]

The reference withdrawal design values for a single wood screw are listed in NDS Table 11.2B and are in units of pounds per inch of threaded penetration into the side grain of the main member (threaded length is approximately two-thirds of the total screw length). The NDS reference design value, W, is multiplied by all applicable adjustment factors (see NDS Table 10.3.1) to obtain the allowable design value, W'. Wood screws cannot be loaded in withdrawal from end grain of wood (see NDS Sec. 11.2.2.2).

NDS Eq 11.2-2, Tbl 11.2B — The reference withdrawal design value is

$$W = 2850G^2D$$

NDS Tbl 10.3.1 — The adjusted withdrawal design value is

$$\begin{aligned} W' &= (\text{reference design values})(\text{product of adjustment factors}) \\ &= WpC_DC_MC_tC_{eg} \end{aligned}$$

C_D	load duration factor (NDS Sec. 2.3.2; NDS App. B)	–
C_M	wet service factor (NDS Table 10.3.3.)	–
p	thread penetration into side grain of main member (thread length is 2/3 the total screw length; see NDS App. Table L3)	in
C_t	temperature factor (NDS Sec. 2.3.3)	–

Example 10.5
Withdrawal Load for Wood Screws

Two no. 12 gage (0.216 in diameter, see NDS App. Table L3), 3 in long wood screws connect a 3/8 in thick steel plate to a 12 × 12 southern pine beam as shown. Assume a wet service condition and normal temperature ($C_t = 1.0$).

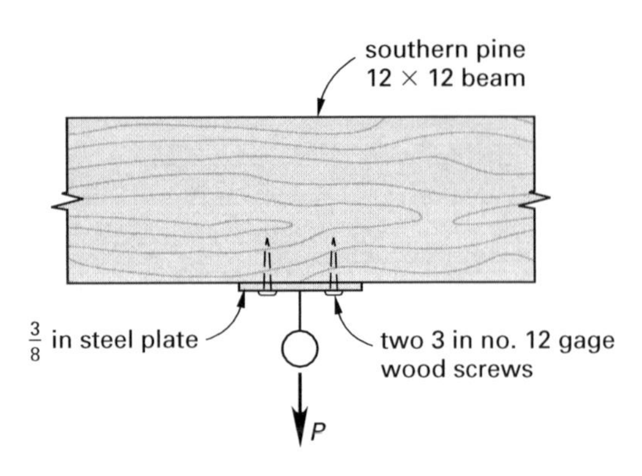

Determine the allowable load for a one-year duration. Assume spacing requirements and edge distance requirements are fully satisfied.

REFERENCE *Solution (ASD Method)*

Calculate the adjusted (allowable) withdrawal design value, W'.

NDS Tbl 11.3.2A — For southern pine, the specific gravity, G, is 0.55.

NDS Tbl 11.2B — The reference withdrawal design value, W, is 186 lbf per inch of threaded penetration length.

The penetration into the beam is

$$3 \text{ in} - \tfrac{3}{8} \text{ in steel plate} = 2.625 \text{ in}$$

As per the NDS Table 11.2B subtitle, the *threaded* length is approximately equal to

$$\left(\frac{2}{3}\right)(3 \text{ in}) = 2 \text{ in} < 2.625 \text{ in}$$

Use $p = 2$ in (thread penetration into side grain of main member).

NDS App B, Fig B1 — $C_D = 1.1$ for one year duration of load.

NDS Tbl 10.3.3 — $C_M = 0.7$

NDS Tbl 10.3.1 — The adjusted withdrawal design value is

$$\begin{aligned} W' &= WpC_DC_MC_t \\ &= \left(186 \ \frac{\text{lbf}}{\text{in}}\right)(2 \text{ in})(1.1)(0.7)(1.0) \\ &= 286.4 \text{ lbf} \end{aligned}$$

The allowable load is

$$\begin{aligned} P_{\text{allow}} &= W'(\text{no. of wood screws}) \\ &= \left(286.4 \ \frac{\text{lbf}}{\text{screw}}\right)(2 \text{ screws}) \\ &= 572.8 \text{ lbf} \end{aligned}$$

6. Lateral Design Values for Wood Screws

[NDS Sec 11.3; NDS Tables 10.3.1, 11L, 11M]

The reference wood screw lateral design value, Z, is the smallest value calculated using NDS yield limit Eqs. 11.3-1 through 11.3-6; the same for dowel type fasteners, nails and spikes, bolts, lag screws, and wood screws.

As an alternative to using these equations, NDS Tables 11L and 11M provide reference wood screw lateral design values, Z, for common connections applications. The yield limit equations must be used for cases not covered in these tables.

These design values, Z, are multiplied by applicable adjustment factors to obtain allowable design values, Z'.

The allowable lateral design value is

$$Z' = ZC_DC_MC_tC_dC_{eg} \quad \text{[NDS Table 10.3.1 and NDS Sec. 11.3]}$$

The end grain factor, C_{eg}, is 0.67 if inserted into the end grain of main member (see NDS Sec. 11.5.2).

Since the reference wood screw lateral design values are based, whether by equations or tables, on a minimum penetration length of $10D$ (ten times the unthreaded shank diameter of the wood screw) into the main member, these values are multiplied by the penetration depth factor, C_d, if the actual penetration, p, is less than $10D$ but greater than or equal to $6D$ (NDS Table 11L, Footnote 2).

When $6D \leq p \leq 10D$,

$$Z' = Z\left(\frac{p}{10D}\right)C_MC_DC_tC_gC_\Delta \quad \text{[NDS Table 10.3.1]}$$

$$C_d = \frac{p}{10D} \quad \text{[NDS Table 11L and Footnotes 2 and 3]}$$

Example 10.6
Lateral Loads on Wood Screws

Eight no. 10 (0.19 in diameter, see NDS App. Table L3), 1.50 in wood screws and two 10-gage ($t_s = 0.134$ in) steel plates, one on each side of the connection, hold a 4 × 4 side member to a 4 × 6 main member. The lumbers are red oak that are exposed to weather in service. Assume normal temperature ($C_t = 1.0$).

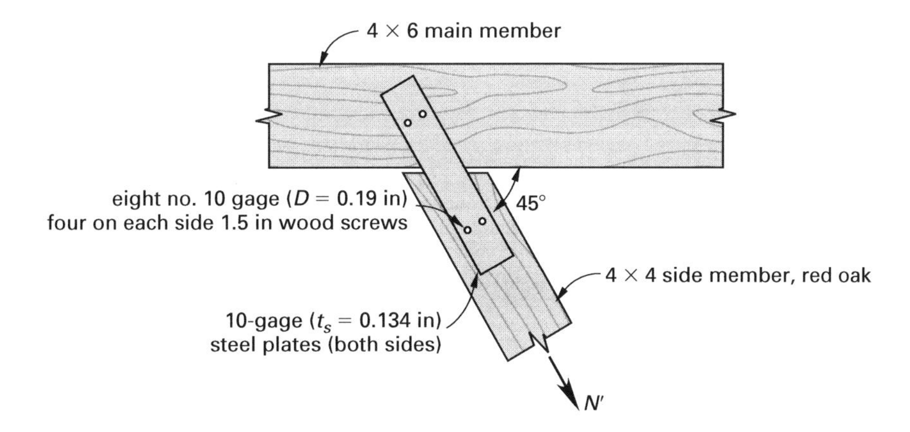

Using NDS Table 11M, determine the allowable load, N', for a one-year load duration. Ignore the capacity of the steel plates.

REFERENCE	*Solution (ASD Method)*
NDS Tbl 11.3.2A	Calculate the adjusted lateral design value, N'. For red oak, the specific gravity, G, is 0.67.
NDS Tbl 11M	The reference lateral design value, Z, is 173 lbf.
NDS Tbl 10.3.3	Because the lumber is exposed to weather, C_M is 0.7.
NDS App B	For a one-year load duration, C_D is 1.10.

The end grain factor, C_{eg}, is not applicable.

Find the penetration depth factor, C_d.

NDS App L

The actual penetration into a 4×6 or a 4×4 member is

$$\begin{aligned} p &= 1.5 \text{ in} - 0.134 \text{ in steel plate} \\ &= 1.37 \text{ in} \end{aligned}$$

The minimum penetration required for full design value is

NDS Tbl 11M, Ftns 2 and 3

$$\begin{aligned} 10D &= (10)(0.19 \text{ in}) \\ &= 1.9 \text{ in} \end{aligned}$$

For the reduced design value,

$$\begin{aligned} 6D &= (6)(0.19 \text{ in}) \\ &= 1.14 \text{ in} < p = 1.37 \text{ in} \end{aligned}$$

$$\begin{aligned} C_d &= \frac{p}{10D} = \frac{1.37 \text{ in}}{1.9 \text{ in}} \\ &= 0.72 \end{aligned}$$

For a 4×4 side member, the allowable load is

$$\begin{aligned} Z' &= ZC_DC_MC_tC_d \\ &= (173 \text{ lbf})(1.10)(0.7)(1.0)(0.72) \\ &= 95.9 \text{ lbf per screw} \\ N' &= Z'(\text{no. of wood screws}) \\ &= \left(95.9 \ \frac{\text{lbf}}{\text{screw}}\right)(4 \text{ screws}) \\ &= 383.6 \text{ lbf} \end{aligned}$$

Note that NDS Sec. 11.4 will apply for determining allowable design values for wood screws loaded at an angle to the wood surface. (See NDS Fig. 11F.)

NDS Sec 11.3.2.1

For a 4×6 main member where $D < 1/4$ in,

$$F_{e\parallel} = F_{e\perp} = F_e$$

Therefore, $Z'_{\parallel} = 383.6 \text{ lbf} = Z'_{\perp}$.

NDS App J

The adjusted lateral design value, calculated using the Hankinson formula is

$$N' = \frac{Z'_{\parallel} Z'_{\perp}}{Z'_{\parallel} \sin^2 45° + Z'_{\perp} \cos^2 45°}$$

$$= \frac{(383.6 \text{ lbf})(383.6 \text{ lbf})}{(383.6 \text{ lbf})(\sin^2 45°) + (383.6 \text{ lbf})(\cos^2 45°)}$$

$$= 383.6 \text{ lbf}$$

Example 10.7

Combined Lateral and Withdrawal Loads

When lag or wood screws are subjected to combined lateral and withdrawal loading, as when the screws are inserted perpendicular to the fiber and the load acts at an angle, α, to the wood surface, the adjusted design value is

$$Z'_{\alpha} = \frac{W'pZ'}{W'p \cos^2 \alpha + Z' \sin^2 \alpha} \quad \text{[NDS Eq. 11.4-1]}$$

p	length of thread penetration in main member	in
α	angle between wood surface and direction of applied load	degree

Determine the allowable load P (combine the lateral and withdrawal load). Assume one 1/8 in thick bracket with one no. 14 × 4 in wood screw, wet service, and $C_D = 1.0$.

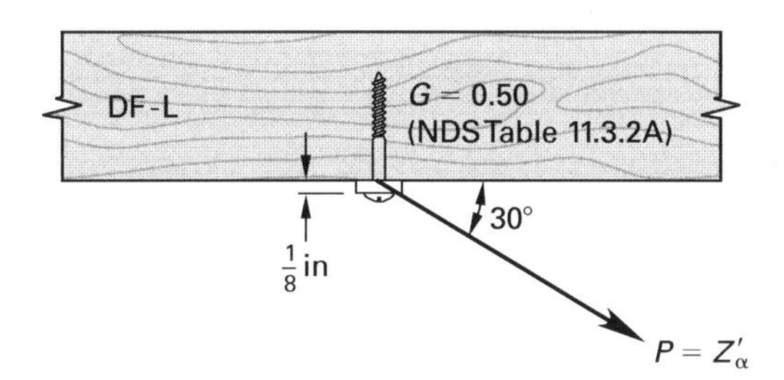

REFERENCE

Solution (ASD Method)

For a wood screw no. 14 × 4 in long,

NDS App Tbl L3

$D = 0.242 \text{ in} \le \frac{1}{4} \text{ in}$

thread length $= \left(\frac{2}{3}\right)(4 \text{ in}) = 2.67 \text{ in}$

NDS Sec 11.5.1.1

The geometry factor, C_{Δ}, is 1.0 when $D < 1/4$ in.

NDS Tbl 11.2B

The reference withdrawal design value is 172 lbf per inch of thread penetration into side grain of main member. The specific gravity, G, is 0.5 for DF-L (see NDS Table 13.2.A).

NDS Tbl 10.3.3

The thread length, p, is 2.67 in.
For wet service conditions, the wet service factor, C_M, is 0.7.

NDS Tbl 10.3.1

The adjusted withdrawal design value is

$$W'p = WpC_DC_MC_t$$

$$= \left(172 \ \frac{\text{lbf}}{\text{in}}\right)(2.67 \text{ in})(1.0)(0.7)(1.0)$$

$$= 321.47 \text{ lbf}$$

NDS Tbl 11M

The side member thickness is $^1/_8$ in.

By interpolation between $t_s = 0.120$ in and $t_s = 0.134$ in, the reference lateral design value is

$$\begin{aligned} Z &= 175\ \text{lbf} + 2\ \text{lbf} \\ &= 177\ \text{lbf} \end{aligned}$$

NDS Sec 11.5.1.1

$C_\Delta = 1.0$ if $D < {}^1/_4$ in.

NDS Tbl 10.3.1

$$\begin{aligned} Z' &= ZC_DC_MC_tC_\Delta \\ &= (177\ \text{lbf})(1.0)(0.7)(1.0)(1.0) \\ &= 123.9\ \text{lbf} \end{aligned}$$

With screws inserted perpendicular to the fiber and the load acting at a 30° angle, α, to the wood surface, the adjusted design value is

$$\begin{aligned} p = Z'_\alpha &= \frac{W'pZ'}{W'p\cos^2\alpha + Z\sin^2\alpha} \\ &= \frac{(321.47\ \text{lbf})(123.9\ \text{lbf})}{(321.47\ \text{lbf})(\cos^2 30°) + (123.9\ \text{lbf})(\sin^2 30°)} \\ &= 146.4\ \text{lbf} \end{aligned}$$

11

Split Rings and Shear Plates

Split rings and shear plates are highly effective mechanical fasteners that can have lateral (shear) load capacities much greater than those of bolts and lag screws. Split rings have a diameter of either $2^1/2$ in or 4 in, while shear plates have a diameter of either $2^5/8$ in or 4 in. Their dimensions are given in NDS App. K and design information is given in NDS Ch. 12.

Split rings are used for wood-to-wood connections. A split ring is fit into a groove cut into the mating surfaces of the wood members being connected. The assembly is held together by a bolt (see Fig. 11.1).

Figure 11.1 Three-Member Connection with Split Rings

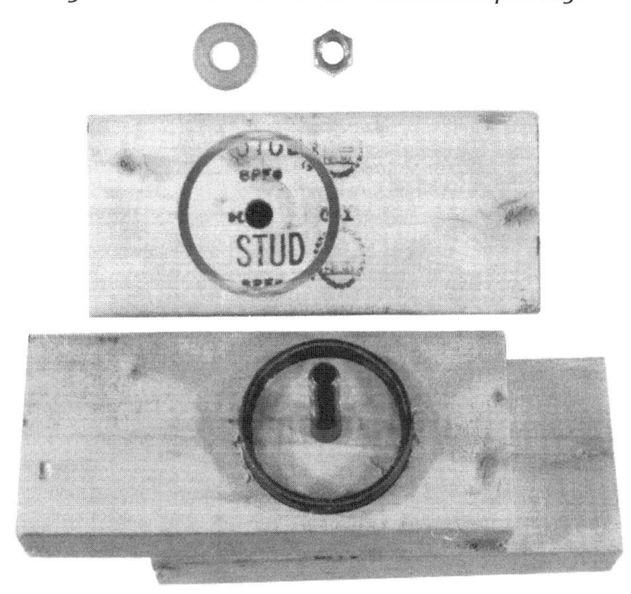

Shear plates can be used for wood-to-metal or wood-to-wood connections. For a three-member wood-to-metal connection, the shear plate cuts into the wood but is flush with the surface of the wood. A bolt holds together the wood, two shear plates, and two steel member plates (see Fig. 11.2b). For a three-member wood-to-wood connection, a total of four shear plates are required to connect the three wood members, and the whole connection is held together by a bolt (see Fig. 11.2a).

Figure 11.2 Three-Member Connections with Shear Plates

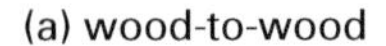

(a) wood-to-wood

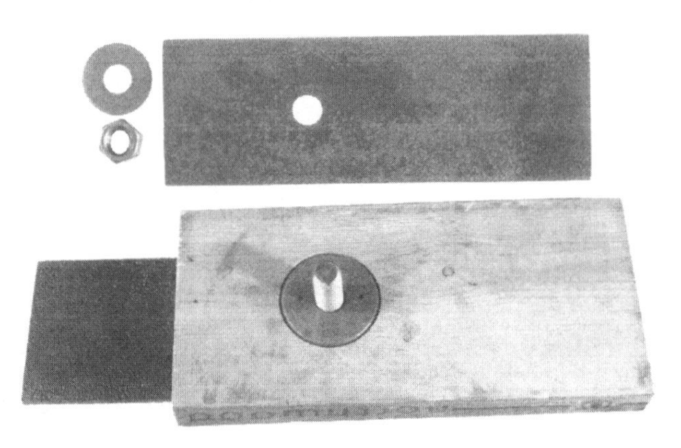

(b) wood-to-metal

1. Lumber Species Group

[NDS Tbl 12A]

Since the density of wood affects the lateral load capacities of split rings and shear plates, the species groupings for these connectors are based on density, as seen in NDS Tables 11.3.2A and 12A.

group A	group B	group C	group D
$G \geq 0.60$	$0.49 \leq G \leq 0.60$	$0.42 \leq G \leq 0.49$	$G < 0.42$

2. Design Values

[NDS Sec 12.2]

NDS Tables 12.2A and 12.2B contain reference design values parallel to grain, P, and perpendicular to grain, Q. These values are multiplied by the applicable adjustment factors specified in NDS Table 10.3.1 to obtain the respective allowable design values, P' and Q', as follows.

$$P' = PC_DC_MC_tC_gC_\Delta C_dC_{st}$$

$$Q' = QC_DC_MC_tC_gC_\Delta C_d$$

P, Q	reference design values (NDS Tables 12.2A and 12.2B)	lbf
C_D	load duration factor (ASD only; NDS Secs. 2.3.2 and 10.3.2)	–

C_M	wet service factor (NDS Table 10.3.3)	–
C_t	temperature factor (NDS Table 10.3.4)	–
C_g	group action factor (NDS Tables 10.3.6B and 10.3.6D)	–
C_Δ	geometry factor for edge distance, end distance, and spacing (NDS Secs. 12.3.1–12.3.7; NDS Table 12.3)	–
C_d	penetration depth factor with lag screws (NDS Table 12.2.3)	–
C_{st}	metal side plate factor (NDS Table 12.2.4)	–

Example 11.1

2½ in Single Split Rings with Parallel-to-Grain and Perpendicular-to-Grain Loadings

One 2×6 and two 2×8s are connected using 2½ in split rings as shown (see NDS App. K for split ring dimensions). Assume that $C_D = C_M = C_t = 1.0$ and that all timbers are douglas fir-larch no. 1 grade.

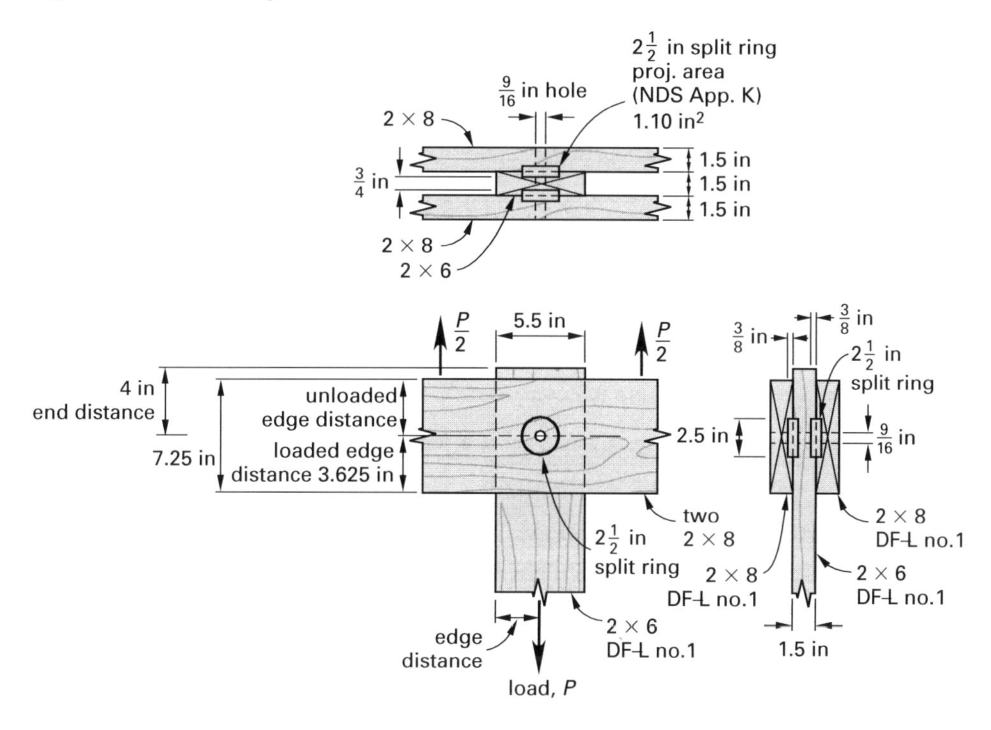

Determine the allowable load capacity for the connection.

REFERENCE

Solution (ASD Method)

1. Connector capacity for 2 × 6 douglas fir-larch no. 1 grade tension member (with split rings on two faces parallel to grain loading)

The net thickness is 1½ in, the load is parallel to the grain, and the rings are in two faces.

The load duration factor, C_D, the wet service factor, C_M, and the temperature factor, C_t, are assumed to be 1.0. Because there is no steel sideplate, group action, or lag screws, C_{st}, C_g, and C_d (no lag screws) are also each 1.0.

NDS Tbls 11.3.2A, 12A

Douglas fir-larch is in species group B for split ring.

NDS Tbl 12.2A

The load, P, parallel to the grain is 2100 lbf for $1^1/_2$ in minimum net member thickness and two faces of the connectors on the same bolt.

NDS Tbl 12.3

Find the end distance factor for the tension member, $2 \times 6, C_\Delta$.

For the full design value of end distance 5.5 in, C_Δ is 1.0.

For the reduced design value of end distance 2.75 in, C_Δ is 0.625.

For the actual end distance of 4 in, interpolate to find C_Δ.

$$\begin{aligned} C_\Delta &= 0.625 + \left(\frac{1.0 - 0.625}{5.5 \text{ in} - 2.75 \text{ in}}\right)(4 \text{ in} - 2.75 \text{ in}) \\ &= 0.795 \quad \text{[controls]} \end{aligned}$$

NDS Tbl 12.3

Find the edge distance factor, C_Δ.

For full the design value of edge distance 1.75 in, C_Δ is 1.0.

For the actual edge distance value, 5.5 in/2 = 2.75 in.

Since the actual edge distance is greater than 1.75 in, the edge distance factor if C_Δ is 1.0.

NDS Tbl 12.3

The spacing factor, C_Δ, is not applicable; therefore, it is 1.0.

NDS Tbl 10.3.1

The allowable load capacity per split ring is

$$\begin{aligned} P' &= PC_DC_MC_tC_gC_\Delta C_dC_{st} \\ &= (2100 \text{ lbf})(1.0)(1.0)(1.0)(1.0)(0.795)(1.0)(1.0) \\ &= 1669.5 \text{ lbf per ring} \end{aligned}$$

Therefore, the total allowable load capacity is

$$\begin{aligned} P_{\text{allow}} &= P'(\text{no. of rings}) = \left(1669.5 \ \frac{\text{lbf}}{\text{ring}}\right)(2 \text{ rings}) \\ &= 3339 \text{ lbf} \end{aligned}$$

2. Connector capacity for 2 × 8 douglas fir-larch no. 1 grade (with two members having split rings perpendicular to loading)

The net thickness is $1^1/_2$ in, the load is perpendicular to the grain, the rings are only in one face, and the loaded edge distance is

$$\frac{7.25 \text{ in}}{2} = 3.625 \text{ in}$$

NDS Tbls 11.3.2A, 12A

Douglas fir-larch is in species group B.

NDS Tbl 12.2A

The load perpendicular to the grain, Q, is 1940 lbf.

NDS Tbl 12.3

Find the edge distance factor for loaded edge, C_Δ.

For the full design value of 2.75 in, C_Δ is 1.0.

For the actual value of 3.625 in, C_Δ is unknown.

Since the actual edge distance is greater than 2.75 in, C_Δ is 1.0. The allowable load capacity per split ring is

NDS Tbl 12.3

$$\begin{aligned} Q' &= QC_DC_MC_tC_gC_\Delta C_dC_{st} \\ &= (1940 \text{ lbf})(1.0)(1.0)(1.0)(1.0)(1.0)(1.0)(1.0) \\ &= 1940 \text{ lbf per ring} \end{aligned}$$

Therefore, the total load capacity, Q_{allow}, is

$$Q_{\text{allow}} = Q'(\text{no. of members}) = \left(1940\ \frac{\text{lbf}}{\text{ring}}\right)(2\ \text{members})$$
$$= 3880\ \text{lbf}$$

3. Member net section load capacities

NDS App K

The ring's projected area is 1.10 in^2 for 2$^1/_2$ in ring in a member.

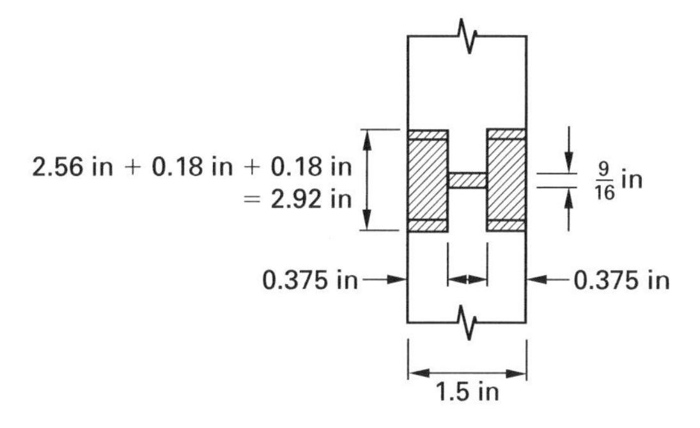

The projected area to be deducted is

$$2.92\ \text{in} \times 0.375\ \text{in} = 1.10\ \text{in}^2$$

The required bolt hole diameter is $^9/_{16}$ in for a 2$^1/_2$ in ring.

The net area of the 2 × 6 (tension load parallel to grain) is

$$A_{\text{net}} = (1.5\ \text{in})(5.5\ \text{in}) - (2)(1.10\ \text{in}^2) - \left(\tfrac{9}{16}\ \text{in}\right)\left(1.5\ \text{in} - (2)(0.375\ \text{in})\right)$$
$$= 5.63\ \text{in}^2$$

The net area of the two 2 × 8s (tension load perpendicular to grain) is

$$A_{\text{net}} = (2)((1.5\ \text{in})(7.25\ \text{in}) - 1.10\ \text{in}^2 - \left(\tfrac{9}{16}\ \text{in}\right)(1.5\ \text{in} - 0.375\ \text{in}))$$
$$= 18.28\ \text{in}^2$$

Find the net section load capacities.

NDS Supp Tbl 4A Adj Fac

For a 2 × 6 douglas fir-larch no. 1 with tension load parallel to grain, the reference tension design value, F_t, is 675 lbf/in^2.

The size factor, C_F, is 1.3.

NDS Tbl 4.3.1

$$F_t' = F_t C_D C_M C_F$$
$$= \left(675\ \frac{\text{lbf}}{\text{in}^2}\right)(1.0)(1.0)(1.0)(1.3)$$
$$= 877.5\ \text{lbf/in}^2$$

The tension load capacity parallel to grain is

$$T_{\|} = F_t' A_{\text{net}} = \left(877.5\ \frac{\text{lbf}}{\text{in}^2}\right)(5.63\ \text{in}^2)$$
$$= 4940.3\ \text{lbf}$$

NDS Supp Tbl 4A

For the two 2 × 8s with compression load perpendicular to grain, the reference compression design value $F_{c\perp}$ is 625.0 lbf/in^2.

NDS Eq 3.10-2

The bearing area factor is

$$C_b = \frac{\ell_{b,\text{ring}} + 0.375 \text{ in}}{\ell_{b,\text{ring}}} = \frac{2.5 \text{ in} + 0.375 \text{ in}}{2.5 \text{ in}}$$
$$= 1.15$$

The adjusted compression design value is

NDS Tbl 4.3.1

$$F'_{c\perp} = F_{c\perp} C_M C_t C_b$$
$$= \left(625.0 \ \frac{\text{lbf}}{\text{in}^2}\right)(1.0)(1.0)(1.15)$$
$$= 718.8 \text{ lbf/in}^2$$

NDS App K

For a 2½ in split ring is ¾ in deep, the compressive dead load capacity is

$$C_{c\perp} = (\text{no. of members})F'_{c\perp}(\text{bearing area})$$
$$= (2 \text{ members})\left(718.8 \ \frac{\text{lbf}}{\text{in}^2}\right)\left((2.5 \text{ in})(0.375 \text{ in})\right)$$
$$= 1347.75 \text{ lbf}$$

4. Summary

The connection load capacities are

$P_{\text{allow}} = 3339$ lbf for the 2 × 6

$Q_{\text{allow}} = 3880$ lbf for the two 2 × 8s

The net section load capacities are

$T_{\|} = 4940.3$ lbf for the 2 × 6

$C_{c\perp} = 1347.75$ lbf for the two 2 × 8s (this value controls the design)

The allowable load capacity is 1347.75 lbf.

Example 11.2
Splice with Multiple 4 in Split Rings

A splice consists of two 2 × 8 side members and a 4 × 8 main member connected with 4 in split rings (see NDS App. K). The splice is fabricated wet and is dry in service. Assume $C_t = C_{di} = C_{st} = 1.0$ and that all lumber is douglas fir-larch no. 1 grade.

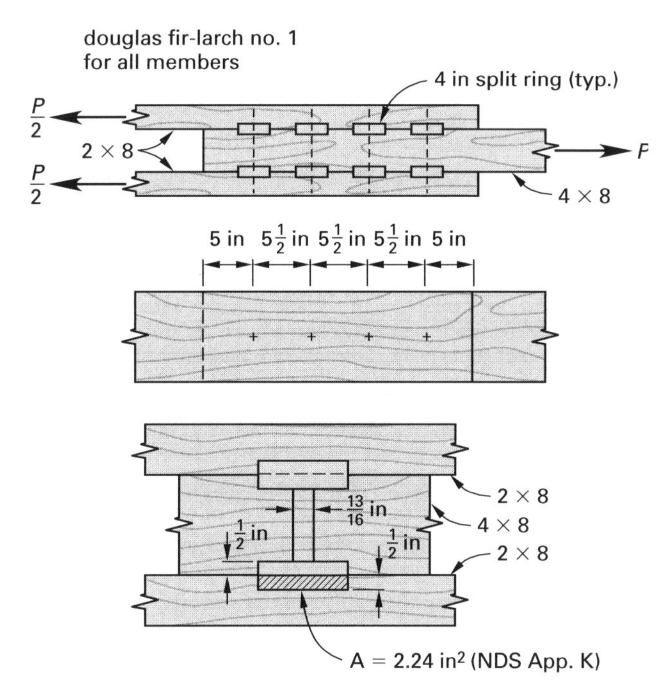

detail for split ring dimensions

Determine the allowable snow load capacity.

REFERENCE

Solution (ASD Method)

1. Connector capacity for 4 × 8 douglas fir-larch no. 1 grade tension member (with one main member)

The net thickness is 3.5 in, the load is parallel to the grain, and the rings are in two faces.

NDS Tbl 2.3.2 — For snow, the load duration factor, C_D, is 1.15.

NDS Tbl 10.3.3 — For fabricated wet and dry service factor, C_M, is 0.80.

NDS Tbls 12.2.3, 12.2.4 — Because there is no lag screw or steel side plate, C_d and C_{st} are each 1.0.

NDS Tbls 11.3.2A, 12A — Douglas fir-larch is in species group B.

NDS Tbl 12.2A — For 3 in or thicker net member thickness (2 faces of 4 × 8 member with split rings on same bolt), the load, P, parallel to the grain is 5260 lbf.

NDS Tbl 12.3 — Find the end distance factor for the tension member, C_Δ.

For the full design value of 7 in, C_Δ 1.0.

For the reduced design value of 3.5 in, C_Δ is 0.625.

For the actual distance of 5 in, interpolate to find C_Δ for 5 in.

$$C_\Delta = 0.625 + \left(\frac{1.0 - 0.625}{7\text{ in} - 3.5\text{ in}}\right)(5\text{ in} - 3.5\text{ in})$$
$$= 0.786$$

NDS Tbl 12.3 — Find the edge distance factor, C_Δ.

For the full design value, the edge distance is 2.75 in (C_Δ is 1.0).

The actual edge distance is

$$\frac{7.25 \text{ in}}{2} = 3.625 \text{ in}$$

Since the actual edge distance is greater than 2.75 in, C_Δ is 1.0.

NDS Tbl 12.3

Find the spacing factor, C_Δ.

For the full design value, spacing is 9 in, C_Δ is 1.0.

For the reduced design value, spacing is 5 in, C_Δ is 0.5.

Interpolate to find C_Δ for the actual distance of 5.5 in.

$$\begin{aligned} C_\Delta &= 0.5 + \left(\frac{1.0 - 0.5}{9 \text{ in} - 5 \text{ in}}\right)(5.5 \text{ in} - 5 \text{ in}) \\ &= 0.562 \quad \text{[controls]} \end{aligned}$$

NDS Tbl 10.3.6B

Find the group action factor, C_g, for four fasteners in a row.

$$\begin{aligned} A_m &= (3.5 \text{ in})(7.25 \text{ in}) \\ &= 25.38 \text{ in}^2 \text{ for one main member} \\ A_s &= (2)(1.5 \text{ in})(7.25 \text{ in}) \\ &= 21.75 \text{ in}^2 \text{ for two side members} \\ \frac{A_s}{A_m} &= \frac{21.75 \text{ in}^2}{25.38 \text{ in}^2} \\ &= 0.857 \end{aligned}$$

By linear interpolation, C_g is 0.858.

NDS Tbl 10.3.1

The allowable load capacity per split ring is

$$\begin{aligned} P' &= PC_DC_MC_tC_gC_\Delta C_dC_{st} \\ &= (5260 \text{ lbf})(1.15)(0.8)(1.0)(0.858)(0.562)(1.0)(1.0) \\ &= 2333.4 \text{ lbf per ring} \end{aligned}$$

Therefore, the total allowable load capacity is

$$\begin{aligned} P_{\text{allow}} &= P'(\text{number of rings}) = \left(2333.4 \ \frac{\text{lbf}}{\text{ring}}\right)(8 \text{ rings}) \\ &= 18{,}667.2 \text{ lbf} \end{aligned}$$

2. Connector capacity for 2 × 8 douglas fir-larch no. 1 grade tension members (with two side members)

The net thickness is 1½ in, the load is parallel to the grain and the rings are only in one face.

NDS Tbl 12A

Douglas fir-larch is in species group B.

The adjustment factors for this step are the same as in the previous step.

NDS Tbl 12.2A

The load, P, parallel to the grain is 5160 lbf/ring.

The allowable load capacity per split ring is

$$\begin{aligned} P' &= PC_DC_MC_tC_gC_\Delta C_dC_{st} \\ &= \left(5160 \ \frac{\text{lbf}}{\text{ring}}\right)(1.15)(0.8)(1.0)(0.858)(0.562)(1.0)(1.0) \\ &= 2289.1 \text{ lbf/ring} \end{aligned}$$

Therefore, the total allowable load capacity is

$$\begin{aligned} P_{\text{allow}} &= P'(\text{no. of rings})(\text{no. of members}) \\ &= \left(2289.1\ \frac{\text{lbf}}{\text{ring}}\right)(4\text{ rings})\ (2\text{ members}) \\ &= 18{,}312.9\text{ lbf} \end{aligned}$$

3. Member net section load capacities

NDS App K

The projected area of 4 in ring is 2.24 in^2 per member. The split ring depth is 1 in.

The required bolt hole diameter is $^{13}/_{16}$ in for a 4 in split ring.

The net area of the 4 × 8 is

$$\begin{aligned} A_{\text{net}} &= (3.5\text{ in})(7.25\text{ in}) - (2)(2.24\text{ in}^2) - \left(\tfrac{13}{16}\text{ in}\right)\left(3.5\text{ in} - (2)(0.5\text{ in})\right) \\ &= 18.86\text{ in}^2 \end{aligned}$$

The net area of the two 2 × 8s is

$$\begin{aligned} A_{\text{net}} &= (2)\left((1.5\text{ in})(7.25\text{ in}) - 2.24\text{ in}^2 - \left(\tfrac{13}{16}\text{ in}\right)(1.5\text{ in} - 0.5\text{ in})\right) \\ &= 15.64\text{ in}^2 \quad \begin{bmatrix}\text{controls since this is} \\ \text{the least net area}\end{bmatrix} \end{aligned}$$

Find the net section load capacities.

NDS Supp Tbl 4A
Adj Fac

The reference tension design value, F_t, is 675 lbf/in^2. For a 2 × 8, the size factor, C_F, is 1.2. For snow, the load duration factor, C_D, is 1.15.

NDS Tbl 4.3.1

The net section load capacity parallel to the grain for the two 2 × 8s is

$$\begin{aligned} F'_t &= F_t C_D C_M C_t C_F \\ &= \left(675.0\ \frac{\text{lbf}}{\text{in}^2}\right)(1.15)(0.8)(1.0)(1.2) \\ &= 745.2\text{ lbf/in}^2 \end{aligned}$$

The tension load capacity is

$$\begin{aligned} P_{\text{allow}} = F'_t A_{\text{net}} &= \left(745.2\ \frac{\text{lbf}}{\text{in}^2}\right)(15.64\text{ in}^2) \\ &= 11{,}654.9\text{ lbf for two } 2 \times 8 \text{ side members} \end{aligned}$$

4. Summary

The connection capacities are

$P_{\text{allow}} = 18{,}667.2$ lbf for the 4 × 8 main member.

$P_{\text{allow}} = 13{,}308.8$ lbf for the two 2 × 8 side members.

The net section load capacities are

$P_{\text{allow}} = 11{,}654.9$ lbf for the two 2 × 8 side members.

The allowable snow load capacity is the lowest capacity, 11,654.9 lbf.

Example 11.3
4 in Split Rings with Lag Screws

A truss connection consists of a 4 in split ring and a $^3/_4$ in by 6 in long lag screw as shown. The wood is seasoned douglas fir-larch no. 1 grade that remains wet in service.

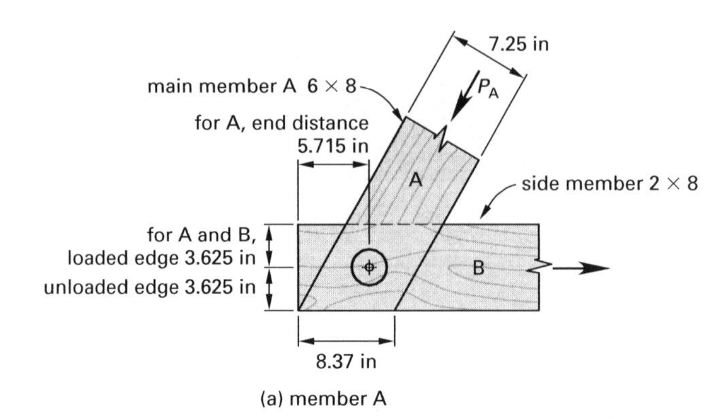

(a) member A

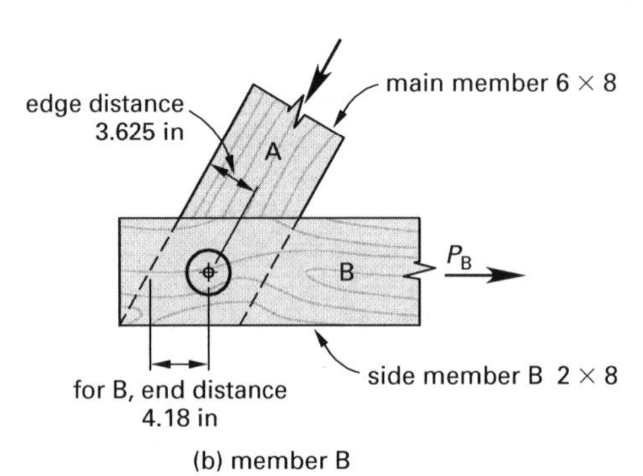

(b) member B

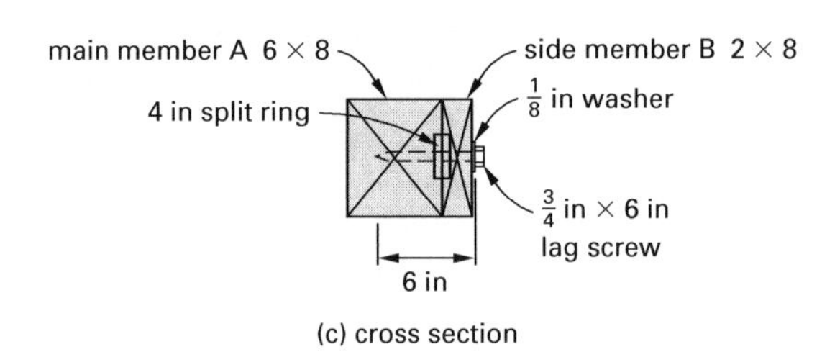

(c) cross section

Determine the wind load capacity, P_{allow}, for the truss connection.

REFERENCE	*Solution (ASD Method)*
NDS Tbl 12A	Douglas fir-larch is in species group B.
NDS Tbl 10.3.3	For wet in service, C_M is 0.7.
NDS Tbl 2.3.2	For wind loads, the load duration factor, C_D is 1.6.
NDS Tbl 2.3.4	For normal temperature, C_t is 1.0.
NDS Tbl 10.3.6B	For one split ring, the group action factor, C_g is 1.0.
NDS Sec 12.2.4	Because there are no steel side plates, C_{st} is 1.0.
NDS Tbl 12.2.3	Find the penetration factor into main member 6 × 8, C_d, for species group B (see NDS Tables 11.3.2A and 12.3).
NDS App L2	The actual penetration of 3/4 in × 6 in lag screw into the 6 × 8 main member is

$$\begin{aligned} p &= L - t_s - t_{\text{washer}} - E \\ &= 6 \text{ in} - 1.5 \text{ in} - 0.125 \text{ in} - 0.5 \text{ in} \\ &= 3.875 \text{ in} \end{aligned}$$

The minimum required penetration for full design value is

$$8D = (8)\left(\tfrac{3}{4}\ \text{in}\right) = 6\ \text{in}$$

The minimum required penetration for reduced design value is

$$3.5D = (3.5)\left(\tfrac{3}{4}\ \text{in}\right) = 2.625\ \text{in}$$

NDS Tbl 12.2.3

$$C_d = 0.75 + \left(\frac{1.0 - 0.75}{6\ \text{in} - 2.625\ \text{in}}\right)(3.875\ \text{in} - 2.625\ \text{in}) = 0.843$$

1. 2 × 8 side member (member B)

The net thickness is 1½ in, the load is parallel to the grain, the ring is only in one face, the loaded edge distance is 3.625 in, and douglas fir-larch is in species group B.

NDS Tbl 12.2A

The load, P, parallel to the grain is 5160 lbf.

NDS Tbl 12.3

For the 6 × 8 main member in compression (member A), find the edge distance factor, C_Δ.

The minimum edge distance for the full design value is 2.75 in and C_Δ is 1.0. The actual value for a loaded or unloaded edge is 3.625 in. Since actual edge distance is greater than 2.75 in, C_Δ is 1.0.

NDS Tbl 12.3

Find the end distance factor, C_Δ, for the 6 × 8 main member in compression.

The minimum end distance for full design value is 5.5 in and C_Δ is 1.0. The actual value is 5.715 in.

Since the actual end distance is greater than 5.5 in, C_Δ is 1.0.

The spacing factor is not applicable. Therefore, $C_\Delta = 1.0$.

NDS Tbl 12.3

For the 2 × 8 side member in tension (member B), find the edge distance factor.

For the full design value of 2.75 in, C_Δ is 1.0.

3.625 in is the actual value.

Since the actual edge distance is greater than 2.75 in, C_Δ is 1.0.

NDS Tbl 12.3

The end distance for the tension member is 7 in for the full design value and C_Δ is 1.0.

For the reduced design value of 3.5 in, C_Δ is 0.625.

For the actual distance of 4.18 in, C_Δ is unknown.

Interpolate to find C_Δ for the actual distance of 4.18 in.

$$C_\Delta = 0.625 + \left(\frac{1.0 - 0.625}{7\ \text{in} - 3.5\ \text{in}}\right)(4.18\ \text{in} - 3.5\ \text{in}) = 0.698$$

Find the allowable wind load capacity of the connector.

$$\begin{aligned} P'_B = P_{\text{allow}} &= P_B C_D C_M C_g C_\Delta C_d C_{st} \\ &= (5160\ \text{lbf})(1.60)(0.7)(1.0)(0.698)(0.843)(1.0) \\ &= 3400.5\ \text{lbf} \end{aligned}$$

2. 2 × 8 in tension (side member B)

NDS App K

For a 4 in split ring, the projected area is 2.24 in^2 in a member.

The required bolt hole diameter is 13/16 in for a 4 in split ring.

The net area of one 2 × 8 is

$$\begin{aligned} A_{\text{net}} &= (1.5\text{ in})(7.25\text{ in}) - 2.24\text{ in}^2 - \left(\tfrac{13}{16}\text{ in}\right)(1.5\text{ in} - 0.5\text{ in}) \\ &= 7.82\text{ in}^2 \end{aligned}$$

The net section load capacity, $T_{||}$, is found as follows.

NDS Supp Tbl 4A

Adj Fac

The reference tension design value, F_t, is 675 lbf/in^2. For a 2 × 8, the size factor, C_F, is 1.2. The load duration factor, C_D, is 1.6. The wet service factor, C_M, is 1.0. The adjusted tension design value is

NDS Tbl 4.3.1

$$\begin{aligned} F'_t &= F_t C_D C_M C_t C_F \\ &= \left(675\ \frac{\text{lbf}}{\text{in}^2}\right)(1.6)(1.0)(1.0)(1.2) \\ &= 1296\ \text{lbf/in}^2 \end{aligned}$$

The tension load capacity is

$$\begin{aligned} P_{\text{allow}} = T_{||} &= F'_t A_{\text{net}} \\ &= \left(1296\ \frac{\text{lbf}}{\text{in}^2}\right)(7.83\text{ in}^2) \\ &= 10{,}147.7\ \text{lbf} \end{aligned}$$

Because $P_{\text{allow}} = T_{||}$, the allowable wind load capacity by net section in tension is 10,147.7 lbf.

The allowable wind load capacity for the connection is controlled by the connector capacity and is 3400.5 lbf.

12

Plywood and Nonplywood Structural Panels

1. Introduction

Plywood has many varied structural applications. It is valuable because of its size and physical properties. A piece of solid-sawn wood 1/2 × 48 × 96 would be very expensive and structurally useless; however, the same size piece of plywood would be much less expensive and, in some ways, stronger than the wood from which it was cut.

Virtually any softwood species can be used in plywood. Plywood is made by peeling logs into veneer and laminating layers of this veneer with glue. The strength results from cutting up and spreading out the knots and other defects, and cross-banding the laminae—that is, placing alternate laminae perpendicular to each other. The result is that plywood has two strong directions instead of one, as in solid timber.

A. Plywood Grades, Wood Species Group, and Exposure Durability

Plywood is available in many forms and grades. The variables in plywood composition are veneer grade and species, veneer configuration, and glue type. Structural plywood veneers are graded for quality from A to D. A is the highest grade and has the tightest grain and the fewest knot holes and other irregularities. Plywood is designated by the veneer grades that appear in the outer, or face, plies. C-D plywood, for example, has a C grade front face with smaller knot holes and a D grade veneer on the back face.

With all the potential variations in plywood composition, it is only through the standards established by The Engineered Wood Association (previously known as the American Plywood Association (APA)) that the end users can know what to expect from a given sheet.[1] The performance-based standards are written as fabrication limits for equivalent plywood designations.

The wood species group system shown in Table 12.1 (APA's *Plywood Design Specification* (PDS) Table 1.5) simplifies the design and identification of a plywood panel. The wood groups listed identify the face and back veneers, and the inner veneers can be of a different group. All plies for marine and structural I grades, however, must be made of a group 1 species. Group 1 includes the strongest varieties of wood that can be used in plywood, and group 4 includes the weakest varieties of wood that can be used in plywood.

There are four durability classifications: exterior (permanently exposed to the weather), exposure 1 (not permanently exposed to the weather), IMG or exposure 2 (protected applications that are not continuously exposed to high humidity conditions), and interior (permanently protected interior applications).

[1]The Engineered Wood Association, 7011 South 19th St., Tacoma, Washington, 98466-5333.

Table 12.1 Classification of Species (APA PDS Table 1.5)

group 1	group 2		group 3	group 4	group 5
Apitong[a,b]	cedar, Port Orford	maple, black	alder, red	aspen	basswood
beech, American	cypress	Mengkulang[a]	birch, paper	bigtooth	poplar, balsam
birch	douglas-fir 2[c]	Meranti, red[a,d]	cedar, Alaska	quaking	
sweet	fir	Mersawa[a]	fir, subalpine	Cativo	
yellow	balsam	pine	hemlock, eastern	cedar	
douglas-fir 1[c]	California red	pond	maple, bigleaf	incense	
Kapur[a]	grand	red	pine	western red	
Keruing[a,b]	noble	Virginia	Jack	cottonwood	
larch, western	Pacific silver	western white	lodgepole	eastern	
maple, sugar	white	spruce	ponderosa	black (western	
pine	hemlock, western	black	spruce	poplar)	
Caribbean	lauan	red	redwood	pine	
Ocote	Almon	Sitka	spruce	eastern white	
pine, southern	Bagtikan	sweetgum	Engelmann	sugar	
loblolly	Mayapis	tamarack	white		
longleaf	red	yellow poplar			
shortleaf	Tangile				
slash	white				
tanoak					

[a]Each of these names represents a trade group of woods consisting of a number of closely related species.

[b]Species from the genus *Dipterocarpus* marketed collectively: Apitong if originating in the Philippines, Keruing if originating in Malaysia or Indonesia.

[c]Douglas-fir from trees grown in the states of Washington, Oregon, California, Idaho, Montana, Wyoming, and the Canadian provinces of Alberta and British Columbia shall be classed as douglas-fir no. 1. Douglas-fir from trees grown in the states of Nevada, Utah, Colorado, Arizona, and New Mexico shall be classed as douglas-fir no. 2.

[d]Red Meranti shall be limited to species having a specific gravity of 0.41 or more based on green volume and oven dry weight.

B. Plywood Structural Applications

The most common plywood applications are flooring, roofing, and siding. The plywood spans the space between joists, rafters, or studs and distributes loads to those members. Engineering calculations can be used to investigate the plywood stresses under these conditions, but it is much simpler to follow the allowable span recommendations found in the certification stamp for most common decking plywoods. Figure 12.1 shows three examples of plywood stamps. The span rating of 32/16 indicates that the panel can be used to span 32 in when used in a roof system and 16 in as floor sheathing.

Figure 12.1 APA Grade-Trademark Stamp

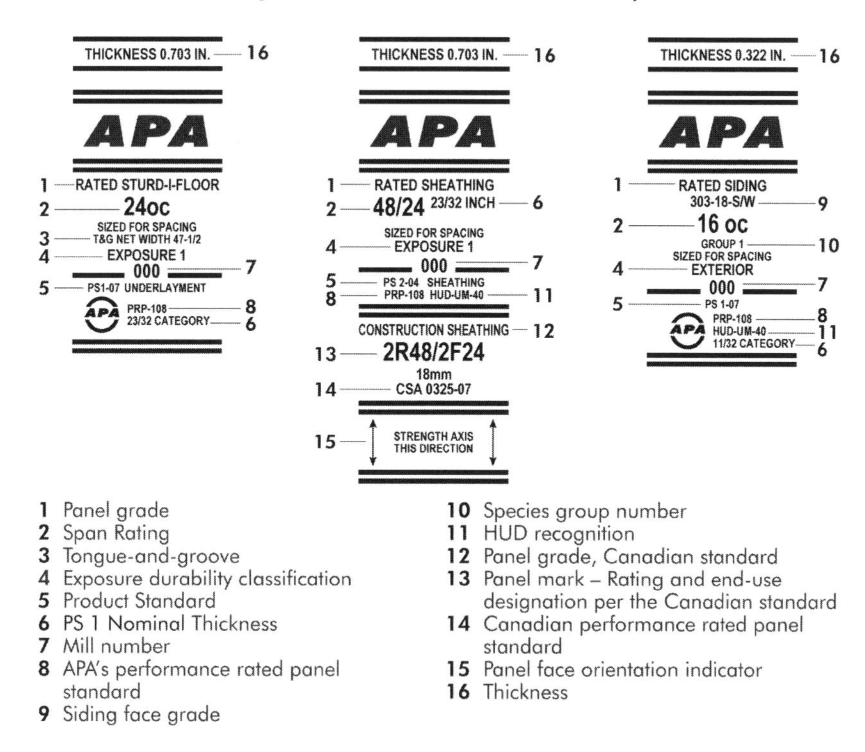

The span rating for sheathing panels is a measure of the panel stiffness and strength parallel to the face grain. The key to span rating in Table 12.2 is used for a roof or floor sheathing without engineering design calculations. The span rating consists of two numbers separated by a slash. The number on the left gives the maximum spacing in inches for roof supports under average loading conditions (good for a live load of 30 lbf/ft^2 or more). The number to the right of the slash shows the maximum spacing in inches for floor supports under average residential loading (maximum allowable uniform loads are 100 lbf/ft^2 or more, depending upon the allowable deflections; PDS Sec. 1.4.1). The span is in the face-grain direction, which is parallel to the 8 ft edge of a standard 4 ft × 8 ft plywood panel.

The more than 70 wood species commonly used in plywood are divided into five groups.

If the veneer species is known, Table 12.1 provides the plywood species classification information. Often, however, the plywood manufacturer will not stamp the plywood with the species. Alternatively, the stamp could include an allowable roof/floor span rating that can be used to indirectly determine the species. Table 12.2 gives the species group as a function of the plywood thickness and span rating. For example, a $^5/_8$ in piece of plywood with group 4 species plies and a $^1/_2$ in sheet of group 1 species plywood will both handle 32 in rafter spacings and 16 in joist spacings.

Table 12.2 Key to Span Rating and Species Group (APA PDS Section 1.4.1)

For panels with "Span Rating" as across top and thickness as at left, use stress for species group given in table.

thickness (in)	span rating (APA rated sheathing grades)							
	12/0	16/0	20/0	24/0	32/16	40/20	48/24	
					span rating (sturd-I-floor grades)			
					16 OC	20 OC	24 OC	48 OC
5/16	4	3	1					
3/8			4	1[a]				
15/32 and 1/2				4	1			
19/32 and 5/8					4	1		
23/32 and 3/4						4	1	
7/8							3[b]	
1 1/8								1

[a]Thicknesses not applicable to APA rated STURD-I-FLOOR.
[b]For APA rated STURD-I-FLOOR 24 OC, use group 4 stresses.

Plywood sheathing on walls, roofs, and floors can resist more than just loads normal to the surface. If the plywood is sized and connected adequately, the walls, roofs, and floors can act as shear diaphragms in resisting lateral loads on a building.

Plywood is very strong and rigid against forces that tend to distort it out of square. Its strength and shape make plywood a logical choice for use as the web of a built-up beam. Box and I-beams can be fabricated with lumber flanges.

It is also possible to make an efficient double use of plywood sheathing in stressed-skin or sandwich panels. In these structural elements, the plywood skin acts as the flanges of a shallow, wide beam. Shear resistance is provided by the lumber webs or sandwich core material. Supplements to the *Plywood Design Specification*, published by the APA, provide the information needed to design and analyze these relatively sophisticated elements.

2. Span Rating

[Table 12.2 (APA PDS Sec 1.4.1)]

The span rating for the sheathing panels is a measure of the panel stiffness and strength parallel to the face grain. It is used as a roof or floor sheathing without engineering design calculations. The span rating consists of two numbers separated by a slash. The number on the left in the span rating gives the maximum spacing for roof supports under average loading conditions (good for 30 lbf/ft^2 live load or better). The number on the right of the slash shows the maximum spacing for floor supports under average residential loading (maximum allowable uniform loads are 100 lbf/ft^2 or more, depending on the allowable deflections).

3. Wood Species Classification

[APA PDS Sec 1.5 and Table 1.5]

The species grouping is divided into four main groups for the seventy-some species of wood from which plywood may be manufactured. Thus, the designer need only be concerned with four species (i.e., four design stresses, instead of seventy). The group classification of a plywood panel is usually determined by the face and back veneer with inner veneers allowed to be of a different group. Group 1 consists of the strongest varieties of wood and Group 4 the weakest varieties of wood.

4. Exposure Durability Classification

There are four durability classifications: Exterior (permanently exposed to the weather); Exposure 1 (not permanently exposed to the weather); IMG or Exposure 2 (protected applications which are not continuously exposed to high humidity conditions); and Interior (permanently protected interior applications).

5. Plywood Section Properties

A. Direction of Face Grain

In plywood, the direction of the grain of alternating plies is usually perpendicular. Since wood has markedly different properties across and along the grain, the direction in which the plies are oriented is important. The standard orientation reference is the grain direction of the visible face plies. The grain in the face plies of a 4 ft × 8 ft plywood panel almost always runs in the 8 ft direction.

The effective section properties of plywood listed in Table 12.3 depend on the orientation of the stresses relative to this face-grain direction. Plywood used as sheathing is strongest against normal loads if the face grain is across the supporting members. Plywood loaded in this strong direction is an example of the "stress applied parallel to face grain" category used in Table 12.3.

B. Thickness for All Properties Except Shear

For all calculations other than those involving shear, the nominal thickness is found in the first column of Table 12.3.

Table 12.3 Effective Section Properties for Plywood (APA PDS Tables 1 and 2)

APA PDS TABLE 1. Face Plies of Different Species Group from Inner Plies (includes all product standard grades except those noted in Table 2)

			stress applied parallel to face grain				stress applied perpendicular to face grain			
nominal thickness (in)	approximate weight (lbf/ft^2)	t_s effective thickness for shear (in)	A area (in^2/ft)	I moment of inertia (in^4/ft)	KS effective section modulus (in^3/ft)	Ib/Q rolling shear constant (in^2/ft)	A area (in^2/ft)	I moment of inertia (in^4/ft)	KS effective section modulus (in^3/ft)	Ib/Q rolling shear constant (in^2/ft)
unsanded panels										
5/16-U	1.0	0.268	1.491	0.022	0.112	2.569	0.660	0.001	0.023	4.497
3/8 -U	1.1	0.278	1.866	0.039	0.152	3.110	0.799	0.002	0.033	5.444
15/32 and 1/2 -U	1.5	0.298	2.292	0.067	0.213	3.921	1.007	0.004	0.056	2.450
19/32 and 5/8 -U	1.8	0.319	2.330	0.121	0.379	5.004	1.285	0.010	0.091	3.106
23/32 and 3/4 -U	2.2	0.445	3.247	0.234	0.496	6.455	1.563	0.036	0.232	3.613
7/8 -U	2.6	0.607	3.509	0.340	0.678	7.175	1.950	0.112	0.397	4.791
1 -U	3.0	0.842	3.916	0.493	0.859	9.244	3.145	0.210	0.660	6.533
1 1/8 -U	3.3	0.859	4.725	0.676	1.047	9.960	3.079	0.288	0.768	7.931
sanded panels										
1/4 -S	0.8	0.267	0.996	0.008	0.059	2.010	0.348	0.001	0.009	2.019
11/32 -S	1.0	0.284	0.996	0.019	0.093	2.765	0.417	0.001	0.016	2.589
3/8 -S	1.1	0.288	1.307	0.027	0.125	3.088	0.626	0.002	0.023	3.510
15/32 -S	1.4	0.421	1.947	0.066	0.214	4.113	1.204	0.006	0.067	2.434
1/2 -S	1.5	0.425	1.947	0.077	0.236	4.466	1.240	0.009	0.087	2.752
19/32 -S	1.7	0.546	2.423	0.115	0.315	5.471	1.389	0.021	0.137	2.861
5/8 -S	1.8	0.550	2.475	0.129	0.339	5.824	1.528	0.027	0.164	3.119
23/32 -S	2.1	0.563	2.822	0.179	0.389	6.581	1.737	0.050	0.231	3.818
3/4 -S	2.2	0.568	2.884	0.197	0.412	6.762	2.081	0.063	0.285	4.079
7/8 -S	2.6	0.586	2.942	0.278	0.515	8.050	2.651	0.104	0.394	5.078
1 -S	3.0	0.817	3.721	0.423	0.664	8.882	3.163	0.185	0.591	7.031
1 1/8 -S	3.3	0.836	3.854	0.548	0.820	9.883	3.180	0.271	0.744	8.428
touch-sanded panels										
1/2 -T	1.5	0.342	2.698	0.083	0.271	4.252	1.159	0.006	0.061	2.746
19/32 and 5/8 -T	1.8	0.408	2.354	0.123	0.327	5.346	1.555	0.016	0.135	3.220
23/32 and 3/4 -T	2.2	0.439	2.715	0.193	0.398	6.589	1.622	0.032	0.219	3.635
1 1/8 -T	3.3	0.839	4.548	0.633	0.977	11.258	4.067	0.272	0.743	8.535

APA PDS TABLE 2. Structural I and Marine

			stress applied parallel to face grain				stress applied perpendicular to face grain			
nominal thickness (in)	approximate weight (lbf/ft^2)	t_s effective thickness for shear (in)	A area (in^2/ft)	I moment of inertia (in^4/ft)	KS effective section modulus (in^3/ft)	Ib/Q rolling shear constant (in^2/ft)	A area (in^2/ft)	I moment of inertia (in^4/ft)	KS effective section modulus (in^3/ft)	Ib/Q rolling shear constant (in^2/ft)
unsanded panels										
5/16-U	1.0	0.356	1.619	0.022	0.126	2.567	1.188	0.002	0.029	6.037
3/8 -U	1.1	0.371	2.226	0.041	0.195	3.107	1.438	0.003	0.043	7.307
15/32 and 1/2 -U	1.5	0.535	2.719	0.074	0.279	4.157	2.175	0.012	0.116	2.408
19/32 and 5/8 -U	1.8	0.707	3.464	0.154	0.437	5.685	2.742	0.045	0.240	3.072
23/32 and 3/4 -U	2.2	0.739	4.219	0.236	0.549	6.148	2.813	0.064	0.299	3.540
7/8 -U	2.6	0.776	4.388	0.346	0.690	6.948	3.510	0.131	0.457	4.722
1 -U	3.0	1.088	5.200	0.529	0.922	8.512	5.661	0.270	0.781	6.435
1 1/8 -U	3.3	1.118	6.654	0.751	1.164	9.061	5.542	0.408	0.999	7.833
sanded panels										
1/4 -S	0.8	0.342	1.280	0.012	0.083	2.009	0.626	0.001	0.013	2.723
11/32 -S	1.0	0.365	1.280	0.026	0.133	2.764	0.751	0.001	0.023	3.397
3/8 -S	1.1	0.373	1.680	0.038	0.177	3.086	1.126	0.002	0.033	4.927
15/32 -S	1.4	0.537	1.947	0.067	0.246	4.107	2.168	0.009	0.093	2.405
1/2 -S	1.5	0.545	1.947	0.078	0.271	4.457	2.232	0.014	0.123	2.725
19/32 -S	1.7	0.709	3.018	0.116	0.338	5.566	2.501	0.034	0.199	2.811
5/8 -S	1.8	0.717	3.112	0.131	0.361	5.934	2.751	0.045	0.238	3.073
23/32 -S	2.1	0.741	3.735	0.183	0.439	6.109	3.126	0.085	0.338	3.780
3/4 -S	2.2	0.748	3.848	0.202	0.464	6.189	3.745	0.108	0.418	4.047
7/8 -S	2.6	0.778	3.952	0.288	0.569	7.539	4.772	0.179	0.579	5.046
1 -S	3.0	1.091	5.215	0.479	0.827	7.978	5.693	0.321	0.870	6.981
1 1/8 -S	3.3	1.121	5.593	0.623	0.955	8.841	5.724	0.474	1.098	8.377
touch-sanded panels										
1/2 -T	1.5	0.543	2.698	0.084	0.282	4.511	2.486	0.020	0.162	2.720
19/32 and 5/8 -T	1.8	0.707	3.127	0.124	0.349	5.500	2.799	0.050	0.259	3.183
23/32 and 3/4 -T	2.2	0.739	4.059	0.201	0.469	6.592	3.625	0.078	0.350	3.596

C. Thickness for Shear

For calculating shear stresses, the effective thicknesses are found in the third column of Table 12.3. For structural plywood grades, this effective thickness can be larger than the actual, nominal thickness.

D. Cross-Sectional Area

The differences in effective areas found in the third and eighth columns of Table 12.3 reflect the differences in grain orientation. Those plies with fibers perpendicular to the direction of stress application are neglected, as they have essentially no stiffness or strength. Note that the effective areas are much larger when the stress is applied parallel to the face.

The effective area is given in square inches per foot. This per-foot basis treats the membrane of plywood as a series of interconnected, 1 ft-wide beams and is common to the rest of the section properties. The foot represents a 12 in-wide strip of the specific piece of plywood, measured perpendicular to the direction of stress application. A 4 ft × 8 ft panel with the stress parallel to the face grain would, therefore, have an effective area resisting that stress of four times the value found in the fourth column of Table 12.3.

E. Moment of Inertia

Only those plies with fibers parallel to the stress direction are included in the values found in the fifth and ninth columns of Table 12.3. The per-foot basis for the unit of the moment of inertia is the same as it is for effective area.

F. Section Modulus

The effective section modulus, KS, includes the empirical factor, K. This KS value should always be used for bending stress calculations when the plywood is installed as sheathing and is loaded normal to the surface. Specifically, the apparently valid I/c value for an effective section modulus should not be used. Figure 12.2 illustrates the difference between loading in the plane of the plywood and loading normal to that plane.

When plywood is used as the web of a plywood-lumber beam, the loads are in the plane of the panel and bending stresses in the plywood are more like axial forces. A different section modulus will be used in those calculations.

Figure 12.2 Plywood in Bending

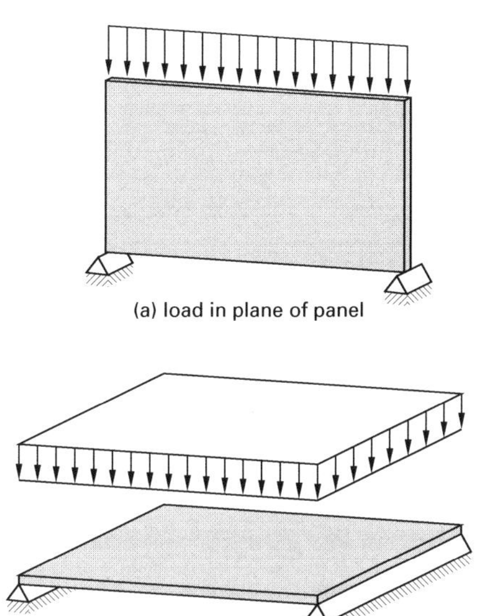

G. Rolling Shear (in the plane of plies)

When plywood is loaded in shear through the thickness, as in Fig. 12.3a, it is tremendously strong because the cross plies are being sheared across the grain. When the shear stresses lie in the plane of the plies, however, plywood is not nearly as strong. This rolling shear that develops, as in Fig. 12.3b, tends to roll the fibers of the cross plies over each other, a tendency against which wood has little resistance. Rolling shear stresses arise in plywood used as sheathing and at the connection between plywood webs and lumber flanges in plywood-lumber beams.

Figure 12.3 Shear Stress Orientations

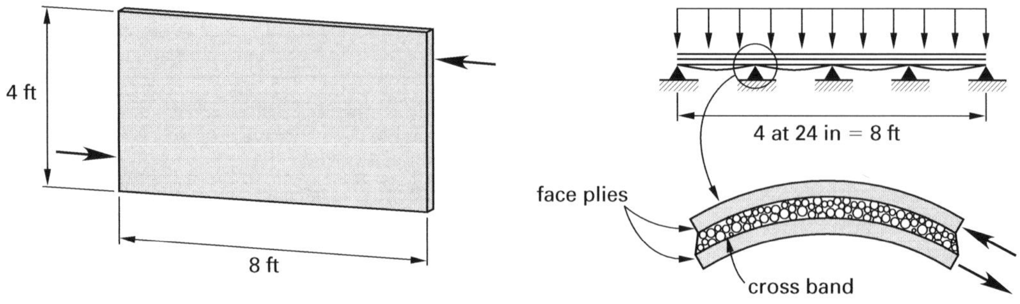

4 ft

8 ft

(a) shear through thickness

4 at 24 in = 8 ft

face plies

cross band

(b) rolling shear

6. Allowable Stresses: Plywood

There are two basic factors that determine the allowable stresses in plywood: the species used in the veneers and the method of assembling those veneers. The allowable stresses are further subjected to modification for load duration and service moisture conditions.

A. Grade Stress Levels

Table 12.4 lists the most common structural plywood types in the first column and gives the stress level (in the sixth column) by plywood type. The fourth column of Table 12.4 (both a and b) lists the grades of the veneer used in the plywood. Plywood is classified into three grade stress levels and is identified in the sixth column of the table as either S-1, S-2, or S-3. The highest stress level, S-1, is associated with plywood that uses only exterior glue and top-grade veneers. The lowest stress level, S-3, is reserved for plywood with interior glue. The intermediate stress level, S-2, incorporates all the other plywoods.

B. Allowable Stress Modification

After the allowable stresses are determined, they are further subjected to modification for load duration and service moisture conditions.

Table 12.5 lists the allowable stresses and stiffnesses that should be used in plywood design.

Since plywood is a wood product, its load capacity is time-dependent. The load duration factors of NDS Table 2.3.2 also apply to the allowable plywood stresses of Table 12.5.

The allowable stresses of Table 12.5 apply to plywood panels that are at least 24 in wide. If the plywood is used in narrower strips, there is an increased possibility of a defect appearing in a critical section. Allowable stresses should be linearly decreased from full strength at 24 in wide to half strength at 8 in wide.

The only two conditions of service that apply to plywood allowable stresses are wet and dry. If the equilibrium moisture content will be less than 16% in service, the dry use values should be used. As long as the plywood is not directly exposed to the weather, this dry condition can be assumed.

Table 12.4a Guide to Use of Allowable Stress and Section Properties Tables (Interior)

INTERIOR OR PROTECTED APPLICATIONS

Plywood Type	Description and Use	Typical Trademarks	Veneer Grade			Common Thicknesses (in)	Grade Stress Level (PDS Table 3)	Species Group	Section Property Table
			Face	Back	Inner				
APA RATED SHEATHING EXP 1 or 2(3)	Unsanded sheathing grade for wall, roof, suflooring, and industrial applications such as pallets and for engineering design, with proper stresses. Manufactured with intermediate and exterior glue.(1) For permanent exposure to weather or moisture only Exterior type plywood is suitable.	APA THE ENGINEERED WOOD ASSOCIATION RATED SHEATHING 32/16 15/32 INCH SIZED FOR SPACING EXPOSURE 1 000 PS 1-95 C-D PRP-108	C	D	D	5/16, 3/8, 15/32, 1/2, 19/32, 5/8, 23/32, 3/4	S-3(1)	See CSTB Table 12.2	Table 1 (unsanded)
APA STRUCTURAL I RATED SHEATHING EXP 1(3)	Plywood grades to use where shear and cross-panel strength propeties are of maximum importance. Made with exterior glue only. Structural I is made from all Group 1 woods.	APA THE ENGINEERED WOOD ASSOCIATION RATED SHEATHING 32/16 15/32 INCH SIZED FOR SPACING EXPOSURE 1 000 PS 1-95 C-D PRP-108	C	D	D	5/16, 3/8, 15/32, 1/2, 19/32, 5/8, 23/32, 3/4	S-2	Group 1	Table 2 (unsanded)
APA RATED STURD-I-FLOOR EXP 1 or 2(3)	For combination subfloor-underlayment. Provides smooth surface for application of carpet and pad. Possesses high concentrated and impact load resitance during construction and occupancy. Manufactured with intermediate and exterior glue. Touch-sanded.(4) Available with tongue-and-groove edges.(5)	APA THE ENGINEERED WOOD ASSOCIATION RATED STURD-I-FLOOR 20 oc 19/32 INCH SIZED FOR SPACING T&G NET WIDTH 47-1/2 EXPOSURE 1 000 PS 1-95 UNDERLAYMENT PRP-108	C plugged	D	C & D	19/32, 5/8, 23/32, 3/4, 1-1/8 (2-4-1)	S-3(1)	See "Key to Span Rating"	Table 1 (touch-sanded)
APA UNDERLAYMENT EXP 1, 2 or INT	For underlayment under carpet and pad. Available with exterior glue. Touch-sanded. Available with tongue-and-groove edges.(5)	APA THE ENGINEERED WOOD ASSOCIATION UNDERLAYMENT GROUP 1 EXPOSURE 1 000 PS 1-95	C plugged	D	C & D	1/2, 19/32, 5/8, 23/32, 3/4	S-3(1)	As specified	Table 1 (touch-sanded)
APA C-D PLUGGED EXP 1, 2 or INT	For built-ins, wall and ceiling tile backing, Not for underlayment. Available with exterior glue. Touch-sanded.(5)	APA THE ENGINEERED WOOD ASSOCIATION C-D PLUGGED GROUP 2 EXPOSURE 1 000 PS 1-95	C plugged	D	D	1/2, 19/32, 5/8, 23/32, 3/4	S-3(1)	As Specified	Table 1 (touch-sanded)
APA APPEARANCE GRADES EXP 1, 2 or INT	Generally applied where a high quality surface is required. Includes APA N-N, N-A, N-B, N-D, A-A, A-B, A-D, B-B, and B-D INT grades.(5)	APA THE ENGINEERED WOOD ASSOCIATION A D GROUP 1 EXPOSURE 1 000 PS 1-95	B or better	D or better	C & D	1/4, 11/32, 3/8, 15/32, 1/2, 19/32, 5/8, 23/32, 3/4	S-3(1)	As Specified	Table 1 (sanded)

(1) When exterior glue is specified, i.e. Exposure 1, stress level 2 (S-2) should be used.
(2) Check local suppliers for availability before specifying Plyform Class II grade, as it is rarely manufactured.
(3) Properties and stresses apply only to APA RATED STURD-I-FLOOR and APA RATED SHEATHING manufactured entirely with veneers.
(4) APA RATED STURD-I-FLOOR 2-4-1 may be produced unsanded.
(5) May be available as Structural I. For such designation use Group 1 stresses and Table 2 section properties.
(6) C face and back must be natural unrepaired; if repaired, use stress level 2 (S-2).

Table 12.4b Guide to Use of Allowable Stress and Section Properties Tables (Exterior)

EXTERIOR APPLICATIONS

Plywood Grade	Description and Use	Typical Trademarks	Veneer Grade			Common Thicknesses	Grade Stress Level (Table 3)	Species Group	Section Property Table
			Face	Back	Inner				
APA RATED SHEATHING EXT(3)	Unsanded sheathing grade with waterproof glue bond for wall, roof, subfloor and industrial applications such as pallet bins.	APA THE ENGINEERED WOOD ASSOCIATION RATED SHEATHING 48/24 23/32 NCH SIZED FOR SPACING EXTERIOR 000 PS 1-95 C-C PRP-108	C	C	C	5/16, 3/8, 15/32, 1/2, 19/32, 5/8 23/32, 3/4	S-1(6)	See "Key to Span Rating"	Table 1 (unsanded)
APA STRUCTURAL I RATED SHEATHING EXT(3)	"Structural" is a modifier for this unsanded sheathing grade. For engineered applications in construction and industry where full Exterior-type panels are required. Structural I is made from Group 1 woods only.	APA THE ENGINEERED WOOD ASSOCIATION RATED SHEATHING STRUCTURAL 24/0 3/8 INCH SIZED FOR SPACING EXPOSURE 1 000 PS 1-95 C-D PRP-108	C	C	C	5/16, 3/8, 15/32, 1/2, 19/32, 5/8, 23/32, 3/4	S-1(6)	Group 1	Table 2 (unsanded)
APA RATED STURD-I-FLOOR EXT(3)	For combination subfloor-underlayment where severe moisture conditions may be present, as in balcony decks. Possesses high concentrated and impact load resistance during construction and occupancy. Touch-sanded.(4) Available with tongue-and-groove edges.(5)	APA THE ENGINEERED WOOD ASSOCIATION RATED STURD-I-FLOOR 20 oc 19/32 NCH SIZED FOR SPACING EXTERIOR 000 PS 1-95 C-C PLUGGED PRP-108	C plugged	C	C	19/32, 5/8, 23/32, 3/4	S-2	See "Key to Span Rating"	Table 1 (touch-sanded)
APA UNDERLAYMENT EXT and APA C-C-PLUGGED EXT	Underlayment for floor where severe moisture conditions may exist. Also for controlled atmosphere rooms and many industrial applications. Touch-sanded. Available with tongue-and-groove edges.(5)	APA THE ENGINEERED WOOD ASSOCIATION C-C PLUGGED GROUP 2 EXTERIOR 000 PS 1-95	C plugged	C	C	1/2, 19/32, 5/8,23/32, 3/4	S-2	As Specified	Table 1 (touch-sanded)
APA B-B PLYFORM CLASS I or II(2)	Concrete-form grade with high reuse factor. Sanded both sides, mill-oiled unless otherwise specified. Available in HDO. For refined design information on this special-use panel see ***APA Design/Construction Guide: Concrete Forming,*** Form No. V345. Design using values from this specification will result in a conservative design.(5)	APA THE ENGINEERED WOOD ASSOCIATION PLYFORM B B CLASS 1 EXTERIOR 000 PS 1-95	B	B	C	19/32, 5/8, 23/32, 3/4	S-2	Class I use Group 1; Class II use Group 3	Table 1 (sanded)
APA MARINE EXT	Superior Exterior-type plywood made only with Douglas-fir or Western Larch. Special solid-core construction. Available with MDO or HDO face. Ideal for boat hull construction.	MARINE · A-A · EXT APA · 000 · PS 1-95	A or B	A or B	B	1/4, 3/8, 1/2, 5/8, 3/4	A face & back use S-1 B face or back use S-2	Group 1	Table 2 (sanded)
APA APPEARANCE GRADES EXT	Generally applied where a high quality surface is required. Includes APA A-A, A-B, A-C, B-B, B-C, HDO and MDO EXT.(5)	APA THE ENGINEERED WOOD ASSOCIATION A C GROUP 1 EXTERIOR 000 PS 1-95	B or better	C or better	C	1/4, 11/32, 3/8, 15/32, 1/2, 19/32, 5/8, 23/32 3/4	A or C face & back use S-1(6) B face or back use S-2	As Specified	Table 1 (sanded)

(1) When exterior glue is specified, i.e. Exposure 1, stress level 2 (S-2) should be used.

(2) Check local suppliers for availability before specifying Plyform Class II grade, as it is rarely manufactured.

(3) Properties and stresses apply only to APA RATED STURD-I-FLOOR and APA RATED SHEATHING manufactured entirely with veneers.

(4) APA RATED STURD-I-FLOOR 2-4-1 may be produced unsanded.

(5) May be available as Structural I. For such designation use Group 1 stresses and Table 2 section properties.

(6) C face and back must be natural unrepaired; if repaired, use stress level 2 (S-2).

Table 12.5 Allowable Stresses for Plywood (APA PDS Table 3)

Allowable Stresses for Plywood (psi) conforming to Voluntary Product Standard PS 1-95 for Construction and Industrial Plywood. Stresses are based on normal duration of load and on common structural applications where panels are 24 in or greater in width. For other use conditions, see NDS Sec. 3.3 for modifications.

type of stress		species group of face ply	grade stress level[a]				
			S-1		S-2		S-3
			wet	dry	wet	dry	dry only
extreme fiber stress in bending (F_b)		1	1430	2000	1190	1650	1650
tension in plane of plies (F_b) face grain parallel or perpendicular to span	F_b and F_t	2, 3	980	1400	820	1200	1200
(at 45° to face grain use $1/6F_t$)		4	940	1330	780	1110	1110
compression in plane of plies		1	970	1640	900	1540	1540
parallel or perpendicular to face grain	F_c	2	730	1200	680	1100	1100
(at 45° to face grain use $1/3F_c$)		3	610	1060	580	990	990
		4	610	1000	580	950	950
shear through the thickness[c]		1	155	190	155	190	160
parallel or perpendicular to face grain	F_v	2, 3	120	140	120	140	120
(at 45° to face grain use $2F_v$)		4	110	130	110	130	115
rolling shear (in the plane of plies) parallel or perpendicular to face grain	F_s	marine and structural I	63	75	63	75	–
(at 45° to face grain use $1^1/3F_s$)		all other[b]	44	53	44	53	48
modulus of rigidity (or shear modulus)		1	70,000	90,000	70,000	90,000	82,000
shear in plane perpendicular to plies (through the thickness)	G	2	60,000	75,000	60,000	75,000	68,000
(at 45° to face grain use $4G$)		3	50,000	60,000	50,000	60,000	55,000
		4	45,000	50,000	45,000	50,000	45,000
bearing (on face)		1	210	340	210	340	340
perpendicular to plane of plies	$F_{c\perp}$	2, 3	135	210	135	210	210
		4	105	160	105	160	160
modulus of elasticity in bending in plane of plies		1	1,500,000	1,800,000	1,500,000	1,800,000	1,800,000
face grain parallel or perpendicular to span	E	2	1,300,000	1,500,000	1,300,000	1,500,000	1,500,000
		3	1,100,000	1,200,000	1,100,000	1,200,000	1,200,000
		4	900,000	1,000,000	900,000	1,000,000	1,000,000

[a]See NDS Tables 12.4a and 12.4b [APA PDS pages 12 and 13] for Guide.
To qualify for stress level S-1, gluelines must be exterior and only veneer grades N, A, and C (natural, not repaired) are allowed in either face or back.
For stress level S-2, gluelines must be exterior and veneer grade B, C-plugged and D are allowed on the face or back.
Stress level S-3 includes all panels with interior or intermediate (IMG) gluelines.

[b]Reduce stresses 25% for three-layer (four- or five-ply) panels over 5/8 in thick. Such layups are possible under PS 1-95 for APA rated sheathing, APA rated sturd-I-floor, underlayment, C-C plugged and C-D plugged grades over 5/8 in through 3/4 in thick.

[c]Shear-through-the-thickness stresses for marine and special exterior grades may be increased 33%. See NDS Sec. 3.8.1 for conditions under which stresses for other grades may be increased.

Example 12.1
Plywood Properties and Allowable Stresses

Roof joists are 24 in on centers and the roof snow load is expected to be 25 lbf/ft^2. 3/8 in APA rated sheathing, exposure 1, 4 ft by 8 ft panels are turned in the strong direction; that is, the face grain is parallel to the 24 in span.

Determine the plywood section properties and allowable stresses for plywood.

REFERENCE

Solution (ASD Method)

Tbl 12.4a (APA PDS pp 14 and 15)

Table 12.4a indicates that unsanded section properties from Table 12.3 should be used in conjunction with the grade stress level S-3. The species group is found using Table 12.2, and determined using the key to span rating of 24/0 for the 3/8 in thickness. The grade stress level for species group 1 in Table 12.5 is to be used for allowable plywood stresses.

For an APA rated sheathing of 3/8 in, exposure 1, and a span rating of 24/0, the section properties and allowable stresses for stresses applied parallel to face grain (across the roof joist supports) are obtained as follows.

1. Section properties

Tbl 12.3 (APA PDS Tbl 1)

From Table 12.3, unsanded 3/8 in plywood (3/8-U),

I	moment of inertia	0.039 in^4/ft
KS	effective section modulus	0.152 in^3/ft
Ib/Q	rolling shear constant	3.110 in^2/ft
	approximate weight	1.1 lbf/ft^2

2. Allowable stresses

Tbl 12.5 (APA PDS Tbl 3)

From Table 12.5, group 1 stresses for S-3 grade stress level for the dry only condition are

F_b	extreme fiber stress in bending	1650 lbf/in^2
F_s	rolling shear	48 lbf/in^2
E	modulus of elasticity in bending of plies	1.8×10^6 lbf/in^2

Example 12.2
Plywood Shelf Design

A 40 in × 30 in shelf is made of 3/4 in APA rated sheathing, with a span rating of 40/20, and exposure 1. The face grain is across supports spanning 40 in.

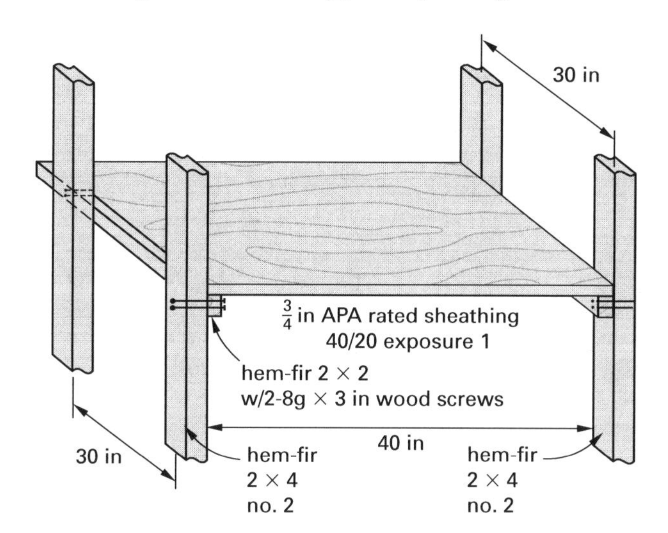

Determine the allowable uniform load for the shelf considering only the 40 in span direction (the strong direction). Assume a load duration of two months, normal temperature conditions, and dry-use conditions. The allowable deflection is $L/360$. Assume $C_D = 1.15$, $C_M = 1.0$, and $C_t = 1.0$.

REFERENCE — *Solution (ASD Method)*

Tbl 12.4a — Table 12.4a shows a stress level of S-3. The species group is determined based on the key to span rating. The section property is determined using Table 12.3 (unsanded).

Tbl 12.2 — The span rating of 40/20 indicates that the species belongs to group 4. The section properties for the 3/4 in unsanded panel are given as follows.

Tbl 12.3

I	moment of inertia	$0.234\ \text{in}^4/\text{ft}$
KS	effective section modulus	$0.496\ \text{in}^3/\text{ft}$
Ib/Q	rolling shear constant	$6.455\ \text{in}^2/\text{ft}$

Tbl 12.5 — The allowable stresses for S-3 grade level with the dry only use condition and species group 4 are

F_b	extreme fiber stress in bending	$1110\ \text{lbf/in}^2$
F_s	rolling shear	$48\ \text{lbf/in}^2$
E	modulus of elasticity in bending in plane of plies	$1.0 \times 10^6\ \text{lbf/in}^2$

1. w, Uniform load based on bending stress, lbf/ft (per foot width of panel)

The maximum calculated bending stress is

$$f_b = \frac{M}{KS} = \frac{\frac{wL^2}{(8\text{ ft})\left(12\ \frac{\text{in}}{\text{ft}}\right)}}{KS} = \frac{wL^2}{(96\text{ in})(KS)}$$

The effective fiber stress in building is

$$F_b' = F_bC_DC_MC_t = \left(1110\ \frac{\text{lbf}}{\text{in}^2}\right)(1.15)(1.0)(1.0) = 1276.5\ \text{lbf/in}^2$$

Set $f_b = F_b'$ and solve for w.

$$w = \frac{F_b'(96\text{ in})(KS)}{L^2} = \frac{\left(1276.5\ \frac{\text{lbf}}{\text{in}^2}\right)(96\text{ in})\left(0.496\ \frac{\text{in}^3}{\text{ft}}\right)}{(40\text{ in} + 1.5\text{ in})^2} = 35\ \text{lbf/ft uniform load per foot width of panel}$$

2. w, Uniform load based on rolling shear (per foot width of panel)

$$V_{\text{max}} = \frac{wL}{2} = \frac{wL}{24\text{ in}}$$

$$f_s = \frac{V_{\text{max}}}{\frac{Ib}{Q}} = \frac{\frac{wL}{24\text{ in}}}{6.455\ \frac{\text{in}^2}{\text{ft}}} = \frac{wL}{154.92\ \frac{\text{in}^3}{\text{ft}}}$$

NDS Tbl 4.3.1

$$F_s' = F_sC_DC_M = \left(48\ \frac{\text{lbf}}{\text{in}^2}\right)(1.15)(1.0) = 55.2\ \text{lbf/in}^2$$

Set $f_s = F_s'$ and solve for w.

$$w = \frac{F_s'\left(154.92\ \frac{\text{in}^3}{\text{ft}}\right)}{L} = \frac{\left(55.2\ \frac{\text{lbf}}{\text{in}^2}\right)\left(154.92\ \frac{\text{in}^3}{\text{ft}}\right)}{40\text{ in} + 1.5\text{ in}} = 206\ \text{lbf/ft uniform load per foot width of panel}$$

3. w, Uniform load based on deflection (per foot width of panel, with w in lbf/ft)

$$\Delta = \frac{5wL^4}{384EI}$$

$$\delta_{\text{max}} = \frac{L}{360}$$

Set $\delta_{\text{max}} = \Delta$ and solve for w.

$$w = \frac{384EI}{(360)(5)L^3} = \frac{(384)\left(1.0 \times 10^6\ \frac{\text{lbf}}{\text{in}^2}\right)(0.234\ \text{in}^4)\left(12\ \frac{\text{in}}{\text{ft}}\right)}{(360)(5)(40\ \text{in} + 1.5\ \text{in})^3}$$

$$= 8\ \text{lbf/ft uniform load per width of panel} \quad \text{[controls]}$$

7. Diaphragms and Shear Walls

Wind and earthquakes are the principal lateral forces a structure must resist. With relatively minor modifications, building walls can act as shear walls to efficiently resist these lateral loads. The floors and roof can also act as diaphragms to redistribute or gather forces and transmit them to the shear walls.

A. Diaphragms

A diaphragm system consists of three parts: the web (diaphragm panels), the chords (edge members), and the connections that hold the web and the chords, as well as transfer loads to the other structural elements and to the foundation.

Just as with other beams, diaphragms are designed to resist the imposed shear and bending stresses. The shear stress is carried by the plywood decking and is assumed to be uniformly distributed across the depth. The nailing schedule and panel splicing details required for a given design shear force can be determined using Table 12.6.

In a *blocked diaphragm*, all panel edges are above and fastened to the common framing lumber. In an *unblocked diaphragm*, only the panel edges in one direction are above and are fastened to common framing lumber (as in a typical roof diaphragm for standard residential construction).

Floors and roofs can be designed to act as very deep, horizontal beams that carry the lateral forces applied to the walls between the floors and roof. These deep beams, or diaphragms, span between shear walls and other structural elements carrying the lateral loads to the building foundation.

In conventional wood frame buildings, the lateral forces due to wind or seismic forces are primarily carried by the wall frames. In other cases, however, the lateral forces are carried by the wall framing to the horizontal diaphragms at the top and then are transferred to the foundation through the vertical wall elements (shear walls). These shear walls used with plywood horizontal diaphragms (roofs and floors) can also be masonry or concrete.

The design of diaphragm panels subjected to the loads normal to the surface of the sheathing must be completed prior to the investigation of diaphragm actions. The diaphragm, shown in Fig. 12.4, is a flat structural member acting like a deep beam, and it acts as a web for resisting shear, while the diaphragm edge members perform the function of flanges for resisting bending stresses. The edge members are called *chords* in the diaphragm system, and they may consist of dimension lumber (top plates), joists, studs, trusses, and so on. Due to the considerable depth of diaphragms in the direction parallel to the applications of lateral loads, shear forces are considered uniform across the depth of the diaphragm (Fig. 12.4b). The diaphragm chords carry all flange forces, acting in compression and tension, to resist bending stresses in the diaphragm system.

The horizontal (in-plane) load-carrying capacity of the diaphragm depends on the nail strength and also on whether the diaphragms are blocked or unblocked. Blocking usually consists of 2×4s or 2×6s fastened between the joists; roof rafters; or other primary structural members; support frames for connecting the edges of the plywood panels. The purpose of blocking is to allow increased shear transfer. Buckling of unsupported panel edges controls unblocked panel loads. In an unblocked diaphragm, two of the diaphragm's four edges are not supported by lumber framing.

Table 12.6 Required Panel Details: Horizontal Diaphragms (APA Design/Construction Guide: Diaphragms and Shear Walls, Table 1)

RECOMMENDED SHEAR (POUNDS PER FOOT) FOR HORIZONTAL APA PANEL DIAPHRAGMS WITH FRAMING OF DOUGLAS-FIR, LARCH OR SOUTHERN PINE[a] FOR WIND OR SEISMIC LOADING

					Blocked Diaphragms				Unblocked Diaphragms	
					Nail Spacing (in.) at diaphragm boundaries (all cases), at continuous panel edges parallel to load (Cases 3 & 4), and at all panel edges (Cases 5 & 6)[b]				Nails Spaced 6" max. at Supported Edges[b]	
					6	4	2-1/2[c]	2[c]		
					Nail Spacing (in.) at other panel edges (Cases 1, 2, 3 & 4)[b] t					
Panel Grade	Common Nail Size	Minimum Nail Penetration in Framing (inches)	Minimum Nominal Panel Thickness (inch)	Minimum Nominal Width of Framing Member (inches)	6	6	4	3	Case 1 (No unblocked edges or continuous joints parallel o load)	All other configurations (Cases 2, 3, 4, 5 & 6)
APA STRUCTURAL I grades	6d[e]	1-1/4	5/16	2	185	250	375	420	165	125
				3	210	280	420	475	185	140
	8d	1-3/8	3/8	2	270	360	530	600	240	180
				3	300	400	600	675	265	200
	10d[d]	1-1/2	15/32	2	320	425	640	730	285	215
				3	360	480	720	820	320	240
APA RATED SHEATHING APA RATED STURD-I-FLOOR and other APA grades except Species Group 5	6d[e]	1-1/4	5/16	2	170	225	335	380	150	110
				3	190	250	380	430	170	125
			3/8	2	185	250	375	420	165	125
				3	210	280	420	475	185	140
	8d	1-3/8	3/8	2	240	320	480	545	215	160
				3	270	360	540	610	240	180
			7/16	2	255	340	505	575	230	170
				3	285	380	570	645	255	190
			15/32	2	270	360	530	600	240	180
				3	300	400	600	675	265	200
	10d[d]	1-1/2	15/32	2	290	385	575	655	255	190
				3	325	430	650	735	290	215
			19/32	2	320	425	640	730	285	215
				3	360	480	720	820	320	240

(a) For framing of other species: (1) Find specific gravity for species of lumber in the AFPA National Design Specification. (2) Find shear value from table above for nail size for actual grade. (3) Multiply value by the following adjustment factor: Specific Gravity Adjustment Factor = [1 – (0.5 – SG)], where SG = specific gravity of the framing. This adjustment shall not be greater than 1.

(b) Space nails maximum 12 inches o.c. along intermediate framing members (6 in. o.c. when supports are spaced 48 in. o.c. or greater). Fasteners shall be located 3/8 inch from panel edges.

(c) Framing at adjoining panel edges shall be 3-in. nominal or wider, and nails shall be staggered where nails are spaced 2 inches o.c. or 2-1/2 inches o.c.

(d) Framing at adjoining panel edges shall be 3-in. nominal or wider, and nails shall be staggered where 10d nails having penetration into framing of more than 1-5/8 inches are spaced 3 inches o.c.

(e) 8d is recommended minimum for roofs due to negative pressures of high winds.

Notes: Design for diaphragm stresses depends on direction of continuous panel joints with reference to load, not on direction of long dimension or strength axis of sheet. Continuous framing may be in either direction for blocked diaphragms.

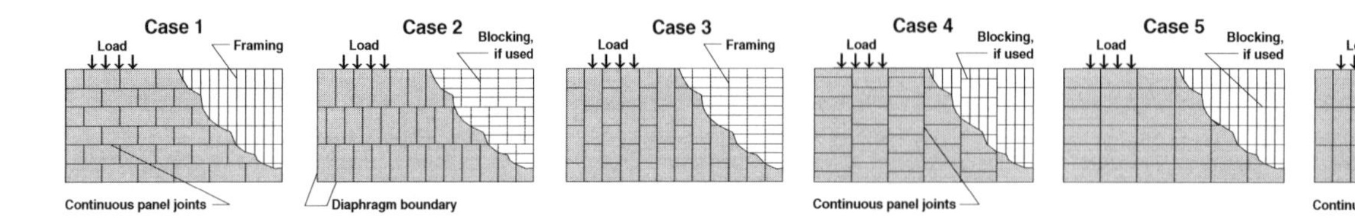

Figure 12.4 Plywood Sheathing Panels for Diaphragms

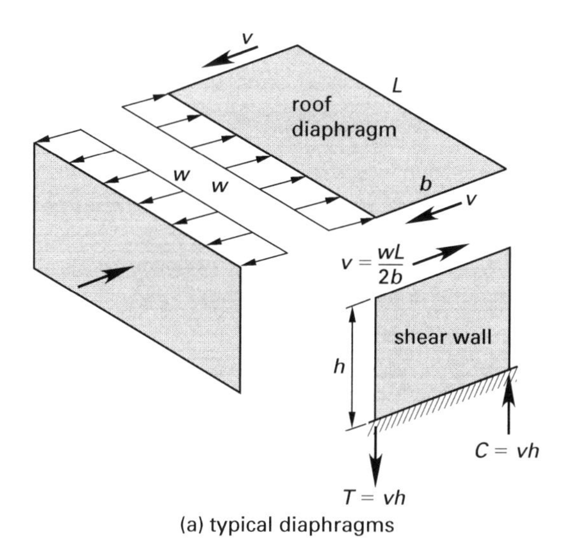

(a) typical diaphragms

strut
chord in
tension
deflected
shapes
chord in
compression
vertical diaphragm
(shear wall)

(b) diaphragm chords

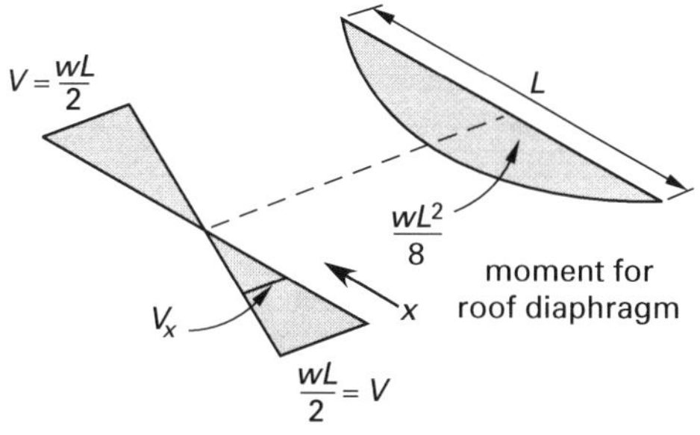

shear for roof diaphragm

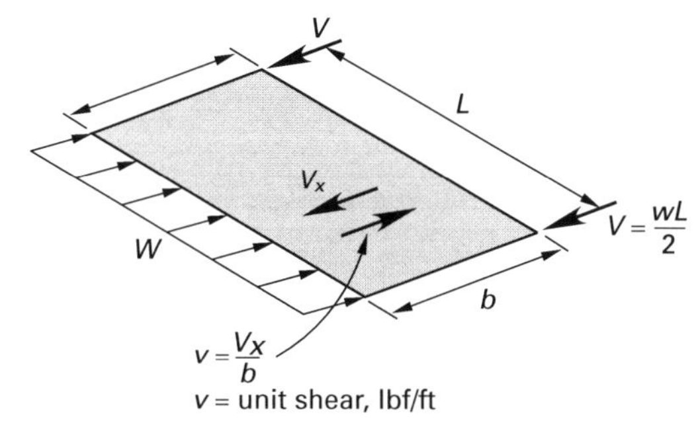

(c) diaphragm shear and moments

Table 12.7 APA Panel Shear Wall Capacities (APA Design/Construction Guide: Diaphragms, Table 2)

RECOMMENDED SHEAR (POUNDS PER FOOT) FOR APA PANEL SHEAR WALLS WITH FRAMING OF DOUGLAS-FIR, LARCH, OR SOUTHERN PINE[a] FOR WIND OR SEISMIC LOADING[b]

Panel Grade	Minimum Nominal Panel Thickness (in.)	Minimum Nail Penetration in Framing (in.)	Panels Applied Direct to Framing: Nail Size (common or galvanized box)	Nail Spacing at Panel Edges (in.): 6	4	3	2[e]	Panels Applied Over 1/2" or 5/8" Gypsum Sheathing: Nail Size (common or galvanized box)	Nail Spacing at Panel Edges (in.): 6	4	3	2[e]
APA STRUCTURAL I grades	5/16	1-1/4	6d	200	300	390	510	8d	200	300	390	510
	3/8	1-3/8	8d	230[d]	360[d]	460[d]	610[d]	10d	280	430	550[f]	730
	7/16			255[d]	395[d]	505[d]	670[d]					
	15/32			280	430	550	730					
	15/32	1-1/2	10d	340	510	665[f]	870		—	—	—	—
APA RATED SHEATHING; APA RATED SIDING[g] and other APA grades except species Group 5	5/16 or 1/4[c]	1-1/4	6d	180	270	350	450	8d	180	270	350	450
	3/8			200	300	390	510		200	300	390	510
	3/8	1-3/8	8d	220[d]	320[d]	410[d]	530[d]	10d	260	380	490[f]	640
	7/16			240[d]	350[d]	450[d]	585[d]					
	15/32			260	380	490	640					
	15/32	1-1/2	10d	310	460	600[f]	770	—	—	—	—	—
	19/32			340	510	665[f]	870	—	—	—	—	—
APA RATED SIDING[g] and other APA grades except species Group 5			**Nail Size (galvanized casing)**					**Nail Size (galvanized casing)**				
	5/16[c]	1-1/4	6d	140	210	275	360	8d	140	210	275	360
	3/8	1-3/8	8d	160	240	310	410	10d	160	240	310[f]	410

(a) For framing of other species: (1) Find specific gravity for species of lumber in the AFPA National Design Specification. (2) For common or galvanized box nails, find shear value from table above for nail size for actual grade. (3) Multiply value by the following adjustment factor: Specific Gravity Adjustment Factor = [1 – (0.5 – SG)], where SG = specific gravity of the framing. This adjustment shall not be greater than 1.

(b) All panel edges backed with 2-inch nominal or wider framing. Install panels either horizontally or vertically. Space nails maximum 6 inches o.c. along intermediate framing members for 3/8-inch and 7/16-inch panels installed on studs spaced 24 inches o.c. For other conditions and panel thicknesses, space nails maximum 12 inches o.c. on intermediate supports. Fasteners shall be located 3/8 inch from panel edges.

(c) 3/8-inch or APA RATED SIDING 16 oc is minimum recommended when applied direct to framing as exterior siding.

(d) Shears may be increased to values shown for 15/32-inch sheathing with same nailing provided (1) studs are spaced a maximum of 16 inches o.c., or (2) if panels are applied with strength axis across studs.

(e) Framing at adjoining panel edges shall be 3-inch nominal or wider, and nails shall be staggered where nails are spaced 2 inches o.c. Check local code for variations of these requirements.

(f) Framing at adjoining panel edges shall be 3-inch nominal or wider, and nails shall be staggered where 10d nails having penetration into framing of more than 1-1/2 inches are spaced 3 inches o.c. Check local code for variations of these requirements.

(g) Values apply to all-veneer plywood APA RATED SIDING panels only. Other APA RATED SIDING panels may also qualify on a proprietary basis. APA RATED SIDING 16 oc plywood may be 11/32 inch, 3/8 inch or thicker. Thickness at point of nailing on panel edges governs shear values.

TYPICAL LAYOUTS FOR SHEAR WALLS

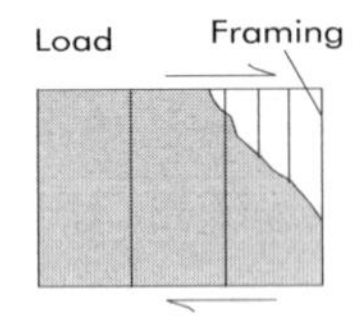

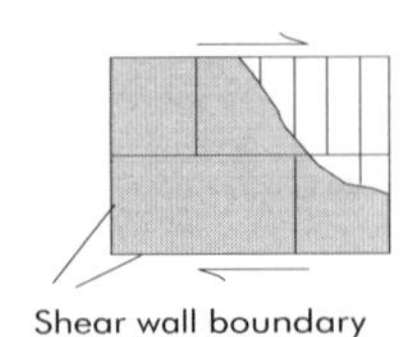

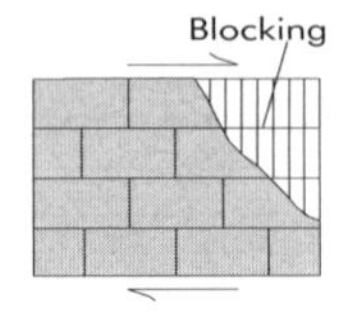

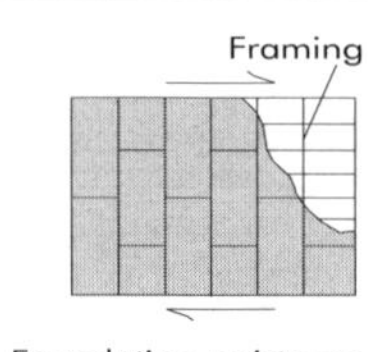

Used with permission of APA—The Engineered Wood Association.

Blocking the unsupported plywood edges between the principal framing members significantly increases the shear capacity, which is evidenced from the data in Tables 12.6 and 12.7. Blocking means installing short pieces of lumber to which all abutting free edges are attached.

The bending forces in diaphragms are usually resisted by the roof or floor perimeter framing, which acts as the diaphragm chords (comparable to the chords of a truss). The chord is sized to resist the calculated axial and bending forces. Since diaphragms are commonly much longer than available or manageable lumber lengths, the chords must be spliced adequately.

Example 12.3
Horizontal Roof Diaphragm

A wood frame building is shown in the following illustration.

Design data is as follows.

L	length of the south wall	80 ft
b	width of the east wall	38 ft
h	average height	16 ft
	wind pressure on vertical surface	33.25 lbf/ft^2
	framing members	2×6
	roof diaphragm selected (based on the load perpendicular to the plywood panel surface)	15/32 in

Sheathing is APA rated with 8d common wire nails. Determine the diaphragm requirements.

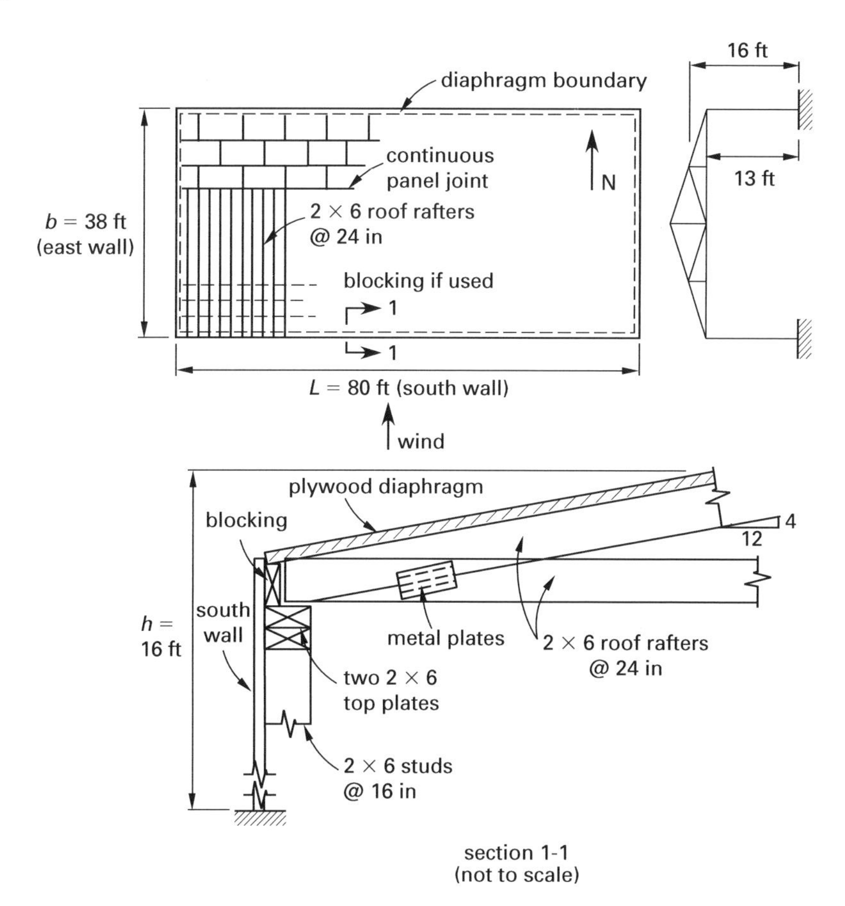

REFERENCE

Solution (ASD Method)

1. Wind pressure on south wall

Tbl 12.6

The lateral force is perpendicular to the continuous panel joint and perpendicular to the unblocked panel edge. Therefore, according to Table 12.6, this is an example of case 1.

The horizontal load acting on the diaphragm is

$$w = (\text{wind pressure})\left(\frac{\text{wall height}}{2}\right) = \left(33.25\ \frac{\text{lbf}}{\text{ft}^2}\right)\left(\frac{16\ \text{ft}}{2}\right)$$
$$= 266\ \text{lbf/ft}$$

The shear at the east and west walls is

$$V = \frac{wL}{2} = \frac{\left(266\ \frac{\text{lbf}}{\text{ft}}\right)(80\ \text{ft})}{2}$$
$$= 10{,}640\ \text{lbf}$$

The unit shear stress along the east and west walls is

$$v = \frac{V}{b} = \frac{10{,}640\ \text{lbf}}{38\ \text{ft}}$$
$$= 280\ \text{lbf/ft}$$

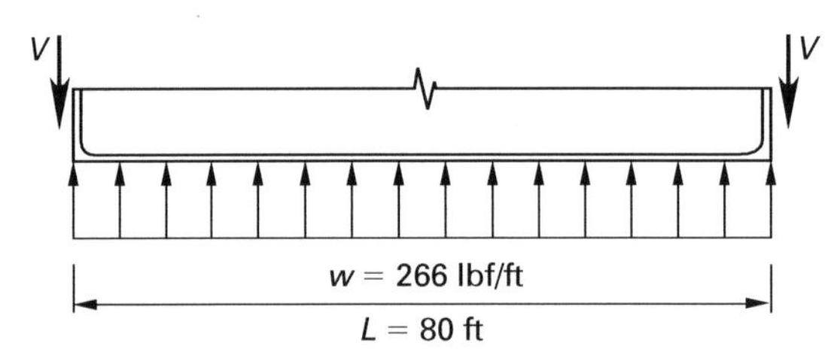

(a) horizontal load on diaphragm

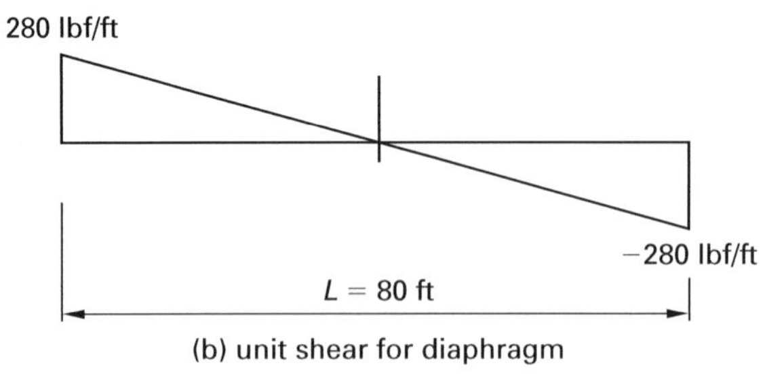

(b) unit shear for diaphragm

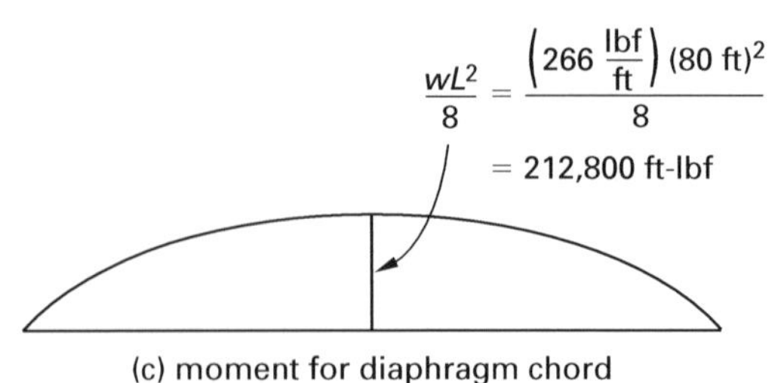

(c) moment for diaphragm chord

Tbl 12.6

The allowable unit shears from Table 12.6 are as follows.

For unblocked diaphragms,

$$v = 240\ \text{lbf/ft} < 280\ \text{lbf/ft} \quad \text{[no good]}$$

For blocked diaphragms,

$$v = 360\ \text{lbf/ft} > 280\ \text{lbf/ft} \quad \text{[OK]}$$

Thus, along the east or west wall, use the blocked diaphragm with nail spacings at 4 in along the diaphragm boundary, 6 in along continuous panel joints as well as along other plywood edges, and 12 in along intermediate framing members.

If reduced unit shears toward the center of the diaphragm (away from east or west wall) are taken into consideration, the nail spacings can be increased and/or blocking can be omitted.

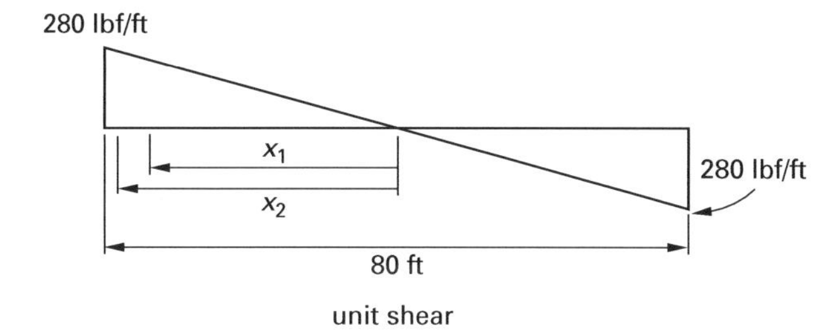

unit shear

Tbl 12.6

For a nail spacing of 6 in at the diaphragm boundary and at other panel edges (see Table 12.6), the allowable shear for unblocked diaphragm is 240 lbf/ft, and allowable shear for blocked diaphragm is 270 lbf/ft.

The blocking along the continuous panel joints can be omitted at a distance from the center of the diaphragm (with nails spaced 6 in at supported edges) of

$$x_1 = \left(\frac{240\ \frac{\text{lbf}}{\text{ft}}}{280\ \frac{\text{lbf}}{\text{ft}}} \right) (40\ \text{ft})$$
$$= 34.3\ \text{ft}$$

Nails spaced at 6 in intervals at the diaphragm boundary for which blocking is provided can begin at a distance from the diaphragm center as follows.

$$x_2 = \left(\frac{270\ \frac{\text{lbf}}{\text{ft}}}{280\ \frac{\text{lbf}}{\text{ft}}} \right) (40\ \text{ft})$$
$$= 38.6\ \text{ft}$$

The axial forces in the diaphragm chord (80 ft in length) are determined from the maximum diaphragm moment, which is resolved into a couple.

The maximum moment is

$$M_{max} = \frac{wL^2}{8} = \frac{\left(266\ \frac{\text{lbf}}{\text{ft}}\right)(80\ \text{ft})^2}{8} = 212{,}800\ \text{ft-lbf}$$

The tension or compression force in the diaphragm is

$$\frac{M_{max}}{b} = \frac{212{,}800\ \text{ft-lbf}}{38\ \text{ft}} = 5600\ \text{lbf}$$

The chord member can be top plates at the top of the stud wall, masonry wall with reinforcement, and so on, and they will be designed for 5600 lbf tension and compression forces.

2. Wind pressure on east wall

Tbl 12.6

The lateral force is parallel to the continuous panel joint and parallel to the unblocked panel edge. Therefore, according to Table 12.6, this is case 3.

The horizontal load acting on the diaphragm is

$$w = \left(33.25\ \frac{\text{lbf}}{\text{ft}^2}\right)\left(\frac{16\ \text{ft}}{2}\right) = 266\ \text{lbf/ft}$$

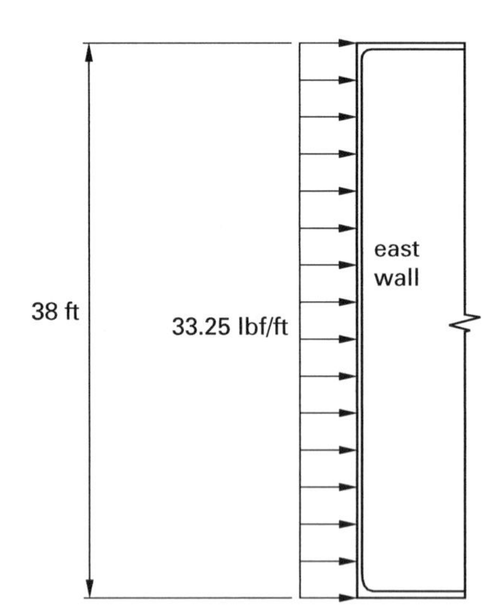

The shear at the east wall is

$$V = \frac{wb}{2} = \frac{\left(266\ \frac{\text{lbf}}{\text{ft}}\right)(38\ \text{ft})}{2} = 5054\ \text{lbf}$$

The unit shear stress along north and south walls is

$$v = \frac{V}{L} = \frac{5054\ \text{lbf}}{80\ \text{ft}} = 63.2\ \text{lbf/ft}$$

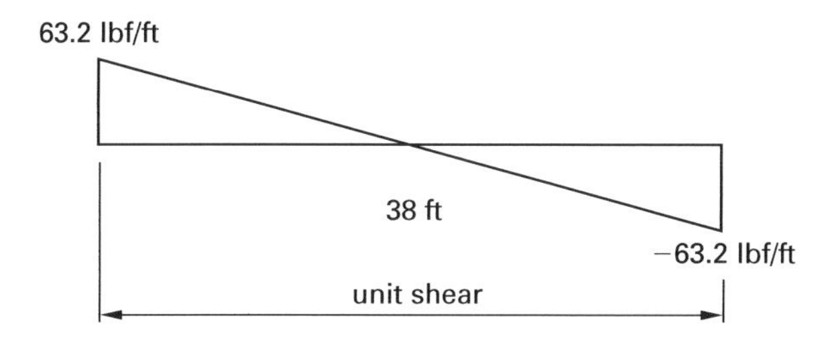

Tbl 12.6

The allowable unit shears from Table 12.6 are as follows.

For unblocked diaphragms with nails spaced 6 in apart at the supported edge,

$v = 180$ lbf/ft

For blocked diaphragms with nails spaced 4 in at boundaries and spaced 6 in at other panel edges,

$v = 360$ lbf/ft

Tbl 12.6

Since the allowable unit shear for unblocked diaphragms is 180 lbf/ft, and greater than the required unit shear, 63.2 lbf/ft, unblocked diaphragms are allowed as seen in case 3 in Table 12.6.

Therefore, use $^{15}/_{32}$ in APA-rated with 8d nails at 4 in on center boundary, 6 in on center all along plywood edges (includes continuous panel joints), and 12 in on center as "field" nailing (i.e., along intermediate framing members). Blocking required: Use 2 × 4 (on edge).

B. Shear Walls

Building walls that are parallel to an applied lateral force can carry that force down to the foundation. Once the shear forces along the shear wall/diaphragm intersection are determined, Table 12.7 can be used to determine the plywood thickness, panel layout, and nailing schedule required to provide the design capacity.

C. Design Methods: Shear Walls and Diaphragms

The basic design procedure for shear walls and diaphragms is to determine the applied loads and detail the respective elements to carry the loads.

step 1: Calculate the applied loads as shears (in lbf/ft) along the supported edge of diaphragms or the loaded edge of shear walls.

step 2: Determine panel layout, plywood thickness, and nailing schedule from Table 12.6 or 12.7.

step 3: Determine diaphragm chord size and detail splices.

step 4: Check deflections by comparing length-width ratios to allowable ones.

step 5: Detail connection between elements and to the foundation.

Example 12.4
Diaphragm and Shear Wall

Determine the design shear on the diaphragms and shear walls of the building shown. Design and detail the roof as a diaphragm and the first-floor interior wall as a shear wall. Assume a wind pressure of 25 lbf/ft^2 and consider only wind against the long side of the building. Use southern pine no. 3 grade for the framing.

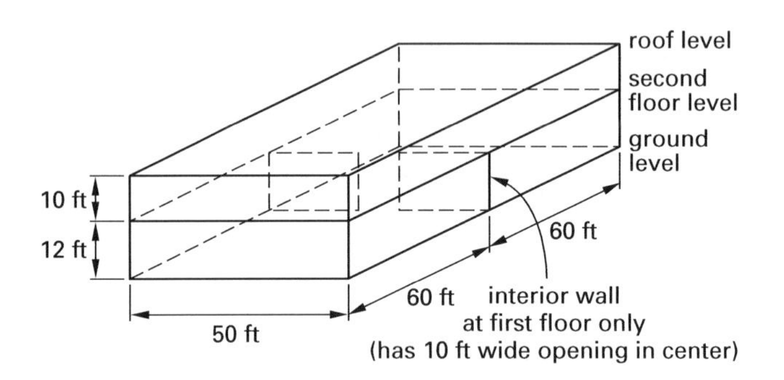

REFERENCE

Solution (ASD Method)

1. Wind loads

Determine the wind loads applied to the long edges of the roof and floor from the wall that frames between them. Assume a tributary area distribution of the uniform load.

The wind load on the 120 ft length of the roof is

$$w_{\text{roof}} = P(\text{tributary height}) = \left(25\ \frac{\text{lbf}}{\text{ft}^2}\right)\left(\frac{10\ \text{ft}}{2}\right)$$
$$= 125\ \text{lbf/ft}$$

The wind load on the 120 ft length of the second floor is

$$w_{\text{second floor}} = P(\text{tributary height}) = \left(25\ \frac{\text{lbf}}{\text{ft}^2}\right)\left(\frac{10\ \text{ft} + 12\ \text{ft}}{2}\right)$$
$$= 275\ \text{lbf/ft}$$

Note that the wind load on the lower 6 ft of the wall is carried directly to the foundation and represents a relatively small shear force.

2. Shears along supported edges

Determine the shears along the supported edges of the diaphragms (where they are connected to the shear walls). The end walls act as shear walls.

Use case 3A from this book's appendix to find the roof-to-end walls shear.

$$V_{\text{roof-to-end wall}} = \frac{w_{\text{roof}}\frac{L}{2}}{\text{end wall length}} = \frac{\left(125\ \frac{\text{lbf}}{\text{ft}}\right)\left(\frac{120\ \text{ft}}{2}\right)}{50\ \text{ft}}$$
$$= 150\ \text{lbf/ft}$$

Use case 4A from this book's appendix to find the floor-to-end walls shear.

$$V_{\text{floor-to-end walls}} = \frac{w_{\text{second floor}}\left(\frac{3}{8}\right)\frac{L}{2}}{\text{end wall length}}$$

$$= \frac{\left(275\ \frac{\text{lbf}}{\text{ft}}\right)\left(\frac{3}{8}\right)\left(\frac{120\ \text{ft}}{2}\right)}{50\ \text{ft}}$$

$$= 124\ \text{lbf/ft}$$

Use case 4A from this book's appendix to find the floor-to-centerline walls shear.

$$V_{\text{floor-to-centerline walls}} = \frac{w_{\text{second floor}}\left(\frac{10}{8}\right)\frac{L}{2}}{\text{interior wall length}}$$

$$= \frac{\left(275\ \frac{\text{lbf}}{\text{ft}}\right)\left(\frac{10}{8}\right)\left(\frac{120\ \text{ft}}{2}\right)}{40\ \text{ft}}$$

$$= 516\ \text{lbf/ft}$$

Note that the 10 ft opening is considered by loading only 40 ft of the interior wall.

Tbl 12.6

3. Roof design

To design the roof as a horizontal diaphragm, first find an adequate configuration in Table 12.6.

Because installing complete blocking is expensive, first try to find an option with unblocked edges. The lower group of panel grades is less expensive, so also try to find a layout in that section of Table 12.6. Several layouts meet the design requirement of 150 lbf/ft. In practice, the choice might be influenced by snow loads, local practice, or other architectural considerations. For this case, assume that 2 in nominal framing is adequate and that a 3/8 in panel thickness is adequate for the wind load. A 160 lbf/ft capacity (requirement is 150 lbf/ft) is provided by the following configuration: APA rated sheathing, 3/8 in panel thickness, 8d nails on 6 in centers along the supported edge (i.e., end wall) and on 12 in centers along the other framing. The edges can be unblocked, and 2 in nominal framing members are adequate. The panel layout can be case 2, 3, 4, 5, or 6.

The chord force is determined by finding the maximum bending moment and dividing by the diaphragm depth—the moment arm of the chord forces.

The maximum moment is

$$M_{\text{max}} = \frac{wL^2}{8} = \frac{\left(125\ \frac{\text{lbf}}{\text{ft}}\right)(120\ \text{ft})^2}{8}$$

$$= 225{,}000\ \text{ft-lbf}$$

Therefore, the bending force in the chords is

$$\frac{M_{\text{max}}}{d} = \frac{225{,}000 \text{ ft-lbf}}{50 \text{ ft}} = 4500 \text{ lbf}$$

NDS Supp Tbl 4B

Since one chord will be in tension and the other in compression, the smaller of the two allowable axial stresses will control the chord size. $F_c = 925 \text{ lbf/in}^2$ and $F_t = 425 \text{ lbf/in}^2$ for southern pine no. 3 grade, so the tension chord controls the design size for an assumed 6 in deep member. The required chord area is, therefore,

$$A_{\text{chord}} = \frac{\frac{M_{\text{max}}}{d}}{F_t(\text{load duration factor})} = \frac{4500 \text{ lbf}}{\left(425 \frac{\text{lbf}}{\text{in}^2}\right)(1.60)} = 6.62 \text{ in}^2$$

NDS Tbl 2.3.2

1.60 is the load duration factor for wind loads. Assuming a doubled 2 in nominal chord, splices will have a 1.5 in wide continuous member. The required depth of the chord is

$$d_{\text{chord}} = \frac{A_{\text{chord}}}{\text{chord width}} = \frac{6.62 \text{ in}^2}{1.5 \text{ in}} = 4.41 \text{ in}$$

Even after removing some of the chord section for the splice fasteners, any chord of doubled 2 × 6s will be adequate. The splices should be designed to handle the maximum chord force.

The diaphragm is only as good as the connection supporting it at the shear walls. This connection should have a design capacity of 160 lbf/ft, the capacity of the diaphragm—not the 150 lbf/ft design load. There are several ways to achieve this capacity. One efficient way is to have the panel wall sheathing overlap the roof perimeter framing.

The *International Building Code* (and most other applicable codes) specifies a maximum length:width ratio of 4:1 for horizontal diaphragms sheathed with plywood. This ratio is intended to eliminate excessive deflection. The ratio in this example is 120 ft:50 ft = 2.4:1, so the criterion is satisfied.

Tbl 12.7

The interior wall has a shear load of 516 lbf/ft, applied on the top by the second floor and transmitted down to the foundation. Table 12.7 distinguishes between walls with the plywood applied directly to the framing and plywood applied over 1/2 in gypsum sheathing intended as fire walls. Assuming this is not a fire wall, a 530 lbf/ft load (requirement is 516 lbf/ft) can be carried in a shear wall with the following specifications: APA rated sheathing, 3/8 in panel thickness, with 8d nails 2 in on center at the boundary framing members and 12 in on center at all other framing. Any of the panel layouts illustrated in Table 12.7 are acceptable.

Tbl 12.7

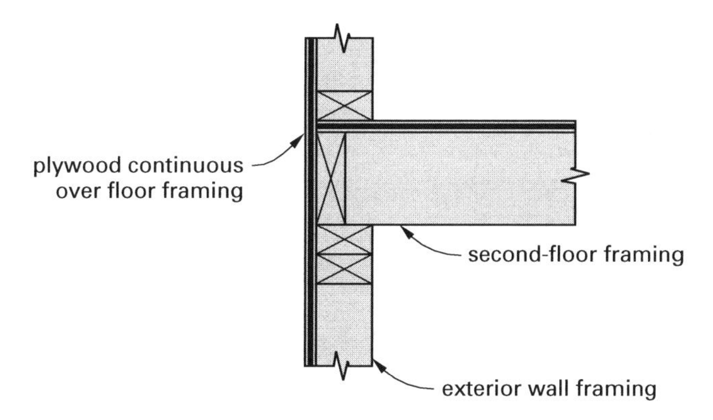

The chords of shear walls are the vertical framing members at the corners and around openings. Since standard framing practice results in triple members at corners and double members at openings, the chords at the openings control. The load at the top of each half of the shear wall is

$$\begin{aligned}(\text{shear at diaphragm edge})(\text{diaphragm length}) &= \left(516\ \frac{\text{lbf}}{\text{ft}}\right)(20\ \text{ft}) \\ &= 10{,}320\ \text{lbf}\end{aligned}$$

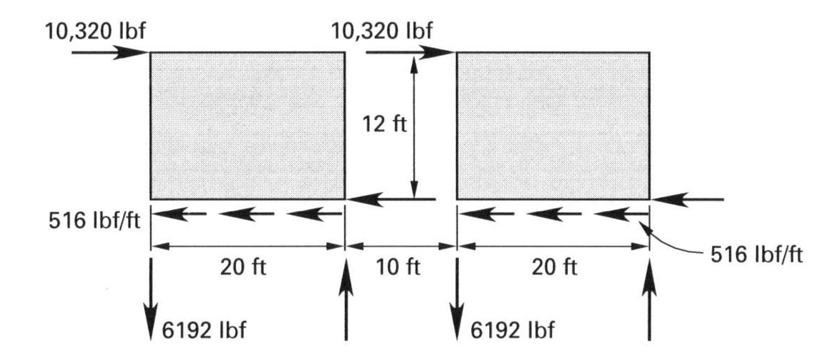

This force creates a bending moment that is maximum at the bottom of the wall, where it is equal to

$$\begin{aligned}M_{\text{bending}} &= v_{\text{top of shear wall}}h = (10{,}320\ \text{lbf})(12\ \text{ft}) \\ &= 123{,}840\ \text{ft-lbf}\end{aligned}$$

The height of the wall is 12 ft. This causes a bending force in the vertical chords equal to the applied moment divided by the chord separation, or shear wall length. The chord force is

$$\begin{aligned}V_{\text{chord}} &= \frac{M_{\text{bending}}}{L_{\text{shear wall}}} = \frac{123{,}840\ \text{ft-lbf}}{20\ \text{ft}} \\ &= 6192\ \text{lbf}\end{aligned}$$

With the allowable tension stress calculated for the chord for the roof diaphragm, this chord force requires an area of

$$\begin{aligned}A_{\text{chord force}} &= \frac{6192\ \text{lbf}}{C_D F_t} = \frac{6192\ \text{lbf}}{(1.60)\left(425\ \frac{\text{lbf}}{\text{in}^2}\right)} \\ &= 9.1\ \text{in}^2\end{aligned}$$

Assuming the verticals will be short enough to be one piece, the full 3 in width will be available. This requires a vertical framing member depth of

$$\begin{aligned} d_{\text{member}} &= \frac{A_{\text{chord force}}}{w} = \frac{9.1 \text{ in}^2}{3 \text{ in}} \\ &= 3.03 \text{ in} \end{aligned}$$

This is less than the 3.5 in of a 2 × 4, so two 2 × 4 walls are adequate.

The 2 × 4 wall must be connected to the foundation to resist both the shear load of 530 lbf/ft (requirement is 516 lbf/ft) and the uplift load in the chords of 6192 lbf at each corner and side of the opening. Anchor bolts at closer than standard spacings or special steel fittings can be used to provide this connection capacity.

The height:width ratio for this shear wall is 12 ft:40 ft, or 0.30:1, less than the 1:1 ratio for which deflections should be calculated. Therefore, in the field, this would be considered a poor design.

---

8. Plywood-Lumber Built-Up Beams

[*PDS Supplement Two: Design and Fabrication of Glued Plywood-Lumber Beams,* 1992]

With its alternating grain directions, plywood is very strong in shear. Its panel configuration also makes plywood convenient to use for webs of built-up beams. Lumber, with all its fibers oriented in the same direction, can efficiently resist the axial forces found in the flanges of built-up beams. For spans longer or loads heavier than dimension lumber beams can handle, plywood-lumber beams can be cheaper and easier to obtain than glued laminated members.

There are many design variables involved in designing a plywood-lumber beam. The beam depth and configuration can be changed. The size, number, species, and grade of the lumber flanges are also variables. Finally, the webs can have any number of plies of any thickness and type of plywood.

Along with all these design variables, there are two basic design methodologies. The first involves tables of plywood-lumber beam capacities, spans, and details, published by several manufacturers' groups and government agencies.

The other methodology is to pick a trial section based on experience and available material. The shear stresses, bending stresses, and deflections are then checked. If the trial section is adequate without being unreasonably understressed, the design is completed by detailing the connections.

Trying to achieve an optimal built-up beam design with all the design variables can be time consuming. Unless the beams are to be mass-produced, it is generally uneconomical to spend a lot of time trying to optimize design. A more important consideration is material availability.

The following design procedure is based on *PDS Supplement Two: Design and Fabrication of Glued Plywood-Lumber Beams.* This publication is a valuable resource on fabrication detailing, asymmetrical sections, and detailing.

A. Design Considerations

The two basic configurations for plywood-lumber beams are the I-beam and the box beam.

The allowable stresses for the lumber flanges are found in the *NDS Supplement*.

Deflections are calculated through standard mechanics procedures and should be limited to the same values as solid beams. Since these beams tend to be long and/or heavily loaded, and are fabricated instead of sawn, camber is a design consideration. The recommended camber is 1.5 times the dead load deflection.

B. Shape

Loads, spans, allowable stresses, and desired appearance determine a beam's proportions. Typical cross sections are shown in Fig. 12.5.

Figure 12.5 Typical Plywood-Lumber Beam Cross Sections
(APA PDS Supplement Two, Design and Fabrication of Glued Plywood-Lumber Beams, Figure 1.2)

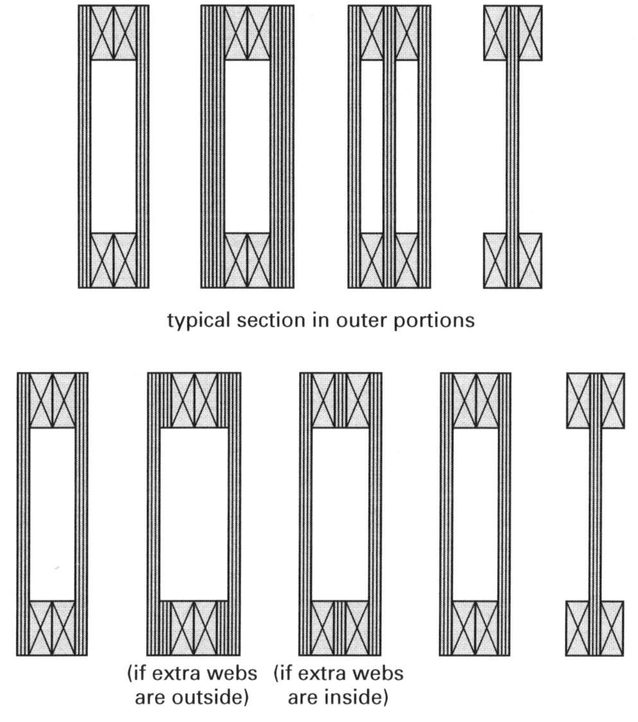

C. Trial Section

To determine a trial section, choose a beam depth that is one-eighth to one-twelfth of the span and that makes efficient use of the 4 ft module of commercially available plywood.

Table 12.8 gives very approximate values for the shear and moment capacities of various cross sections. These are only preliminary cross sections, and they must be checked against the actual design and allowable stresses.

D. Lumber Flanges

Since the lumber flanges are primarily loaded axially by the bending stresses, F_c and F_t are the considered allowable stresses. If the beam is symmetrical, the smaller of the two allowable axial stresses will control. The usual bending stress equation is used with the net moment of inertia, I_n, which is calculated neglecting those longitudinal fibers in the plywood webs and lumber flanges that are interrupted by butt splices.

Table 12.8 Preliminary Capacities of Plywood-Lumber Beam Cross Sections (APA PDS Supplement Two, Design and Fabrication of Glued Plywood-Lumber Beams, Appendix B)

Plywood webs, butt joints staggered 24" minimum, spliced per PDS Section 5.6.3.2

Continuous lumber flanges (no butt joints), resurfaced, for gluing per Part 1, Section 4.1.2

		Max. Moment,(1)(2) M (ft-lb)			Max. Shear,(1) V_h (lb)
Depth	Flange	M_{flange}	M_{web}(3)(4)	M_{total}	$V_{horizontal}$(5)(6)
12"	1-2x4	3563	538	4101	1145
	2-2x4	7126	538	7664	1159
	3-2x4	10690	538	11227	1165
16"	1-2x4	5657	971	6629	1612
	2-2x4	11314	971	12286	1648
	3-2x4	16971	971	17943	1663
	1-2x6	5863	842	6704	1486
	2-2x6	11725	842	12567	1497
	3-2x6	17588	842	18430	1502
20"	1-2x4	7826	1533	9358	2073
	2-2x4	15652	1533	17184	2135
	3-2x4	23477	1533	25010	2162
	1-2x6	8639	1328	9968	1950
	2-2x6	17279	1328	18607	1978
	3-2x6	25918	1328	27247	1990
	1-2x8	8632	1226	9858	1845
	2-2x8	17264	1226	18490	1856
	3-2x8	25896	1226	27122	1861
24"	1-2x4	9831	2197	12028	2515
	2-2x4	19662	2197	21859	2606
	3-2x4	29493	2197	31690	2647
	1-2x6	11389	1904	13294	2407
	2-2x6	22779	1904	24683	2457
	3-2x6	34168	1904	36073	2477
	1-2x8	11820	1758	13578	2295
	2-2x8	23639	1758	25397	2321
	3-2x8	35459	1758	37217	2331
	1-2x10	11452	1611	13064	2183
	2-2x10	22905	1611	24516	2192
	3-2x10	34357	1611	35969	2196
30"	2-2x4	26229	3463	29691	3317
	3-2x4	39343	3463	42806	3381
	4-2x4	52457	3463	55920	3418
	2-2x6	31666	3001	34667	3189
	3-2x6	47498	3001	50500	3226
	4-2x6	63331	3001	66332	3246
	2-2x8	34101	2770	36871	3051
	3-2x8	51151	2770	53921	3074
	4-2x8	68201	2770	70972	3086
	2-2x10	34397	2539	36936	2898
	3-2x10	51595	2539	54134	2910
	4-2x10	68793	2539	71332	2917
	2-2x12	32692	2309	35001	2768
	3-2x12	49039	2309	51347	2773
	4-2x12	65385	2309	67693	2776
36"	2-2x4	32842	5015	37856	4014
	3-2x4	49262	5015	54277	4104
	4-2x4	65683	5015	70698	4156
	2-2x6	40721	4346	45067	3915
	3-2x6	61081	4346	65427	3971
	4-2x6	81441	4346	85787	4003
	2-2x8	44930	4012	48942	3785
	3-2x8	67395	4012	71407	3823
	4-2x8	89860	4012	93872	3844
	2-2x10	46605	3677	50283	3626
	3-2x10	69908	3677	73585	3650
	4-2x10	93210	3677	96888	3663
	2-2x12	45487	3343	48831	3475
	3-2x12	68231	3343	71574	3489
	4-2x12	90975	3343	94318	3497
42"	2-2x6	49871	5939	55810	4632
	3-2x6	74806	5939	80746	4711
	4-2x6	99742	5939	105681	4755
	2-2x8	55968	5482	61450	4515
	3-2x8	83952	5482	89434	4571
	4-2x8	111936	5482	117418	4602
	2-2x10	59220	5026	64245	4360
	3-2x10	88830	5026	93855	4398
	4-2x10	118440	5026	123465	4419
	2-2x12	58956	4569	63525	4202
	3-2x12	88434	4569	93003	4227
	4-2x12	117912	4569	122481	4241
48"	2-2x6	59080	7781	66861	5340
	3-2x6	88621	7781	96402	5443
	4-2x6	118161	7781	125942	5502
	2-2x8	67135	7182	74318	5240
	3-2x8	100703	7182	107885	5316
	4-2x8	134270	7182	141453	5358
	2-2x10	72087	6584	78670	5093
	3-2x10	108130	6584	114714	5147
	4-2x10	144173	6584	150757	5177
	2-2x12	72843	5985	78829	4935
	3-2x12	109265	5985	115250	4973
	4-2x12	145687	5985	151672	4994

(Continued)

Table 12.8 Preliminary Capacities of Plywood-Lumber Beam Cross Sections (PDS Supplement Two, Design and Fabrication of Glued Plywood-Lumber Beams, Appendix B) (continued)

Bases and Adjustments:

(1) Basis: Normal duration of load (C_D): 1.00

Adjustments: 0.90 for permanent load (over 50 years)
1.15 for 2 months, as for snow
1.25 for 7 days
1.6 for 10 minutes, as for wind or earthquake
2.00 for impact

(2) Basis: F_t of flange = 1,000 psi, corrected by C_F (douglas fir-larch Select Structural, 1997 NDS)

2 x 4 = 1,000 x 1.5 = 1,500 psi
2 x 6 = 1,000 x 1.3 = 1,300 psi
2 x 8 = 1,000 x 1.2 = 1,200 psi
2 x 10 = 1,000 x 1.1 = 1,100 psi
2 x 12 = 1,000 x 1.0 = 1,000 psi

Adjustment: $\frac{F_t}{1,000}$ for other tabulated tension stresses.

(C_F in numerator and denominator cancel when flanges are same width.) Also see PDS Section 5.7.3 for adjustments due to butt joints.

(3) Basis: One web effective in bending because web joints are assumed to be unspliced.

Adjustment: 2.0 for web splices per PDS 5.6.1.

(4) Basis: $A_{||}$ of webs for 15/32" or 1/2" APA RATED SHEATHING EXP 1 (CDX). See PDS Tables 1 and 2.

Adjustments: 0.81 for 3/8"
0.97 for 3/8" STRUCTURAL I
1.19 for 15/32" or 1/2" STRUCTURAL I
1.02 for 19/32" or 5/8"
1.51 for 19/32" or 5/8" STRUCTURAL I
1.42 for 23/32" or 3/4"
1.84 for 23/32" or 3/4" STRUCTURAL I

(5) Basis: t_s of webs for 15/32" or 1/2" APA RATED SHEATHING EXP 1 (CDX). See PDS Tables 1 and 2.

Note: Adjustments below may in some cases cause rolling shear to control final design.

Adjustments: 0.93 for 3/8"
1.24 for 3/8" STRUCTURAL I
1.80 for 15/32" or 1/2" STRUCTURAL I
1.07 for 19/32" or 5/8"
2.37 for 19/32" or 5/8" STRUCTURAL I
1.49 for 23/32" or 3/4"
2.48 for 23/32" or 3/4"" STRUCTURAL I

(6) Basis: Plywood edges parallel to face grain glued to continuous framing per PDS 3.8.1

Adjustments: 1.12 for all plywood edges glued to framing per PDS 3.8.1 (1.33/1.19)
0.84 for non-glued conditions (1.00/1.19)

Used with permission of APA—The Engineered Wood Association.

When calculating I_n, three reductions in the lumber flange area must be considered. The first is due to the resurfacing required for a competent glue joint at the web/flange connection. Assume this is a 1/8 in reduction in width across the beam cross section. The second area reduction is due to resurfacing the depth of the assembled beam and is intended to smooth out irregularities in the fit along the top and bottom of the beam. Assume this reduction in beam and flange depth to be 3/8 in for beams less than 24 in deep and 1/2 in for beams 24 in deep or deeper.

The last flange area reduction results from butt-splicing the lumber to achieve the required beam length. The reduction is a function of the butt-joint spacings. The butt-jointed flange members are neglected in any I_n calculation. If two flange pieces are butt-jointed at a spacing less than 10 times the thickness of the flange pieces, both flange pieces are neglected at the more critical of the two locations. Unless the splices are spaced more than 50 times the flange member width, the unjointed flange members also have their areas reduced according to the factor from Table 12.9. When checking flange tension stresses, the jointed members are neglected, the unjointed members are reduced according to Table 12.9, and the allowable stress is further reduced by 20%.

The plywood webs contribute some bending resistance. This is accounted for in the I_n calculation by including only the plywood plies parallel to the longitudinal beam axis. Butt-spliced web members must be neglected in this calculation unless they are spliced full depth with a plywood scab.

Table 12.9 Effective Area of Unspliced Flange Members (PDS Section 5.7.3)

butt-joint spacing (t = lamination thickness)	effective laminae area
$30t$	90%
$20t$	80%
$10t$	60%

Used with permission of APA—The Engineered Wood Association.

E. Plywood Webs

The major stress in the plywood web is through-the-panel shear stress. When calculating the maximum value of the shear stress (at the neutral axis), the standard equation, VQ/Ib, is used. In this calculation, however, the Q and I section properties are calculated including all longitudinal fibers, regardless of butt splicing. The thickness, b, is the sum of all shear thicknesses of plywood present at the section.

F. Flange to Web Connection

The shear connection between the web(s) and flange(s) causes rolling shear in the plywood. The shear stress is evaluated at the glue line with the standard equation. This stress is compared with an allowable rolling shear from Table 12.5. The table value should be reduced by 50% to account for the stress concentrations that arise at the connection.

G. Deflections

PDS *Supplement Two* provides a refined way to calculate the separate bending and shear components of deflection in plywood-lumber beams. An approximate method is to calculate the bending component with standard deflection equations and increase it to account for shear deflection. The magnitude of the increase is a function of the span-to-depth ratio, as indicated in Table 12.10. The values of Table 12.10 may be linearly interpolated for actual span-to-depth ratios.

Table 12.10 Bending Deflection Increase to Account for Shear (PDS Supplement Two, Section 7.1)

span/depth	magnitude of increase
10	1.5 (150%)
15	1.2 (120%)
20	1.0 (100%)

Used with permission of APA—The Engineered Wood Association.

H. Details

Any splices must be detailed to transmit the resultant design forces. In addition, stiffeners are added wherever a point load or support reaction is applied to the beam. These stiffeners fit vertically between the flanges to reinforce the web and distribute the point load. They are sized to not crush the flanges at the bearing points and to not induce rolling shear failure in the web as the load is transferred from flange to web.

I. Lateral Stability

Plywood-lumber beams can be long and subject to lateral buckling. The American Plywood Association suggests considering the ratio of cross-sectional stiffnesses about the vertical and horizontal axes, $\sum I_x$, as a measure of the beam's tendency to buckle laterally, $\sum I_y$. The moments of inertia include all longitudinal fibers, regardless of splicing. The bracing recommendations, as a function of that ratio, are found in Table 12.11.

Table 12.11 Lateral Bracing Required for Plywood-Lumber Beams (PDS Supplement Two, Section 9)

$\frac{\sum I_x}{\sum I_y}$	provision for lateral bracing
up to 5	none required
5 to 10	ends held in position at bottom flanges at supports
10 to 20	beams held in line at ends (both top and bottom flanges restrained from horizontal movement in planes perpendicular to beam axis)
20 to 30	one edge (either top or bottom) held in line
30 to 40	beam restrained by bridging or other bracing at intervals of not more than 8 ft
more than 40	compression flanges fully restrained and forced to deflect in a vertical plane, as with a well-fastened joist and sheathing, or stressed-skin panel system

Used with permission of APA—The Engineered Wood Association.

Example 12.5
Plywood-Lumber Beam

A plywood-lumber beam is 24 in deep (nominally) with a 25 ft clear span. The webs are $^5/_8$ in 32/16 APA rated sheathing exterior with exterior glue. The webs are not spliced at the butt joints. The flanges are three douglas fir-larch, select structural 2 × 6s that are butt-jointed on staggered 48 in centers. The standard resurfacing has been done. The beam is subjected to normal duration and dry loading. Deflection is limited to span/360.

What is the beam's uniform load capacity as limited by (a) flange stresses, (b) web stresses, (c) web/flange connection stresses, and (d) deflection?

REFERENCE

Solution (ASD Method)

The first step is to evaluate the section properties. Butt-jointed members are considered intermittently effective in the net section properties. Once the section properties are known, the design equations are solved for the allowable uniform load.

The members are resurfaced during beam fabrication. The flange members are sanded $^1/_8$ in thinner for better gluing. After assembly, the beam's depth is resurfaced and reduced by $^1/_2$ in. This depth reduction is assumed to reduce the depth of each flange by $^1/_4$ in.

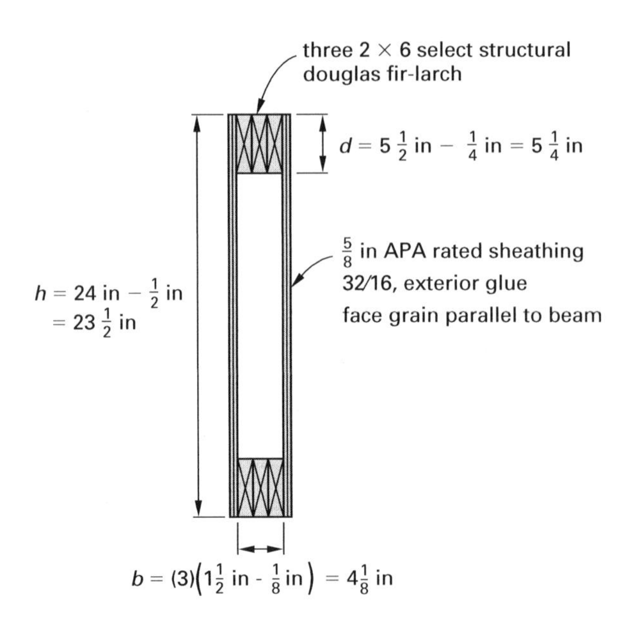

1. Section properties

PDS Supp 2 App A4

Moments of inertia, I (net and total), are as follows.

$$I_{\text{flanges,net}} = \frac{b\left(h^3 - (h-2d)^3\right)}{12}$$

$$= \frac{\begin{array}{c}(0.90\text{ in})(2)(1.5\text{ in} - 0.125\text{ in})\\ \times\left((23.5\text{ in})^3 - \left(23.5\text{ in} - (2)(5.25\text{ in})^3\right)\right)\end{array}}{12}$$

$$= 2224\text{ in}^4$$

This is the net moment of inertia for the flanges, where one of the veneers is neglected (leaving two pieces) and the remaining members are only 90% effective. The butt-joint spacing is

$$48\text{ in}/1.5\text{ in} \approx 30t$$

Tbl 12.9

Table 12.9 indicates that the effective laminae area for $30t$ butt-joint spacing is 90%. The total moment of inertia includes all parallel fibers, regardless of butt-splicing.

$$I_{\text{flanges,total}} = \frac{(2224\text{ in}^4)\left(\dfrac{3\text{ pieces}}{2\text{ pieces}}\right)}{0.9} = 3707\text{ in}^4$$

$$I_{\text{per web}} = t_{\parallel}\left(\frac{h^3}{12}\right) = (0.194\text{ in})\left(\frac{(23.5\text{ in})^3}{12}\right) = 210\text{ in}^4$$

Tbl 12.3

From Table 12.3, the area of a 5/8 in unsanded panel is 2.33 in^2. Using this information, calculate the effective thickness.

$$t_{\parallel} = \frac{2.33\text{ in}^2}{12\text{ in}} = 0.194\text{ in}$$

Tbls 12.2, 12.4b

Only the plywood fibers that are parallel to the beam axis contribute to section properties. Table 12.2 gives the group 4 species, and Table 12.4b gives the unsanded finish and S-1 stress rating (APA rated sheathing exterior).

The composite section properties are a combination of the plywood webs and the lumber flanges.

$$\begin{aligned} I_n = I_{\text{net}} &= I_{\text{flanges}} + I_{\text{web}} \quad \text{[contributing web]} \\ &= 2224 \text{ in}^4 + 210 \text{ in}^4 \\ &= 2434 \text{ in}^4 \quad \text{[only one web is considered]} \end{aligned}$$

$$\begin{aligned} I_t &= I_{\text{flanges}} + I_{\text{all parallel web plys}} \\ &= 3707 \text{ in}^4 + (4)(210 \text{ in}^4) \\ &= 4547 \text{ in}^4 \end{aligned}$$

The statical moment, Q_{total}, is found as follows.

$$\begin{aligned} Q_{\text{flanges}} &= bd\left(\frac{h}{2} - \frac{d}{2}\right) \\ &= \big((3)(1.5 \text{ in} - 0.125 \text{ in})\big)(5.25 \text{ in})\left(\frac{23.5 \text{ in}}{2} - \frac{5.25 \text{ in}}{2}\right) \\ &= 197.6 \text{ in}^3 \end{aligned}$$

$$\begin{aligned} Q_{\text{webs}} &= t_{\parallel}\left(\frac{h}{2}\right)\left(\frac{h}{4}\right)(\text{no. of webs}) \\ &= (0.194 \text{ in})\left(\frac{23.5 \text{ in}}{2}\right)\left(\frac{23.5 \text{ in}}{4}\right)(4) \\ &= 53.6 \text{ in}^3 \end{aligned}$$

$$\begin{aligned} Q_{\text{total}} &= Q_{\text{flange}} + Q_{\text{webs}} = 197.6 \text{ in}^3 + 53.6 \text{ in}^3 \\ &= 251.2 \text{ in}^3 \end{aligned}$$

Note that only the total Q value is used. A net value of Q is not calculated. In both I and Q calculations, the web contribution is much smaller than that of the flanges. This is one justification for not correcting for the different stiffnesses of the lumber flanges and plywood webs before combining them in their composite section.

2. Allowable loads

A. The allowable load is limited by the bending stresses in the flanges. For the lumber flanges, the allowable tensile stress is

$$\begin{aligned} F_t' = F_t C_D C_M &= \left(1000 \ \frac{\text{lbf}}{\text{in}^2}\right)(1.0)(1.0) \\ &= 1000 \text{ lbf/in}^2 \end{aligned}$$

NDS Supp Tbl 4A; PDS Sec 5.7.3

This controls over the allowable stress in the compression flange, F_c', 1700 lbf/in^2. F_t is further reduced by 20% at the butt-jointed sections. The allowable moment is

$$\begin{aligned} \frac{F_t I_n}{0.5h} &= \left(\frac{(0.8)\left(1000 \ \frac{\text{lbf}}{\text{in}^2}\right)(2434 \text{ in}^4)}{(0.5)(23.5 \text{ in})}\right)\left(\frac{1 \text{ ft}}{12 \text{ in}}\right) \\ &= 13{,}810 \text{ ft-lbf} \end{aligned}$$

The allowable load is

$$\begin{aligned} \frac{8M_{\text{allow}}}{(\text{span})^2} &= \frac{(8)(13{,}810 \text{ ft-lbf})}{(25 \text{ ft})^2} \\ &= 176.8 \text{ lbf/ft} \end{aligned}$$

PDS Sec 3.8.1; Tbl 12.5

B. The *Plywood Design Specification* allows a 33% increase in allowable shear through the thickness of plywood if the plywood panel is rigidly glued to continuous framing around its edges. Table 12.5 gives the grade stress level as 130 lbf/in^2 for S-1, species group 4, dry use. For this application,

$$F_v = (1.33)(130) = 172.9 \text{ lbf/in}^2$$

The allowable horizontal shear for S-1 species group 4, dry use, is

Tbl 12.3

$$t_s = 0.319 \text{ in}$$

$$V_h = \frac{F_v I_t t_s}{Q} = \frac{\left(172.9 \ \frac{\text{lbf}}{\text{in}^2}\right)(4547 \text{ in}^4)(0.319 \text{ in})(4)}{251.2 \text{ in}^3} = 3994 \text{ lbf}$$

Neglecting the uniform load within a beam depth of the supports, the allowable uniform load for this allowable horizontal shear is

$$w_{\text{allow}} = \frac{2V}{L-2h} = \frac{(2)(3994 \text{ lbf})}{25 \text{ ft} - \frac{(2)(23.5 \text{ in})}{12 \ \frac{\text{in}}{\text{ft}}}} = 379 \text{ lbf/ft}$$

Tbl 12.5

C. From Table 12.5, for S-1 dry use conditions, the rolling shear stress level, F_s, is 53 lbf/in^2. Using this information, the rolling shear at the glue line between web and flange, V_s, is found as follows. F_s is reduced 50% for the flange-to-web joints in box beams.

$$V_s = \frac{2F_s d I_t}{Q_{\text{flanges}}} = \frac{(2)\left(\frac{53 \frac{\text{lbf}}{\text{in}^2}}{2}\right)(5.25 \text{ in})(4547 \text{ in}^4)}{197.6 \text{ in}^3} = 6403 \text{ lbf}$$

PDS Sec 3.8.2

This allowable shear translates into the following allowable uniform load.

$$w_{\text{allow}}\left(\frac{V_s}{V_h}\right) = \left(379 \ \frac{\text{lbf}}{\text{ft}}\right)\left(\frac{6403 \text{ lbf}}{3994 \text{ lbf}}\right) = 608 \text{ lbf/ft}$$

Tbl 12.10

D. A simple method of calculating deflections in plywood-lumber beams is to increase calculated bending deflections by a factor to account for the neglected, but significant, shear deflections. Table 12.10 gives this factor as a function of the span-to-depth ratio.

$$\frac{\text{span}}{\text{depth}} = \frac{25 \text{ ft}}{\frac{23.5 \text{ in}}{12 \ \frac{\text{in}}{\text{ft}}}} = 12.8$$

Interpolating between ratios of 10 and 15 yields a shear deflection factor of 1.33.

The allowable deflection, span/360, is

$$\frac{(25\ \text{ft})\left(12\ \frac{\text{in}}{\text{ft}}\right)}{360} = 0.83\ \text{in}$$

Use the maximum bending deflection equation for uniform load on simple spans to calculate the allowable uniform load.

$$\Delta_{\text{allow}} = \left(\frac{5w_{\text{allow}}\ell^4}{384EI_t}\right)(\text{shear deflection factor})$$

$$\begin{aligned} w_{\text{allow}} &= \frac{\Delta_{\text{allow}}384EI_t}{(\text{shear deflection factor})5\ell^4} \\ &= \frac{(0.83\ \text{in})(384)\left(1{,}900{,}000\ \frac{\text{lbf}}{\text{in}^2}\right)(4547\ \text{in}^4)}{(1.33)(5)(25\ \text{ft})^4\left(12\ \frac{\text{in}}{\text{ft}}\right)^3} \\ &= 613\ \text{lbf/ft} \end{aligned}$$

E. The uniform load capacity of the plywood-lumber beam is limited by bending stresses to 176.8 lbf/ft and by shear in the plywood webs to 379 lbf/ft. There are several ways to increase the capacity: use thicker plywood; use a higher grade plywood—such as structural 1, with its species group 1 allowable stresses; or simply use more webs of the same plywood at the beam ends where shear stresses are maximum. It is also possible to increase the flanges in number or size, or to specify a higher stress species or grade.

9. Other Nonplywood Structural Panels

These panels may be broadly divided into two categories: particle boards and composite plywood panels.

Particle boards such as oriented strand board and waferboard are nonveneer panels consisting of wood particles in the form of flakes, strands, wafers, chips, shavings, and so on, that are bonded with resins under heat and pressure.

Oriented strand board (OSB; Fig. 12.6) is a nonveneer panel consisting of strand-like particles that are compressed and bonded with resin. Three to five layers of the strand fibers in a panel are arranged at right angles to one another in the same manner as in plywood. These panels may be rated under the APA performance specifications for structural sheathings.

Figure 12.6 Oriented Strand Board (OSB)

OSB is considered a major part of structural applications and is interchangeable with plywood (veneer). APA maintains design information on its website in the section called "Oriented Strand Board." For APA rated sheathing (see Fig. 12.1), the span rating looks like a fraction, such as $^{48}/_{24}$. The numerator denotes the maximum spacing of supports (in inches) when the panel is used for roof sheathing, and the denominator denotes the maximum spacing of supports when the panel is used for subflooring.

Waferboard (Fig. 12.7) is a nonveneer structural panel that is manufactured with wood flakes (wafer-like particles) of $1^1/_4$ in or longer lengths. These flakes are randomly arranged and compressed and bonded with an adhesive, although they can be arranged in certain manners to increase the strength and stiffness properties of a panel. Waferboard is the predecessor to oriented strand board, but it is not recognized by the NDS as a structural panel. These panels are rated by the Structural Board Association of Canada (there is no rating in the United States).

Figure 12.7 Waferboard

The composite plywood panel (Fig. 12.8) has a veneer face and back with an inner corestock of oriented strand material or veneer crossband. The grain direction of face and back veneers is in the long direction of the panel. These panels are rated under the APA performance specifications.

Figure 12.8 Composite Plywood Panel

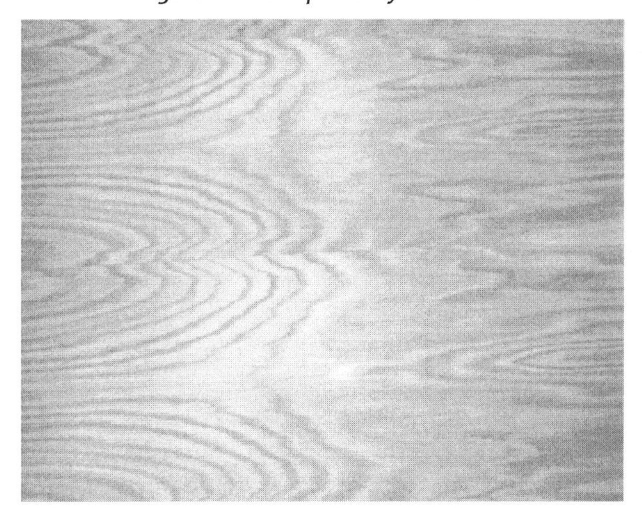

13

Practice Problems

Practice Problem 1: Timber Formwork and Shoring

During curing, timber formwork and shoring support a cast-in-place concrete slab. The 9 in concrete slab is placed on the $^3/_4$ in, APA B-B Plyform Class II (see Table 12.4b) supported by 2×4 joists spaced 16 in apart as shown. 4×6 stringers spaced 36 in apart support these joists. The shores directly under the stringers are 4×4 and are spaced at 5 ft intervals along the stringers. The lumber is no. 2 douglas fir-larch. The concrete has a specific weight of 150 lbf/ft^3 and a construction live load of 50 lbf/ft^2. Limit all deflections to $L/360$.

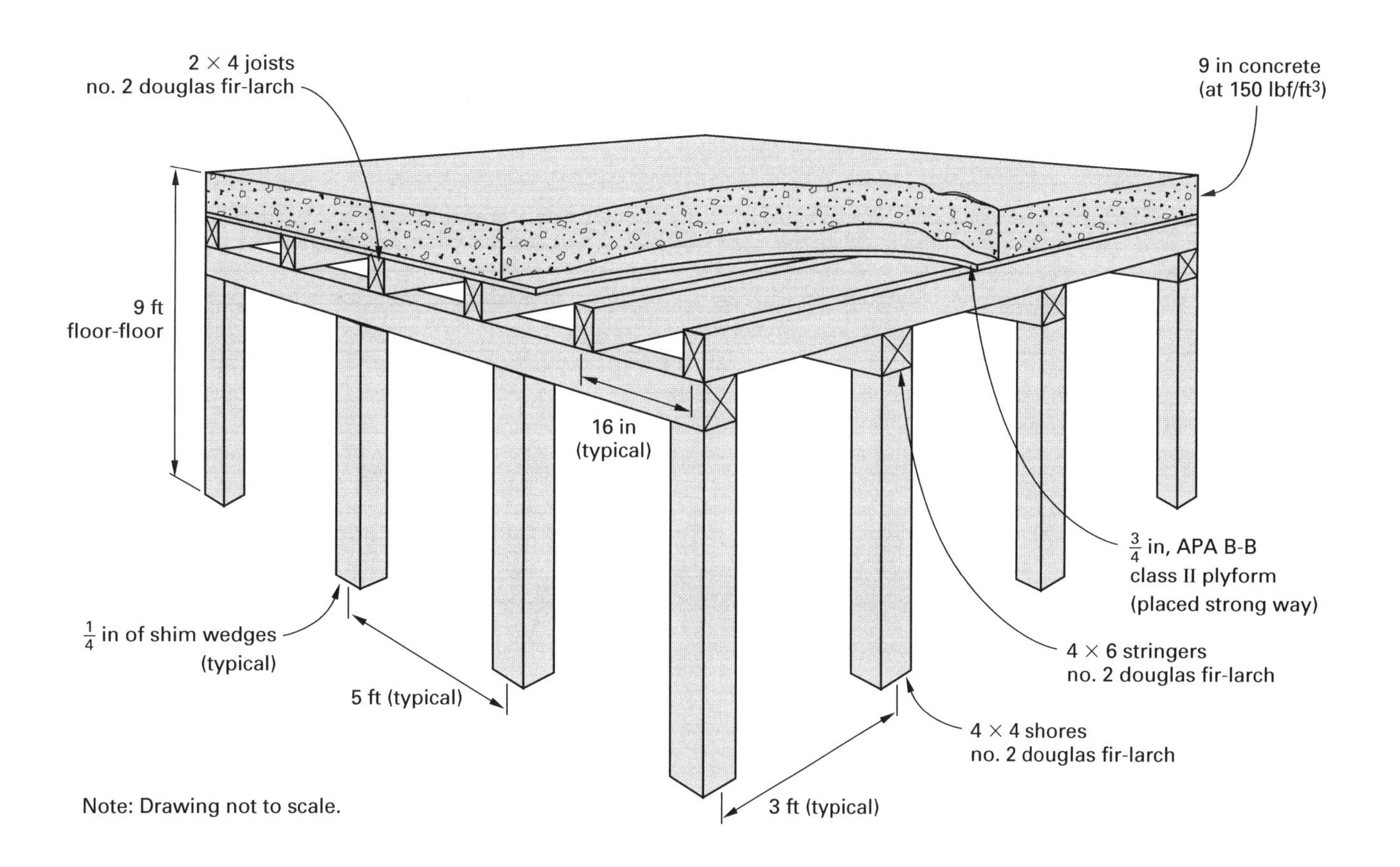

Use the NDS and the PDS or Ch. 12 of this book to do the following. References to Ch. 12 refer to this book and not the NDS.

1. Check the joist spacing of 16 in.
2. Check the stringer spacing of 36 in.
3. Check the shore spacing of 5 ft along each stringer.
4. Determine the allowable shore capacity.
5. Check bearing stresses at the joints connecting shore to stringer and also connecting stringer to joist.

REFERENCE *Solution (ASD Method)*

1. Joist spacing

The total applied load, w, is found as follows.

$$
\begin{aligned}
w &= \text{concrete slab load} + \text{construction live load} + \text{estimated dead load} \\
&= \left(150\ \frac{\text{lbf}}{\text{ft}^3}\right)\left(\frac{9\ \text{in}}{12\ \frac{\text{in}}{\text{ft}}}\right) + 50\ \frac{\text{lbf}}{\text{ft}^2} + 4\ \frac{\text{lbf}}{\text{ft}^2} \\
&= 166\ \text{lbf/ft}^2
\end{aligned}
$$

w corresponds to 166 lbf/ft per 1 ft width of plyform board.

Check joist spacing (i.e., determine the maximum allowable plyform span length).

On 1 ft widths of the plyform, check the plyform for shear, bending, and deflection.

Chap 12
Tbl 12.4b

Class II plyform is in the group 3 species, S-2 grade stress level and sanded.

The following illustration shows shear and moment diagrams for a three-span beam.

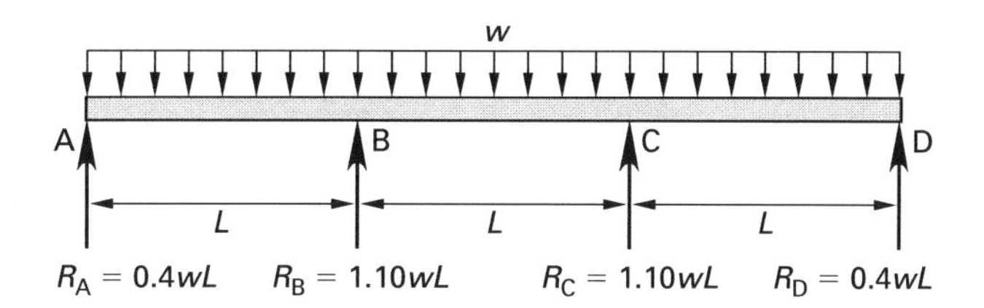

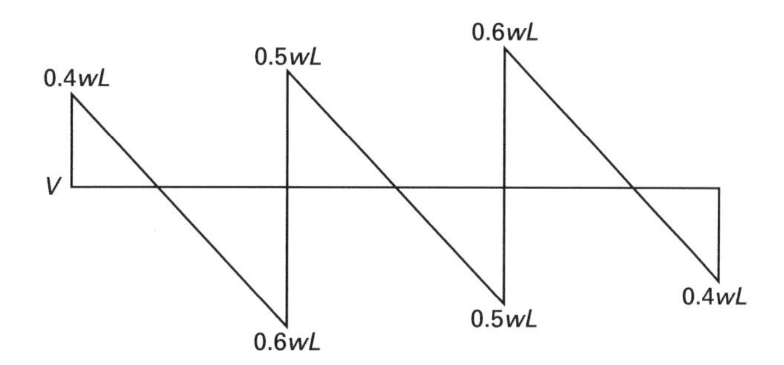

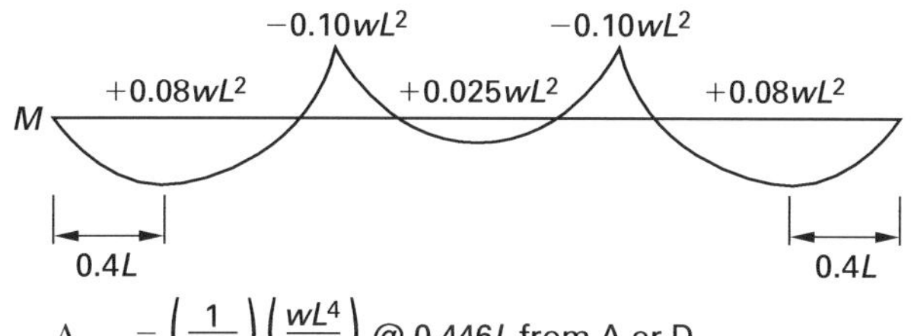

$$\Delta_{max} = \left(\frac{1}{145}\right)\left(\frac{wL^4}{EI}\right) \text{ @ } 0.446L \text{ from A or D}$$

Assume that the plyform is supported in the strong direction, which is along the 8 ft side of the plyform crossing the joists. Also assume that the plyform is continuous over at least three joists (in this case, the plyform is actually continuous over more than three joists).

Chap 12
Tbl 12.5

For wet service (accounting for C_M), the rolling shear, F_s, is 44 lbf/in^2 (APA PDS Table 3).

NDS Tbl 2.3.2

For a seven-day construction load duration, $C_D = 1.25$.

The allowable rolling shear stress is

$$F'_s = F_s C_D = \left(44\ \frac{\text{lbf}}{\text{in}^2}\right)(1.25)$$
$$= 55\ \text{lbf/in}^2$$

The rolling shear, f_v, is

$$f_v = \frac{VQ}{Ib} = V\left(\frac{Q}{Ib}\right)$$

Chap 12
Tbl 12.3

The rolling shear constant for sanded panel 3/4-S is

$$\frac{Ib}{Q} = 6.762\ \text{in}^2\text{/ft panel width}$$

The maximum shear for a three-span beam is

$$V_{max} = V = 0.60wL$$

Setting F'_s equal to f_v,

$$55\ \frac{\text{lbf}}{\text{in}^2} = V_{\text{max}}\left(\frac{Q}{Ib}\right) = 0.60wL\left(\frac{1}{6.762\ \frac{\text{in}^2}{\text{ft}}}\right)$$

Rearranging,

$$L = \frac{\left(55\ \frac{\text{lbf}}{\text{in}^2}\right)\left(6.762\ \frac{\text{in}^2}{\text{ft}}\right)}{(0.60)\left(166\ \frac{\text{lbf}}{\text{ft}^2}\right)\left(\frac{1\ \text{ft}}{12\ \text{in}}\right)} = 44.8\ \text{in}$$

Therefore, the maximum allowable joist spacing as limited by rolling shear stress is 44.8 in.

Chap 12
Tbl 12.5

For wet service, the allowable bending stress, F_b, is 820 lbf/in^2. The adjusted bending stress is

$$F'_b = F_bC_D = \left(820\ \frac{\text{lbf}}{\text{in}^2}\right)(1.25) = 1025\ \text{lbf/in}^2$$

The maximum calculated bending stress is

$$f_b = \frac{M}{S} = \frac{0.10wL^2}{KS}$$

Chap 12
Tbl 12.3

The effective section modulus, KS, for sanded panel 3/4-S is 0.412 in^3. Setting F'_b to f_b,

$$1025\ \frac{\text{lbf}}{\text{in}^2} = \frac{0.01wL^2}{0.412\ \frac{\text{in}^3}{\text{ft}}}$$

Rearranging,

$$L = \sqrt{\frac{\left(1025\ \frac{\text{lbf}}{\text{in}^2}\right)\left(0.412\ \frac{\text{in}^3}{\text{ft}^3}\right)}{(0.10)\left(166\ \frac{\text{lbf}}{\text{ft}^2}\right)\left(\frac{1\ \text{ft}}{12\ \text{in}}\right)}} = 17.5\ \text{in}$$

Therefore, the maximum allowable joist spacing as limited by bending is 17.5 in.

The calculated maximum deflection for a three-span beam is given by

$$\Delta_{\text{max}} = \frac{wL^4}{145EI}$$

The maximum allowable deflection is

$$\Delta_{\text{allow}} = \frac{L}{360}$$

Chap 12
Tbl 12.5

For group 3 species and wet service, E is 1,100,000 lbf/in^2.

Chap 12
Tbl 12.3

The moment of inertia, I, is 0.197 in^4/ft.

Setting the maximum deflection equal to the maximum allowable deflection and rearranging,

$$\Delta_{\text{max}} = \Delta_{\text{allow}} \quad \text{or} \quad \frac{wL^4}{145EI} = \frac{L}{360}$$

$$\begin{aligned} L &= \sqrt[3]{\frac{(145)\left(1{,}100{,}000\ \frac{\text{lbf}}{\text{in}^2}\right)\left(0.197\ \frac{\text{in}^4}{\text{ft}}\right)}{(360)\left(166\ \frac{\text{lbf}}{\text{ft}^2}\right)\left(\frac{1\ \text{ft}}{12\ \text{in}}\right)}} \\ &= 18.5\ \text{in} \end{aligned}$$

Therefore, the maximum allowable joist spacing as limited by deflection is 18.5 in.

The given joist spacing of 16 in is less than the maximum allowable spacing of 17.5 in as limited by bending.

2. Check stringer spacing (i.e., maximum allowable joist span length).

For 2 × 4 joists of no. 2 douglas fir-larch, the NDS values for seven days' load duration and wet service are

NDS Supp Tbl 4A
Adj Fac

	C_F	C_M
$F_b = 900\ \text{lbf/in}^2$	1.5	0.85
$F_v = 180\ \text{lbf/in}^2$	–	0.97
$E = 1{,}600{,}000\ \text{lbf/in}^2$	–	0.90

NDS Tbl 2.3.2

For a load duration of seven days, C_D is 1.25.

The allowable bending stress (excluding C_t, C_{fu}, and C_i) is

NDS Tbl 4.3.1

$$\begin{aligned} F_b' &= F_bC_DC_MC_LC_FC_r = \left(900\ \frac{\text{lbf}}{\text{in}^2}\right)(1.25)(0.85)(1.0)(1.5)(1.0) \\ &= 1434.3\ \text{lbf/in}^2 \end{aligned}$$

Since the joist span is only 16 in, C_L is 1.0. For C_r, use 1.0 (this is conservative), since load-sharing among joists probably does not occur from simultaneous loading on all joists with fresh concrete.

Simplification from "point loads" to distributed loading is assumed.

The maximum calculated bending stress for a three-span beam is

$$f_b = \frac{M}{S} = \frac{0.10wL^2}{S}$$

The section modulus for 2 × 4 joists is

$$\begin{aligned} S &= \frac{bd^2}{6} = \frac{(1.5\ \text{in})(3.5\ \text{in})^2}{6} \\ &= 3.062\ \text{in}^3 \end{aligned}$$

The joist load is

$$w = \left(166\ \frac{\text{lbf}}{\text{ft}^2}\right)(16\ \text{in})\left(\frac{1\ \text{ft}}{12\ \text{in}}\right)$$
$$= 221\ \text{lbf/ft}$$

Setting F_b' equal to f_b and rearranging,

$$1434.3\ \frac{\text{lbf}}{\text{in}^2} = \frac{0.10wL^2}{S}$$

$$L = \sqrt{\frac{\left(1434.3\ \frac{\text{lbf}}{\text{in}^2}\right)(3.062\ \text{in}^2)}{(0.10)\left(221\ \frac{\text{lbf}}{\text{ft}}\right)\left(\frac{1\ \text{ft}}{12\ \text{in}}\right)}}$$
$$= 48.8\ \text{in}$$

The allowable shear stress is

NDS Tbl 4.3.1

$$F_v' = F_v C_D C_M$$
$$= \left(180\ \frac{\text{lbf}}{\text{in}^2}\right)(1.25)(0.97)$$
$$= 218.3\ \text{lbf/in}^2$$

The maximum calculated shear stress for a three-span beam is

$$f_v = \frac{3V}{2A} = \frac{(3)(0.6wL)}{2A}$$

The cross-sectional area is

$$A = (1.5\ \text{in})(3.5\ \text{in})$$
$$= 5.25\ \text{in}^2$$

Setting F_v' equal to f_v and rearranging,

$$218.3\ \frac{\text{lbf}}{\text{in}^2} = \frac{(3)(0.6wL)}{2A}$$

$$L = \frac{(2)(5.25\ \text{in}^2)\left(218.3\ \frac{\text{lbf}}{\text{in}^2}\right)}{(3)(0.6)\left(221\ \frac{\text{lbf}}{\text{ft}}\right)\left(\frac{1\ \text{ft}}{12\ \text{in}}\right)}$$
$$= 69.2\ \text{in}$$

The maximum calculated deflection for a three-span beam is

$$\Delta_{\text{max}} = \frac{wL^4}{145E'I}$$

$$I = \frac{(1.5\ \text{in})(3.5\ \text{in})^3}{12}$$
$$= 5.359\ \text{in}^4$$

$$E' = EC_M = \left(1{,}600{,}000\ \frac{\text{lbf}}{\text{in}^2}\right)(0.90)$$
$$= 1{,}440{,}000\ \text{lbf/in}^2$$

The maximum allowable deflection is

$$\Delta_{\text{allow}} = \frac{L}{360}$$

Setting the maximum calculated deflection equal to the maximum allowable deflection and rearranging,

$$\begin{aligned}\Delta_{\text{max}} &= \Delta_{\text{allow}} \\ \frac{wL^4}{145E'I} &= \frac{L}{360} \\ L &= \sqrt[3]{\frac{145E'I}{360w}} \\ &= \sqrt[3]{\frac{(145)\left(1{,}440{,}000\ \frac{\text{lbf}}{\text{in}^2}\right)(5.359\ \text{in}^4)}{(360)\left(221\ \frac{\text{lbf}}{\text{ft}}\right)\left(\frac{1\ \text{ft}}{12\ \text{in}}\right)}} \\ &= 55.3\ \text{in}\end{aligned}$$

Bending controls the maximum allowable joist span length. Therefore, the maximum allowable stringer spacing is 48.8 in.

3. Shore spacing

The NDS values for 4 × 6 stringers, which are no. 2 douglas fir-larch, are

NDS Supp Tbl 4A

	C_F	C_M
$F_b = 900\ \text{lbf/in}^2$	1.30	1.00
$F_v = 180\ \text{lbf/in}^2$	–	0.97
$E = 1{,}600{,}000\ \text{lbf/in}^2$	–	0.90

For seven days, $C_D = 1.25$.

The section properties of the 4 × 6 lumber are

$$\begin{aligned}S &= \tfrac{1}{6}bh^2 = \left(\frac{1}{6}\right)(3.5\ \text{in})(5.5\ \text{in})^2 \\ &= 17.65\ \text{in}^3\end{aligned}$$

$$\begin{aligned}I &= \tfrac{1}{12}bh^3 = \left(\frac{1}{12}\right)(3.5\ \text{in})(5.5\ \text{in})^3 \\ &= 48.53\ \text{in}^4\end{aligned}$$

$$\begin{aligned}A &= bh = (3.5\ \text{in})(5.5\ \text{in}) \\ &= 19.25\ \text{in}^2\end{aligned}$$

The stringer loading is

$$\begin{aligned}w &= \left(166\ \frac{\text{lbf}}{\text{ft}}\right)(36\ \text{in})\left(\frac{1\ \text{ft}}{12\ \text{in}}\right) \\ &= 498\ \text{lbf/ft}\end{aligned}$$

NDS Sec 3.3.3

The allowable bending stress (where C_L is 1.0) is

$$F'_b = F_b C_D C_M C_L C_F = \left(900\ \frac{\text{lbf}}{\text{in}^2}\right)(1.25)(1.0)(1.0)(1.3)$$
$$= 1462.5\ \text{lbf/in}^2$$

The maximum calculated bending stress is

$$f_b = \frac{M}{S} = \frac{0.1wL^2}{S}$$

Set F'_b equal to f_b and rearrange to find the maximum allowable stringer span length.

$$1462.5\ \frac{\text{lbf}}{\text{in}^2} = \frac{0.10wL^2}{S}$$

$$L = \sqrt{\frac{\left(1462.5\ \frac{\text{lbf}}{\text{in}^2}\right)(17.65\ \text{in}^3)}{(0.10)\left(498\ \frac{\text{lbf}}{\text{ft}}\right)\left(\frac{1\ \text{ft}}{12\ \text{in}}\right)}}$$
$$= 78.9\ \text{in}$$

The maximum allowable shear stress is

$$F'_v = F_v C_D C_M = \left(180.0\ \frac{\text{lbf}}{\text{in}^2}\right)(1.25)(0.97)$$
$$= 218.3\ \text{lbf/in}^2$$

The maximum calculated shear stress for a three-span beam (where $V = 0.6wL$) is

$$f_v = \frac{3V}{2A}$$

Setting the maximum allowable shear stress equal to the maximum calculated shear stress and rearranging,

$$F'_v = f_v$$

$$218.3\ \frac{\text{lbf}}{\text{in}^2} = \frac{(3)(0.6wL)}{2A}$$

$$L = \frac{(2)(19.25\ \text{in}^2)\left(218.3\ \frac{\text{lbf}}{\text{in}^2}\right)}{(3)(0.6)\left(498\ \frac{\text{lbf}}{\text{ft}}\right)\left(\frac{1\ \text{ft}}{12\ \text{in}}\right)}$$
$$= 112.5\ \text{in}$$

Check deflections.

The maximum calculated deflection is

$$\Delta_{\text{max}} = \frac{wL^4}{145E'I}$$

$$E' = EC_M C_t = \left(1{,}600{,}000\ \frac{\text{lbf}}{\text{in}^2}\right)(0.90)(1.0)$$
$$= 1{,}440{,}000\ \text{lbf/in}^2$$

The maximum allowable deflection is

$$\Delta_{\text{allow}} = \frac{L}{360}$$

Setting the maximum calculated deflection equal to the maximum allowable deflection and rearranging,

$$\Delta_{\text{max}} = \Delta_{\text{allow}}$$

$$\frac{wL^4}{145E'I} = \frac{L}{360}$$

$$L = \sqrt[3]{\frac{145E'I}{360w}}$$

$$= \sqrt[3]{\frac{(145)\left(1{,}440{,}000\ \frac{\text{lbf}}{\text{in}^2}\right)(48.53\ \text{in}^4)}{(360)\left(498\ \frac{\text{lbf}}{\text{ft}}\right)\left(\frac{1\ \text{ft}}{12\ \text{in}}\right)}}$$

$$= 87.9\ \text{in}$$

Bending stress limits the maximum allowable stringer span length to 78.9 in. The actual stringer span length is 5 ft (60 in), which is within the maximum limit.

4. Shore capacity

The shore load is

$$P = (\text{applied loads})(\text{area}) = wA = \left(166\ \frac{\text{lbf}}{\text{ft}^2}\right)\left((3\ \text{ft})(5\ \text{ft})\right)$$

$$= 2490\ \text{lbf}$$

The shore is a solid column with a total length calculated as follows.

9 ft	floor to floor
−9 in	slab
−3/4 in	plyform
−3 1/2 in	2 × 4 joist
−5 1/2 in	4 × 6 stringer
−1/4 in	shims

total ℓ = 89.0 in

For 4 × 4 shores of no. 2 douglas fir-larch, the NDS values for seven days' load duration and wet service are

NDS Supp Tbl 4A Adj Fac

	C_F	C_M
$F_c = 1350\ \text{lbf/in}^2$	1.15	0.8
$E_{\text{min}} = 580{,}000\ \text{lbf/in}^2$	–	0.9

The allowable compression design value is

$$F_c' = F_c C_D C_M C_t C_F C_P = F_c^* C_P$$

$$F_c^* = \left(1350\ \frac{\text{lbf}}{\text{in}^2}\right)(1.25)(0.8)(1.0)(1.15)$$

$$= 1552.5\ \text{lbf/in}^2$$

$$E'_{\text{min}} = E_{\text{min}} C_M C_t = \left(580{,}000\ \frac{\text{lbf}}{\text{in}^2}\right)(0.9)(1.0)$$

$$= 522{,}000\ \text{lbf/in}^2$$

NDS Sec 3.7

Determine the column stability factor, C_P.
$c = 0.80$ for sawn lumber.

NDS App. G

$K_e = 1.0$

NDS Sec 3.7.1

$$\left(\frac{\ell_e}{d}\right)_x = \left(\frac{\ell_e}{d}\right)_y = \frac{K_e \ell_u}{d} = \frac{(1.0)(89.0 \text{ in})}{3.5 \text{ in}} = 25.4$$

$$F_{cE} = \frac{0.822E'_{\text{min}}}{\left(\frac{\ell_e}{d}\right)^2} = \frac{(0.822)\left(522{,}000 \ \frac{\text{lbf}}{\text{in}^2}\right)}{(25.4)^2} = 665.1 \text{ lbf/in}^2$$

$$\frac{F_{cE}}{F_c^*} = \frac{665.1 \ \frac{\text{lbf}}{\text{in}^2}}{1552.5 \ \frac{\text{lbf}}{\text{in}^2}} = 0.428$$

NDS Eq 3.7-1

$$C_P = \frac{1+\frac{F_{cE}}{F_c^*}}{2c} - \sqrt{\left(\frac{1+\frac{F_{cE}}{F_c^*}}{2c}\right)^2 - \frac{\frac{F_{cE}}{F_c^*}}{c}}$$

$$= \frac{1+0.428}{(2)(0.8)} - \sqrt{\left(\frac{1+0.428}{(2)(0.8)}\right)^2 - \frac{0.428}{0.8}}$$

$$= 0.381$$

The allowable compressive stress is

$$F'_c = F_c^* C_P = \left(1552.5 \ \frac{\text{lbf}}{\text{in}^2}\right)(0.381) = 591.5 \text{ lbf/in}^2$$

The calculated compressive stress is

$$f_c = \frac{2490 \text{ lbf}}{(3.5 \text{ in})(3.5 \text{ in})} = 203.3 \text{ lbf/in}^2 < F'_c = 591.5 \text{ lbf/in}^2 \quad \text{[OK]}$$

The shore capacity is

$$P_{\text{allow}} = F'_c(4 \times 4 \text{ shore area}) = \left(591.5 \ \frac{\text{lbf}}{\text{in}^2}\right)\left((3.5 \text{ in})(3.5 \text{ in})\right) = 7245.9 \text{ lbf} > \text{shore load of } 2490 \text{ lbf} \quad \text{[OK]}$$

NDS Secs 2.3.10, 3.10

5. Bearing stresses

The NDS value for compression perpendicular to grain for no. 2 douglas fir-larch is

NDS Supp 4A

$$F_{c\perp} = 625 \text{ lbf/in}^2$$
$$C_M = 0.67$$
$$C_t = 1.0 \text{ for normal temperature}$$

The allowable compressive (bearing) stress design value is

NDS Tbl 4.3.1; NDS Secs 3.10.2, 3.10.4

$$F'_{c\perp} = F_{c\perp} C_M C_t C_b$$

C_b can be used for $\ell_b < 6$ in and when bearing is at least 3 in away from member end.

$$C_b = \frac{\ell_b + \frac{3}{8} \text{ in}}{\ell_b}$$

The following illustration shows bearing stresses on 4 × 4 shores with 4 × 6 stringers.

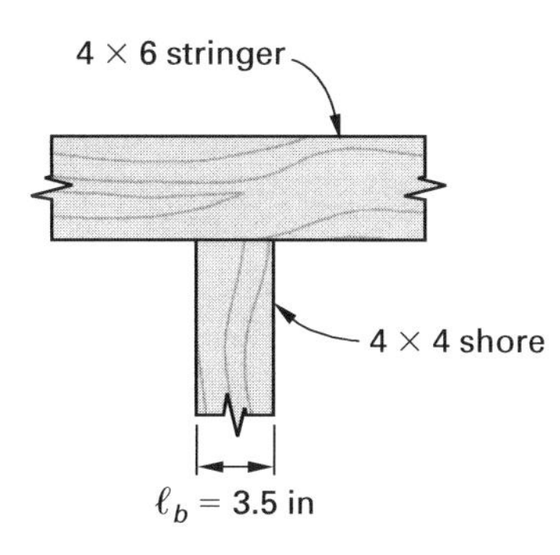

$$C_b = \frac{\ell_b + \frac{3}{8} \text{ in}}{\ell_b} = \frac{3.5 \text{ in} + \frac{3}{8} \text{ in}}{3.5 \text{ in}} = 1.107$$

$$F'_{c\perp} = F_{c\perp} C_M C_t C_b = \left(625 \ \frac{\text{lbf}}{\text{in}^2}\right)(0.67)(1.0)(1.107) = 463.6 \text{ lbf/in}^2$$

Checking bearing stress gives

$$f_{c\perp} = \frac{\text{shore load}}{\text{bearing area}} = \frac{2490 \text{ lbf}}{(3.5 \text{ in})(3.5 \text{ in})} = 203.3 \text{ lbf/in}^2 < F'_{c\perp} = 463.6 \text{ lbf/in}^2 \quad \text{[OK]}$$

The following illustration shows bearing stresses on a 4 × 6 stringer with 2 × 4 joist.

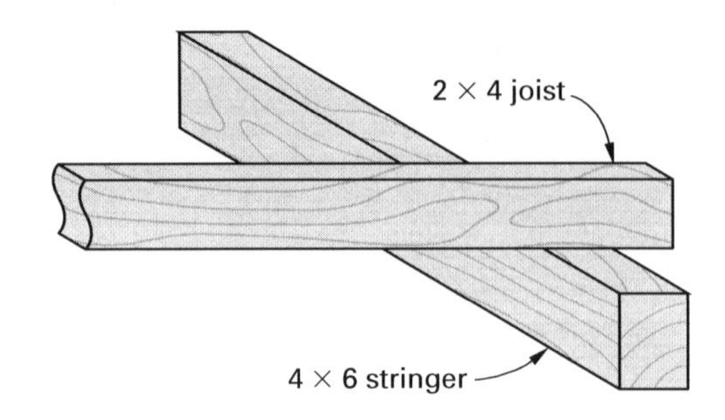

Since both members are compressed perpendicular to grain, the following allowable value is applicable, with a factor of 1/2.

$$C_b = \frac{\ell_b + \frac{3}{8}\text{ in}}{\ell_b} = \frac{3.5\text{ in} + \frac{3}{8}\text{ in}}{3.5\text{ in}} = 1.107$$

$$\begin{aligned} F'_{c\perp} &= \tfrac{1}{2} F_{c\perp} C_M C_t C_b = \left(\frac{1}{2}\right)\left(625\ \frac{\text{lbf}}{\text{in}^2}\right)(0.67)(1.0)(1.107) \\ &= 231.8\text{ lbf/in}^2 \end{aligned}$$

The bearing load from 2 × 4 joists to 4 × 8 stringers is

$$\begin{aligned} P &= wA \\ &= \left(166\ \frac{\text{lbf}}{\text{in}^2}\right)(16\text{ in})(36\text{ in})\left(\frac{1\text{ ft}}{12\text{ in}}\right)^2 \\ &= 664\text{ lbf} \end{aligned}$$

The actual bearing stress is

$$\begin{aligned} f_{c\perp} &= \frac{P}{\text{bearing area}} = \frac{664\text{ lbf}}{(1.5\text{ in})(3.5\text{ in})} \\ &= 126.5\text{ lbf/in}^2 < F_{c\perp} = 231.8\text{ lbf/in}^2 \quad \text{[OK]} \end{aligned}$$

Therefore, the shores have sufficient load capacity.

Practice Problem 2: Bridge Stringers and Deck

A section through the deck of a 30 ft long single-span timber bridge is shown.

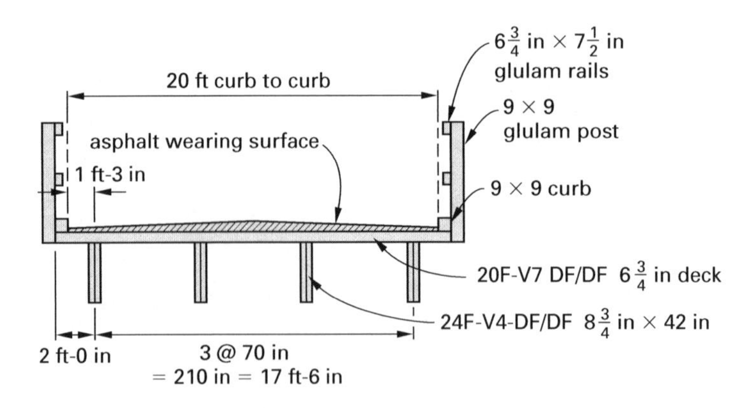

The bridge width is 20 ft curb-to-curb.

The bridge span is 30 ft.

The asphalt wearing surface average thickness is 3 in.

The deck is made of glulam douglas fir 20F-V7, 6³/₄ in thick with bonded edge joint.

The stringers are made of glulam douglas fir 24F-V4, and are 8³/₄ in × 42 in.

The applicable specification is *AASHTO Standard Specifications for Highway Bridges*, 17th Ed., 2002.

The live load is HS20-44.

The timber unit weight is 50 lbf/ft^3.

The allowable live load deflection is $L/500$. The maximum dead load plus live load moments are 221.84 ft-kips for each interior stringer and 12.25 ft-kips for a 21.75 in deck width. The maximum dead plus live load shears are 27.763 kips for each interior stringer and 7.422 kips for a 21.75 in deck width.

Check whether the interior stringers and deck are sufficiently structurally designed. Assume that the deck-stringer connections provide continuous lateral support to the stringers. Checking deck deflection is not required for this problem.

REFERENCE *Solution (ASD Method)*

1. Interior stringers (8³/₄ in × 42 in, glulam douglas fir 24F-V4 DF/DF)

NDS Supp Tbl 1C

The section properties of an 8³/₄ in × 42 in glulam beam are

$$A = 367.5 \text{ in}^2$$

$$S_x = 2573 \text{ in}^3$$

$$I_x = 54{,}020 \text{ in}^4$$

NDS Supp Tbl 5A

Design values for glulam combination 24F-V4 douglas fir are

	C_M
$F_{bx} = 2400 \text{ lbf/in}^2$	0.80
$F_{vx} = 265 \text{ lbf/in}^2$	0.875
$E_x = 1{,}800{,}000 \text{ lbf/in}^2$	0.833

NDS Supp Tbl 5A Adj Fac, NDS Sec 5.3.6

The volume factor is

$$C_V = \left(\frac{21}{L}\right)^{1/x} \left(\frac{12}{d}\right)^{1/x} \left(\frac{5.125}{b}\right)^{1/x} \le 1.0$$

$x = 10$ for wood other than southern pine.

Thus,

$$C_V = \left(\frac{21}{30 \text{ ft}}\right)^{1/10} \left(\frac{12}{42 \text{ in}}\right)^{1/10} \left(\frac{5.125}{8.75 \text{ in}}\right)^{1/10} = 0.807$$

NDS Sec 3.3.3.3

The beam stability factor, C_L, is 1.0 since the compression edge of the stringer is continuously laterally supported by the deck, assuming that the deck is fastened appropriately. Therefore, $C_V = 0.807$ controls.

NDS Tbl 2.3.2

C_D is 1.0 for a 10-year load duration.

NDS Tbl 5.3.1

The allowable design values are

$$F_b' = F_b C_D C_M C_V = \left(2400 \ \frac{\text{lbf}}{\text{in}^2}\right)(1.0)(0.8)(0.807)$$
$$= 1549.4 \text{ lbf/in}^2$$

$$F_v' = F_v C_D C_M = \left(265 \ \frac{\text{lbf}}{\text{in}^2}\right)(1.0)(0.875)$$
$$= 231.9 \text{ lbf/in}^2$$

$$E' = E_x C_M = \left(1{,}800{,}000 \ \frac{\text{lbf}}{\text{in}^2}\right)(0.833)$$
$$= 1{,}499{,}400 \text{ lbf/in}^2$$

Check the bending stress.

$M_{\text{max}} = 221.84$ ft-kips [given]

The calculated bending stress is

$$f_b = \frac{M_{\text{max}}}{S_x} = \left(\frac{221.84 \text{ ft-kips}}{2573 \text{ in}^3}\right)\left(12 \ \frac{\text{in}}{\text{ft}}\right)\left(1000 \ \frac{\text{lbf}}{\text{kip}}\right)$$
$$= 1034.6 \text{ lbf/in}^2 < F_b' = 1594.4 \text{ lbf/in}^2 \quad \text{[OK]}$$

Check the shear stress.

$V_{\text{max}} = 27.763$ kips [given]

The calculated shear stress is

$$f_v = \frac{3V_{\text{max}}}{2A} = \frac{(3)(27.763 \text{ kips})\left(1000 \ \frac{\text{lbf}}{\text{kip}}\right)}{(2)(367.5 \text{ in}^2)}$$
$$= 113.3 \text{ lbf/in}^2 < F_v' = 231.9 \text{ lbf/in}^2 \quad \text{[OK]}$$

Check the live load deflection.

AASHTO Tbl 3.23.1

The wheel load multiplied by the stringer distribution factor is

$$P = (16{,}000 \text{ lbf})(1.167) = 18{,}672 \text{ lbf}$$

$$\Delta = \frac{PL^3}{48E'I}$$
$$= \frac{(18{,}672 \text{ lbf})(360 \text{ in})^3}{(48)\left(1{,}499{,}400 \ \frac{\text{lbf}}{\text{in}^2}\right)(54{,}020 \text{ in}^4)}$$
$$= 0.224 \text{ in}$$

$$\Delta_{\text{allow}} = \frac{L}{500} = \frac{360 \text{ in}}{500}$$
$$= 0.72 \text{ in}$$

$0.72 \text{ in} > \Delta = 0.224 \text{ in}$ [OK]

Use 8¾ in × 42 in glulam, 24F-V4 douglas fir stringer.

2. Deck thickness

AASHTO 3.25.1

The 6¾ in thick section properties are based on a 21.75 in deck width and a 68 in deck span.

$$A = (21.75 \text{ in})(6.75 \text{ in}) = 146.81 \text{ in}^2$$

$$S_x = \frac{(21.75 \text{ in})(6.75 \text{ in})^2}{6} = 165.16 \text{ in}^3$$

NDS Supp Tbl 5A Adj Fac

Design values for glulam combination 20F-V7 DF/DF are

		C_M	
F_{by}	$= 1450 \text{ lbf/in}^2$	0.80	
F_{vy}	$= 230 \text{ lbf/in}^2$	0.875	
E_y	$= 1{,}600{,}000 \text{ lbf/in}^2$	0.833	
C_{fu}	flat use factor	1.07	for 6¾ in thickness
C_D	load duration factor	1.0	for 10-year load duration

NDS Secs 3.3.3, 5.3.2

Volume factor, C_V, and beam stability factor, C_L, are both equal to 1.0 because of the deck thickness and geometry.

The allowable design values are

$$F_b' = F_{by}C_DC_MC_{fu}(\text{the smaller of } C_L \text{ and } C_V) = \left(1450 \ \frac{\text{lbf}}{\text{in}^2}\right)(1.0)(0.80)(1.07)(1.0) = 1241.2 \text{ lbf/in}^2$$

$$F_v' = F_{vy}C_DC_M = \left(230 \ \frac{\text{lbf}}{\text{in}^2}\right)(1.0)(0.875) = 201.3 \text{ lbf/in}^2$$

Check bending and shear stresses.

$$f_b = \frac{M}{S_x} = \left(\frac{12.25 \text{ ft-kips}}{165.16 \text{ in}^3}\right)\left(12 \ \frac{\text{in}}{\text{ft}}\right)\left(1000 \ \frac{\text{lbf}}{\text{kip}}\right) = 890.0 \text{ lbf/in}^2 < F_b' = 1241.2 \text{ lbf/in}^2 \quad [\text{OK}]$$

$$f_v = \frac{3V}{2A} = \frac{(3)(7.422 \text{ kips})\left(1000 \ \frac{\text{lbf}}{\text{kip}}\right)}{(2)(146.81 \text{ in}^2)} = 75.8 \text{ lbf/in}^2 < F_v' = 201.3 \text{ lbf/in}^2 \quad [\text{OK}]$$

Use 6¾ in thick glulam 20F-V7 DF/DF deck.

Note that a deflection check for this deck is not necessary for this problem.

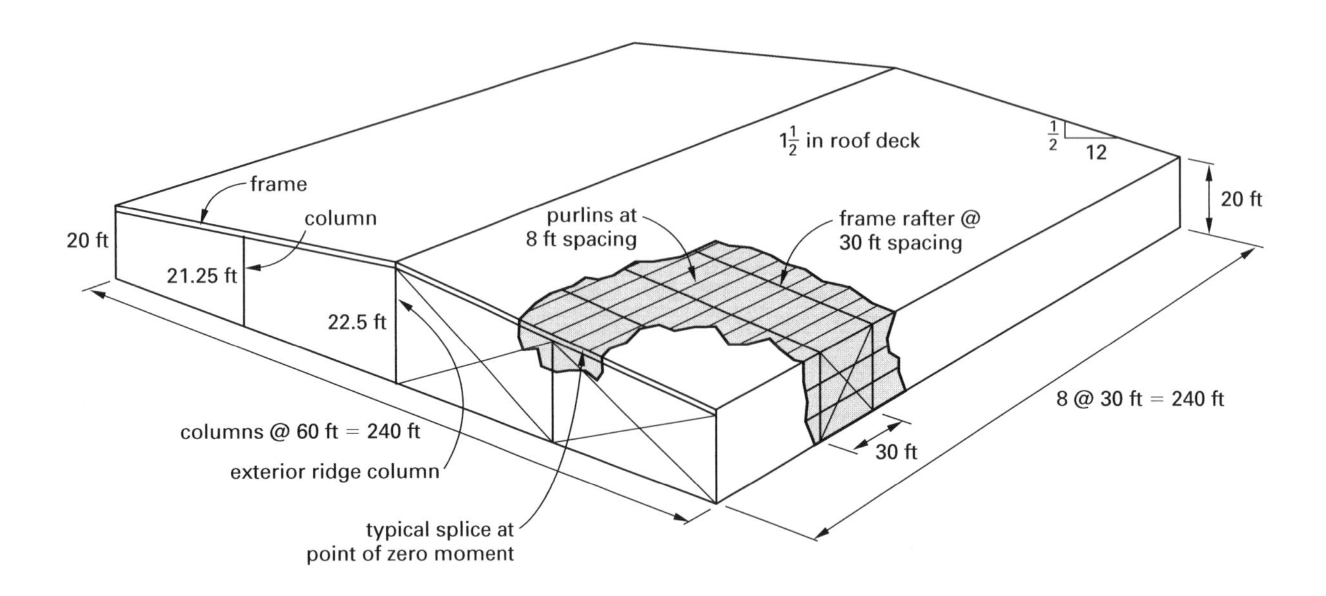

Practice Problem 3: Commercial Building

Snow loading and wind pressures on a large one-story commercial building have been determined in accordance with ASCE/SEI7-05 provisions. The design loads are dead loads plus snow loads for the roof structure and dead loads plus wind loads for the columns. The design snow load is 20 lbf/ft^2 and the wind pressure on the windward wall is 10 lbf/ft^2. The roof structure consists of a $1^1/2$ in thick timber plank deck supported by purlins spaced at 8 ft on centers. The purlins rest on frame rafters spaced at 30 ft on centers as shown. The maximum allowable total deflection is $L/180$. Assume timber unit weights of 36 lbf/ft^3 for the deck and 40 lbf/ft^3 for the other timber components. Use douglas fir-larch.

1. Design the roof deck.
2. Design the purlins.
3. Design the frame rafters (assume two-span cantilever beam system).
4. Design the connections between the purlins and frame rafters.
5. Design the end column.

REFERENCE

Solution (ASD Method)

1. Roof deck design

The deck span is 8 ft between roof purlins and is continuous over two purlins. The deck consists of 2 × 6 tongue-and-groove douglas fir-larch planks, Commercial Dex grade.

The total roof load, w, is found as follows.
The dead load is

$$w_D = \begin{array}{c}\text{roofing, felt, and}\\ \text{vapor barrier load}\end{array} + \begin{array}{c}\text{insulation}\\ \text{load}\end{array} + \begin{array}{c}\text{2 in nominal tongue-}\\ \text{and-groove plank load}\end{array}$$
$$= 5.0\ \frac{\text{lbf}}{\text{ft}^2} + 4.0\ \frac{\text{lbf}}{\text{ft}^2} + \left(36\ \frac{\text{lbf}}{\text{ft}^3}\right)(1.5\ \text{in})\left(\frac{1\ \text{ft}}{12\ \text{in}}\right)$$
$$= 13.5\ \text{lbf/ft}^2$$

The snow load, w_S, is 20 lbf/ft^2.

The total load is

$$w = w_D + w_S = 13.5\ \frac{\text{lbf}}{\text{ft}^2} + 20\ \frac{\text{lbf}}{\text{ft}^2}$$
$$= 33.5\ \text{lbf/ft}^2$$

NDS Supp Tbl 4E
Adj Fac

Design values for douglas fir-larch Commercial Dex grade are as follows.

$C_F =$ 1.10 for 2 in nominal thickness
$F_{c\perp} = 625\ \text{lbf/in}^2$
$E = 1{,}700{,}000\ \text{lbf/in}^2$
$F_v =$ not available in the NDS tables
$C_{fu} =$ flat use factor already considered in the NDS design values

NDS Tbl 2.3.2

$C_D =$ 1.15 for snow loads

NDS Sec 4.1.4

$C_M =$ 1.0 for 19% or less moisture content

NDS Tbl 4.3.1

Allowable design values are

$$F_b' = F_b C_D C_M C_F = \left(1450\ \frac{\text{lbf}}{\text{in}^2}\right)(1.15)(1.0)(1.10)$$
$$= 1834.2\ \text{lbf/in}^2$$
$$F_{c\perp}' = F_{c\perp} C_M = \left(625\ \frac{\text{lbf}}{\text{in}^2}\right)(1.0)$$
$$= 625\ \text{lbf/in}^2$$
$$E' = EC_M = \left(1{,}700{,}000\ \frac{\text{lbf}}{\text{in}^2}\right)(1.0)$$
$$= 1{,}700{,}000\ \text{lbf/in}^2$$

2 in nominal deck section properties per foot width are

$$A = (1.5\ \text{in})(12\ \text{in})$$
$$= 18.0\ \text{in}^2$$
$$S = \frac{bh^2}{6} = \frac{(12\ \text{in})(1.5\ \text{in})^2}{6}$$
$$= 4.5\ \text{in}^3$$
$$I = \frac{bh^3}{12} = \frac{(12\ \text{in})(1.5\ \text{in})^3}{12}$$
$$= 3.375\ \text{in}^4$$

The load per foot on the 2 in nominal deck is

$$\left(33.5\ \frac{\text{lbf}}{\text{ft}^2}\right)(1\ \text{ft}) = 33.5\ \text{lbf/ft}$$

The maximum moment and shear for a two-span deck are

$$M = -\tfrac{1}{8}wL^2 = -\left(\frac{1}{8}\right)\left(33.5\ \frac{\text{lbf}}{\text{ft}}\right)(8\ \text{ft})^2\left(12\ \frac{\text{in}}{\text{ft}}\right)$$
$$= -3216\ \text{in-lbf}$$
$$V = \tfrac{5}{8}wL = \left(\frac{5}{8}\right)\left(33.5\ \frac{\text{lbf}}{\text{ft}}\right)(8\ \text{ft})$$
$$= 167.5\ \text{lbf}$$

The actual maximum bending and shear stresses are

$$f_b = \frac{|M|}{S} = \frac{3216\ \text{in-lbf}}{4.5\ \text{in}^3}$$
$$= 714.7\ \text{lbf/in}^2 < F_b' = 1834.2\ \text{lbf/in}^2 \quad [\text{OK}]$$
$$f_v = \frac{3V}{2A} = \frac{(3)(167.5\ \text{lbf})}{(2)(18\ \text{in}^2)}$$
$$= 13.96\ \text{lbf/in}^2 \quad \begin{bmatrix}\text{very small by} \\ \text{inspection; OK}\end{bmatrix}$$

Check the maximum deflection for the 2 in nominal deck per foot width.

$$\Delta = \frac{wL^4}{185E'I}$$
$$= \frac{\left(33.5\ \frac{\text{lbf}}{\text{ft}}\right)\left(\frac{1\ \text{ft}}{12\ \text{in}}\right)\left((8\ \text{ft})\left(12\ \frac{\text{in}}{\text{ft}}\right)\right)^4}{(185)\left(1{,}700{,}000\ \frac{\text{lbf}}{\text{in}^2}\right)(3.375\ \text{in}^4)}$$
$$= 0.223\ \text{in}$$
$$\Delta_{\text{allow}} = \frac{L}{180} = \frac{(8\ \text{ft})\left(12\ \frac{\text{in}}{\text{ft}}\right)}{180}$$
$$= 0.53\ \text{in} > \Delta = 0.223\ \text{in} \quad [\text{OK}]$$

Use 2 in nominal deck, douglas fir-larch Commercial Dex, continuous two-span, 8 ft each span.

2. Purlin design

Assume glulam 3¹/8 in × 18 in, 20F-V3 douglas fir-larch at 8 ft spacing. Assume timber density of 40 lbf/ft^3. Alternatively, calculate timber densities from the specific gravities tables in NDS Secs. 8, 9, 11, and 12.

The total purlin load, w, is found as follows.

$$\text{roof loads} = \left(33.5\ \frac{\text{lbf}}{\text{ft}^2}\right)(8\ \text{ft}) = 268\ \text{lbf/ft}$$

$$\text{purlin dead load} = \frac{\left(40\ \frac{\text{lbf}}{\text{ft}^3}\right)(3.125\ \text{in})(18\ \text{in})}{\left(12\ \frac{\text{in}}{\text{ft}}\right)^2} = 15.6\ \text{lbf/ft}$$

$$w = 268\ \frac{\text{lbf}}{\text{ft}} + 15.6\ \frac{\text{lbf}}{\text{ft}} = 283.6\ \text{lbf/ft}$$

NDS Supp Tbl 1C

Section properties of 3¹/₈ in × 18 in glulam are as follows. The span is 30 ft, simply supported.

$$A = 56.25\ \text{in}^2$$

$$S_x = 168.8\ \text{in}^3$$

$$I_x = 1519\ \text{in}^4$$

NDS Supp Tbl 5A

Design values for glulam combination 20F-V3 douglas fir-larch are

$$F_{bx} = 2000\ \text{lbf/in}^2$$

$$F_{vx} = 265\ \text{lbf/in}^2$$

$$E_x = 1{,}600{,}000\ \text{lbf/in}^2$$

$C_M = 1.0$ for assumed 15% or less moisture content

$C_D = 1.15$ for snow load

NDS Sec 3.3.3.3

The beam stability factor, C_L, is 1.0 since the joists are continuously supported by the deck above it.

NDS Supp Tbl 5A Adj Fac

The volume factor is

$$C_V = \left(\frac{21}{L}\right)^{1/x}\left(\frac{12}{d}\right)^{1/x}\left(\frac{5.125}{b}\right)^{1/x} \leq 1.0$$

$x = 10$ for other than southern pine

$$C_V = \left(\frac{21}{30\ \text{ft}}\right)^{1/10}\left(\frac{12}{18\ \text{in}}\right)^{1/10}\left(\frac{5.125}{3.125\ \text{in}}\right)^{1/10} = 0.974$$

NDS Tbl 5.3.1

The allowable design values are

$$F_b' = F_{bx}C_DC_M\begin{pmatrix}\text{the smaller of}\\ C_V \text{ and } C_L\end{pmatrix} = \left(2000\ \frac{\text{lbf}}{\text{in}^2}\right)(1.15)(1.0)(0.974) = 2240.2\ \text{lbf/in}^2$$

$$F_v' = F_{vx}C_DC_M = \left(265\ \frac{\text{lbf}}{\text{in}^2}\right)(1.15)(1.0) = 304.8\ \text{lbf/in}^2$$

$$E' = EC_M = \left(1{,}600{,}000\ \frac{\text{lbf}}{\text{in}^2}\right)(1.0) = 1{,}600{,}000\ \text{lbf/in}^2$$

Check the actual maximum stresses.

$$M = \frac{wL^2}{8} = \left(\frac{\left(283.6 \ \frac{\text{lbf}}{\text{ft}}\right)(30 \text{ ft})^2}{8}\right)\left(12 \ \frac{\text{in}}{\text{ft}}\right)$$

$$= 382{,}860 \text{ in-lbf}$$

$$f_b = \frac{M}{S_x} = \frac{382{,}860 \text{ in-lbf}}{168.8 \text{ in}^3}$$

$$= 2268.1 \text{ lbf/in}^2 \approx F'_b = 2240.2 \text{ lbf/in}^2 \quad \text{[consider OK]}$$

Note that when actual values slightly exceed allowable design values, engineering judgment must be used to determine whether or not this is acceptable. However, this may not be the case when taking the engineering registration examination(s) and strict compliance may be required.

NDS Sec 3.4.3

$$V = w\left(\frac{\text{span}}{2} - \text{beam depth}\right)$$

$$= \left(283.6 \ \frac{\text{lbf}}{\text{in}^2}\right)\left((30 \text{ ft})\left(\frac{1}{2}\right) - (18 \text{ in})\left(\frac{1 \text{ ft}}{12 \text{ in}}\right)\right)$$

$$= 3828.6 \text{ lbf}$$

$$f_v = \frac{3V}{2A} = \frac{(3)(3828.6 \text{ lbf})}{(2)(56.25 \text{ in}^2)}$$

$$= 102.1 \text{ lbf/in}^2 < F'_v = 304.8 \text{ lbf/in}^2 \quad \text{[OK]}$$

Check the maximum deflection.

$$\Delta = \frac{5wL^4}{384E'I} = \frac{(5)\left(283.6 \ \frac{\text{lbf}}{\text{ft}}\right)\left(\frac{1 \text{ ft}}{12 \text{ in}}\right)\left((30 \text{ ft})\left(12 \ \frac{\text{in}}{\text{ft}}\right)\right)^4}{(384)\left(1{,}600{,}000 \ \frac{\text{lbf}}{\text{in}^2}\right)(1519 \text{ in}^4)}$$

$$= 2.13 \text{ in}$$

$$\Delta_{\text{allow}} = \frac{L}{180} = \frac{(30 \text{ ft})\left(12 \ \frac{\text{in}}{\text{ft}}\right)}{180}$$

$$= 2.0 \text{ in} \approx \Delta = 2.13 \text{ in} \quad \text{[consider OK]}$$

Note that when actual values slightly exceed allowable design values, engineering judgment must be used to determine whether or not this is acceptable. However, this may not be the case when taking the engineering registration examination(s) and strict compliance may be required.

Use glulam 3¹/8 in × 18 in, 20F-V3 douglas fir-larch for the 30 ft purlins.

3. Frame rafter design

Assume glulam 8³/4 in × 39 in 24F-V8 douglas fir and a two-span cantilever beam system. Assume a timber density of 40 lbf/ft^3.

An optimum design for bending strength is determined by choosing the two-span beam system as shown in Fig. 13.1. The splice is placed at a location where there is no bending moment and is modeled as a hinge. See the Appendix: Beam Formulas, or the AITC's *Timber Construction Manual* for appropriate equations. The equations can also be derived as in this problem solution.

Figure 13.1 Two-Span Cantilever Beam System

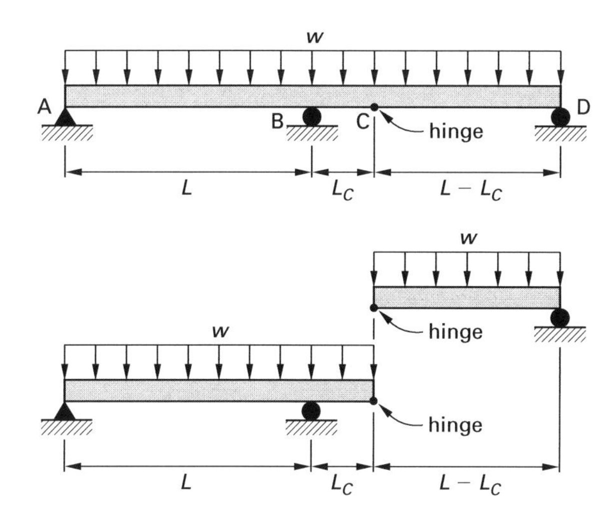

The position of an inflection point, L_c, is determined by making both the negative moment at B and positive moment in span CD equal in absolute value. The negative moment at B is

$$M_{\text{B,neg}} = \frac{wL_c^2}{2} + \frac{w(L - L_c)L_c}{2}$$

The maximum positive moment in span CD is

$$M_{\text{CD,pos}} = \frac{w(L - L_c)^2}{8}$$

Letting $|M_\text{B}| = |M_\text{CD}|$ and rearranging,

$$\frac{wL_c^2}{2} + \frac{w(L - L_c)L_c}{2} = \frac{w(L - L_c)^2}{8}$$

Therefore, the inflection point, L_c, is located at $0.172L$.

Also, the moment, M, is found to be $0.086wL^2$.

The shears, V, are as follows.

At the ends (point A or point D),

$$\begin{aligned} V &= \frac{w(L - L_c)}{2} = \frac{w(L - 0.172L)}{2} \\ &= 0.414wL \end{aligned}$$

At center (at B),

$$\begin{aligned} V &= -0.414wL - wL_c = -0.414wL - w(0.172L) \\ &= -0.586wL \end{aligned}$$

NDS Supp Tbl 1C

Section properties of $8^3/_4$ in × 39 in glulam are

span = 60 ft

$A = 341.3\ \text{in}^2$

$S_x = 2218\ \text{in}^3$

$I_x = 43{,}250\ \text{in}^4$

NDS Supp Tbl 5A

Design values for glulam 24F-V8 douglas fir-larch spaced at 30 ft with 60 ft spans are as follows.

The tension zone stressed in tension is

$$F_{bx}^{+} = F_{bx,t/t} = 2400 \text{ lbf/in}^2$$

The compression zone stressed in tension is

$$\begin{aligned} F_{bx}^{-} &= F_{bx,c/t} = 2400 \text{ lbf/in}^2 \\ F_{vx} &= 265 \text{ lbf/in}^2 \\ E_x &= 1{,}800{,}000 \text{ lbf/in}^2 \\ E_y &= 1{,}600{,}000 \text{ lbf/in}^2 \\ C_M &= 1.0 \text{ for dry condition} \\ C_D &= 1.15 \text{ for snow} \\ E_{x,\min} &= 0.93 \times 10^6 \text{ lbf/in}^2 \\ E_{y,\min} &= 0.83 \times 10^6 \text{ lbf/in}^2 \end{aligned}$$

The rafter load is

$$\begin{aligned} w &= \frac{\text{roof loads}}{(\text{snow} + \text{deck})} + \text{purlin loads} + \text{rafter loads} \\ &= \left(33.5 \ \frac{\text{lbf}}{\text{ft}^2}\right)(30 \text{ ft}) + \left(40 \ \frac{\text{lbf}}{\text{ft}^3}\right)\left(\frac{25 \text{ in}}{8}\right) \\ &\quad \times (18 \text{ in})\left(\frac{1 \text{ ft}}{12 \text{ in}}\right)^2 (30 \text{ ft})\left(\frac{1}{8 \text{ ft}}\right) \\ &\quad + \left(40 \ \frac{\text{lbf}}{\text{ft}^3}\right)\left(\frac{35 \text{ in}}{4}\right)(39 \text{ in})\left(\frac{1 \text{ ft}}{12 \text{ in}}\right)^2 \\ &= 1158.4 \text{ lbf/ft} \quad (1.158 \text{ kips/ft}) \end{aligned}$$

The rafter glulam 24F-V8 douglas fir is loaded with snow loads on both spans as shown in Fig. 13.5. The inflection point is located at distance L_c from B.

$$\begin{aligned} L_c &= 0.172L = (0.172)(60 \text{ ft}) \\ &= 10.32 \text{ ft} \end{aligned}$$

It is noted that this region is where the compression zone of the rafter glulam is stressed in tension. The moments and shears are shown in Fig. 13.2.

Figure 13.2 Behavior of Rafter Under Load (Forces and moments deviate slightly as calculated from Appendix: Beam Formulas because of round-off error.)

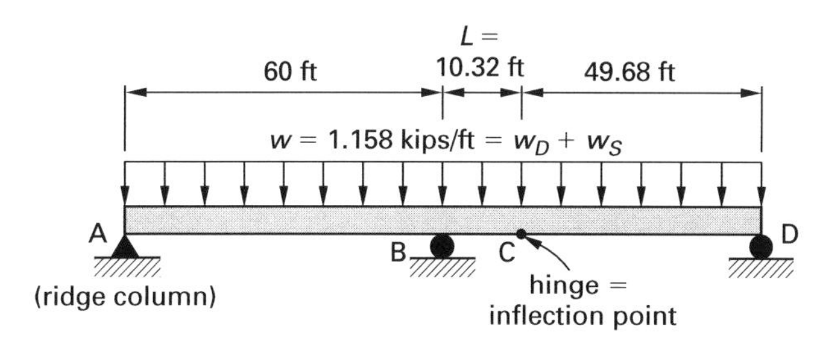

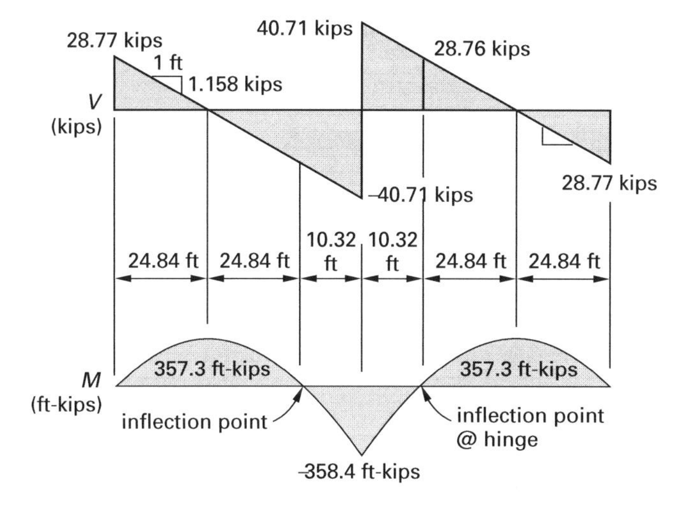

For member AC (see Fig. 13.2):

NDS Supp Tbl 5A

For bending, the glulam 24F-V8 douglas fir-larch combination is "balanced" to provide equal positive and negative moment capacity.

$F_{bx} = 2400\ \text{lbf/in}^2$

(F_{bx} for tension zone stressed in tension $=$ F_{bx} for compression zone stressed in tension.)

App: Beam Formulas; Fig 13.2

$M_{\max} = -M_{\max} = 358.4\ \text{ft-kips}$

The volume effect factor is determined as follows.

NDS Sec 5.3.6

$$C_V = \left(\frac{21}{L}\right)^{1/x} \left(\frac{12}{d}\right)^{1/x} \left(\frac{5.125}{b}\right)^{1/x} \leq 1.0$$

For other than southern pine, $x = 10$.

The largest distance between points of zero moments is

$$L = 60\ \text{ft} - 10.32\ \text{ft} = 49.68\ \text{ft}$$

NDS Sec 5.3.6

$$C_V = \left(\frac{21}{L}\right)^{1/x}\left(\frac{12}{d}\right)^{1/x}\left(\frac{5.125}{b}\right)^{1/x}$$
$$= \left(\frac{21}{49.68 \text{ ft}}\right)^{1/10}\left(\frac{12}{39 \text{ in}}\right)^{1/10}\left(\frac{5.125}{8.75 \text{ in}}\right)^{1/10}$$
$$= 0.773$$

The lateral stability factor, C_L, is determined as follows.

The maximum unbraced length, ℓ_u, is the 8 ft purlin spacing. However, in the negative-moment region it is 10.32 ft.

NDS Sec 3.3.3

$$\ell_u = (10.32 \text{ ft})\left(12 \ \frac{\text{in}}{\text{ft}}\right)$$
$$= 123.8 \text{ in}$$

NDS Tbl 3.3.3

$$\frac{\ell_u}{d} = \frac{123.8 \text{ in}}{39 \text{ in}} = 3.17 < 7$$

$$\ell_e = 2.06\ell_u = (2.06)(123.8 \text{ in})$$
$$= 255 \text{ in}$$

$$R_B = \sqrt{\frac{\ell_e d}{b^2}} = \sqrt{\frac{(255 \text{ in})(39 \text{ in})}{(8.75 \text{ in})^2}}$$
$$= 11.4$$

$$E'_{y,\min} = E_{y,\min}C_M = \left(830{,}000 \ \frac{\text{lbf}}{\text{in}^2}\right)(1.0)$$
$$= 830{,}000 \text{ lbf/in}^2$$

$$F_{bE} = \frac{1.20E'_{y\ \min}}{R_B^2} = \frac{(1.20)\left(830{,}000 \ \frac{\text{lbf}}{\text{in}^2}\right)}{(11.4)^2}$$
$$= 7663.9 \text{ lbf/in}^2$$

$$F_b^* = F_{bx}C_DC_M = \left(2400 \ \frac{\text{lbf}}{\text{in}^2}\right)(1.15)(1.0)$$
$$= 2760 \text{ lbf/in}^2$$

$$\frac{F_{bE}}{F_b^*} = \frac{7663.9 \ \frac{\text{lbf}}{\text{in}^2}}{2760 \ \frac{\text{lbf}}{\text{in}^2}}$$
$$= 2.78$$

NDS Eq 3.3-6

$$C_L = \frac{1 + \frac{F_{bE}}{F_b^*}}{1.9} - \sqrt{\left(\frac{1 + \frac{F_{bE}}{F_b^*}}{1.9}\right)^2 - \frac{\frac{F_{bE}}{F_b^*}}{0.95}}$$
$$= \frac{1 + 2.78}{1.9} - \sqrt{\left(\frac{1 + 2.78}{1.9}\right)^2 - \frac{2.78}{0.95}}$$
$$= 0.973$$

C_V (0.773) is less than C_L (0.973), so use C_V.

Check bending stress. The allowable bending stress is

$$F'_{bx} = F_{bx}C_DC_MC_V = \left(2400\ \frac{\text{lbf}}{\text{in}^2}\right)(1.15)(1.0)(0.773)$$
$$= 2133.48\ \text{lbf/in}^2$$

The maximum bending stress is

$$f_b = \frac{M}{S_x}$$
$$= \frac{(358.4\ \text{ft-kips})\left(12\ \frac{\text{in}}{\text{ft}}\right)\left(1000\ \frac{\text{lbf}}{\text{kip}}\right)}{2218\ \text{in}^3}$$
$$= 1939\ \text{lbf/in}^2 < F'_{bx} = 2133.48\ \text{lbf/in}^2 \quad [\text{OK}]$$

Check the shear stress.

The allowable shear stress is

$$F'_v = F_{vx}C_DC_M = \left(265\ \frac{\text{lbf}}{\text{in}^2}\right)(1.15)(1.0)$$
$$= 304.8\ \text{lbf/in}^2$$

The maximum shear force and shear stress are $V_{\text{max}} = 40.71$ kips (see Fig. 13.2).

$$f_v = \frac{3V}{2A} = \frac{(3)(40.71\ \text{kips})\left(1000\ \frac{\text{lbf}}{\text{kip}}\right)}{(2)(341.3\ \text{in}^2)}$$
$$= 179\ \text{lbf/in}^2 < F'_v = 304.8\ \text{lbf/in}^2 \quad [\text{OK}]$$

For bending, member CD has positive moment throughout the entire beam and has continuous lateral support on the compression edge from the decking, assuming that the decking is fastened appropriately (see Fig. 13.3).

Therefore, $C_L = 1.0$.

$C_V = 0.773$ since both members AC and CD are the same glulam size and have the same distance between the regions of zero moment.

The maximum moments and shears are also approximately the same for both members AC and CD. Therefore, bending and shear stress checks are approximately the same for both members. This is a design judgment, and other situations may require the calculation of stresses in each member.

Check deflection.

The largest deflection in the frame rafter occurs in span AB. This deflection is given by

$$E'_x = E_xC_M = \left(1{,}800{,}000\ \frac{\text{lbf}}{\text{in}^2}\right)(1.0)$$
$$= 1{,}800{,}000\ \text{lbf/in}^2$$

$$I_x = 43{,}253\ \text{in}^4$$

The rafter load was found to be

$$w = 1158.4\ \text{lbf/in}^2$$

$$\Delta_{\text{max}} = \left(\frac{0.370}{48}\right)\left(\frac{wL^4}{E'_x I_x}\right)$$

$$= \left(\frac{0.370}{48}\right)\left(\frac{\left(1158.4\ \frac{\text{lbf}}{\text{ft}}\right)(60\ \text{ft})^4\left(12\ \frac{\text{in}}{\text{ft}}\right)^3}{\left(1{,}800{,}000\ \frac{\text{lbf}}{\text{in}^2}\right)(43{,}250\ \text{in}^4)}\right)$$

$$= 2.57\ \text{in}$$

$$\Delta_{\text{allow}} = \frac{L}{180} = \frac{(60\ \text{ft})\left(12\ \frac{\text{in}}{\text{ft}}\right)}{180}$$

$$= 4\ \text{in} > \Delta_{\text{max}} = 2.57\ \text{in} \quad \text{[OK]}$$

However, camber is also necessary to prevent visible sagging of the frame. Determination of the amount of camber required is beyond the scope of this text. (It is often given in prevailing design specifications.)

Use glulam 8³/₄ in × 39 in, 24F-V8 douglas fir-larch for frame rafter as a two-span cantilever beam system.

4. Connection design

Between the DF-L frame rafter and purlins, the vertical load from each purlin to the rafter is

$$R = (\text{purlin load})\left(\frac{\text{purlin span length}}{2}\right) = \left(283.6\ \frac{\text{lbf}}{\text{ft}}\right)\left(\frac{30\ \text{ft}}{2}\right)$$

$$= 4254\ \text{lbf}$$

Use 1 in bolts and 4 in × 4 in × ¹/₄ in × 12 in long steel angles, as seen in Fig. 13.3, acting at a 90° angle to the grain of the wood.

NDS Tbl 11D

$$t_m = 3^1/_8\ \text{in (DF-L purlin)}$$

$$t_s = \frac{1}{4}\ \text{in}$$

$$Z = Z_\perp$$

$$= (740\ \text{lbf})(2\ \text{bolts}) = 1480\ \text{lbf}$$

NDS Tbl 11I

For t_m = 8³/₄ in (frame rafter for double shear), use $Z_\perp = 3000$ lbf.

$$Z = 1480\ \text{lbf} \quad \text{[controls]}$$

For snow, $C_D = 1.15$.

For dry condition (i.e., moisture content ≤ 16%), $C_M = 1.0$.

For normal temperature, $C_t = 1.0$.

NDS Sec 10.3.6

For group action, $C_g = 1.0$.

NDS Sec 11.5;
NDS Tbl 10.3.1

$$C_\Delta = 1.0$$

$$\begin{aligned} Z' &= ZC_DC_MC_tC_gC_\Delta \\ &= (1480 \text{ lbf})(1.15)(1.0)(1.0)(1.0)(1.0) \\ &= 1702 \text{ lbf per bolt} \end{aligned}$$

Since the rafter is connected to two purlins, each set of bolts actually resists two purlin loads, and the bolts are in double shear. The number of bolts required in the rafter is

$$\begin{aligned} N &= \frac{2R}{Z'} = \frac{(2)(4254 \text{ lbf})}{1702 \frac{\text{lbf}}{\text{bolt}}} \\ &= 5.0 \text{ bolts} \end{aligned}$$

Use six bolts, three in each side of each purlin (see Fig. 13.3).

Figure 13.3 Connection Design

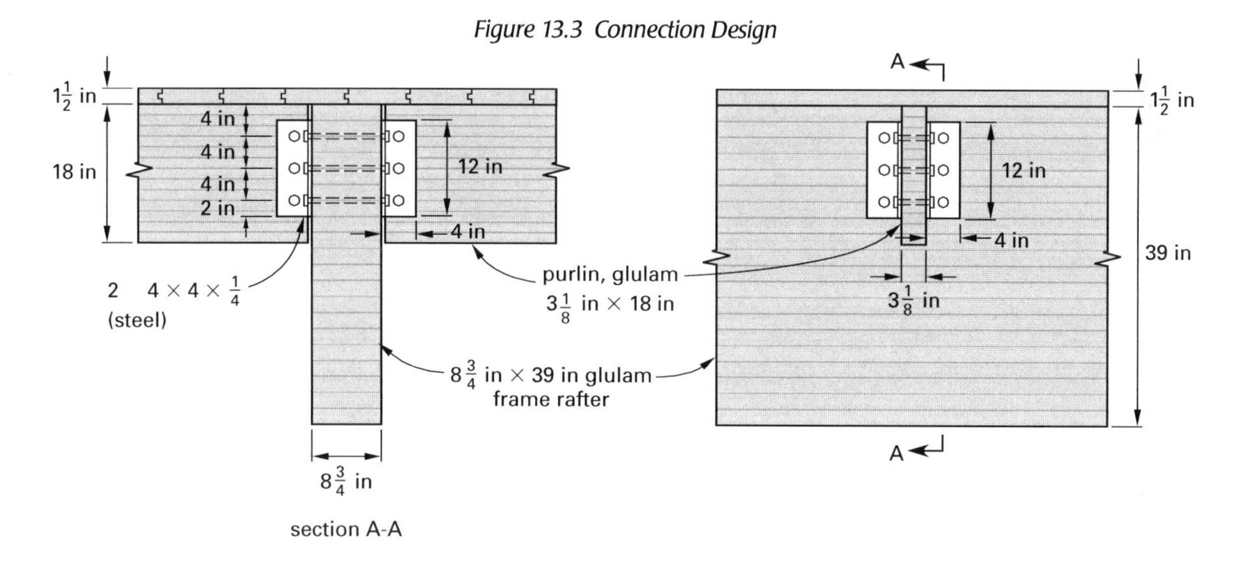

Note that the load capacity of steel hardware must also be checked, although it is not done in this problem solution.

5. Exterior ridge column design

The column is investigated for two load cases.

From Fig. 13.2, the dead load plus snow load for vertical loads is

$$\begin{aligned} w_D + w_S &= \left(\frac{1}{2}\right)(28.77 \text{ kips})(2) \\ &= 28.77 \text{ kips} \end{aligned}$$

(This load is found by noting that each side of the building contributes column load and that exterior columns support one-half the load that corresponding interior columns support.)

The dead load plus wind load for vertical and lateral loads by ratio of w_D to $(w_D + w_S)$ is (see roof deck design for w_D and w_S) as follows.

$$\begin{aligned} \text{vertical load } w_D &= \frac{(28.77 \text{ kips})w_D}{w_D + w_S} = \frac{(28.77 \text{ kips})\left(13.5 \ \frac{\text{lbf}}{\text{ft}^2}\right)}{33.5 \ \frac{\text{lbf}}{\text{ft}^2}} \\ &= 11.59 \text{ kips} \end{aligned}$$

$$\begin{aligned} \text{lateral load } w_W &= (\text{wind pressure})(\text{column spacing}) \\ &= \left(10 \ \frac{\text{lbf}}{\text{ft}^2}\right)(60 \text{ ft}) \\ &= 600 \text{ lbf/ft} \end{aligned}$$

Fig 13.4

The calculated column loads are shown in Fig. 13.4.

The column is assumed to be 8¾ in × 12 in 24F-V10 douglas fir-larch glulam. The wind pressure is assumed to act on the 8¾ in face, which is the wide face of the laminations. It is assumed that the moisture content will not exceed 16%. Normal temperatures (i.e., less than 100°F) apply. Therefore, $C_M = C_t = 1.0$.

NDS Supp Tbl 5A

The design values are

$$\begin{aligned} F_{bx} &= 2400 \text{ lbf/in}^2 \\ F_c &= 1550 \text{ lbf/in}^2 \\ E_{x,\min} &= 0.93 \times 10^6 \text{ lbf/in}^2 \\ E_{y,\min} &= 0.78 \times 10^6 \text{ lbf/in}^2 \end{aligned}$$

Figure 13.4 Column Loads for Both Load Cases

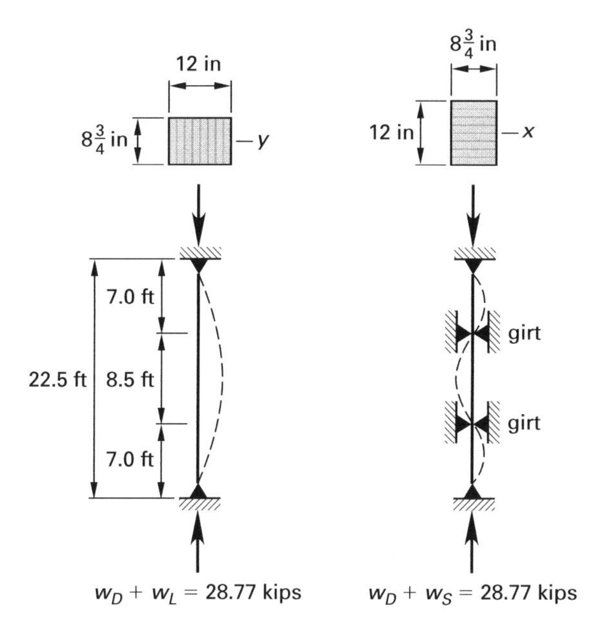

(a) $w_D + w_S$

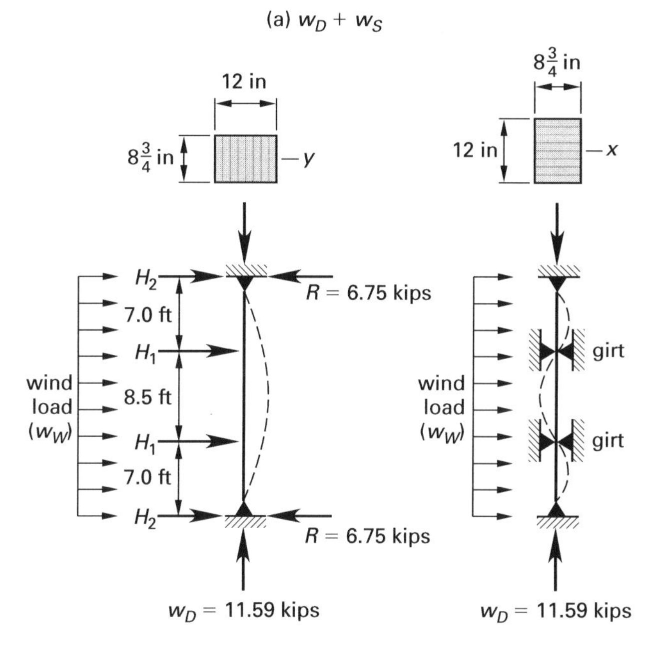

$$w_W = 0.6 \text{ kip/ft}$$

$$H_1 = \left(0.6 \ \frac{\text{kip}}{\text{ft}}\right)(7 \text{ ft} + 8.5 \text{ ft})\left(\frac{1}{2}\right) = 4.65 \text{ kips}$$

$$H_2 = \left(0.6 \ \frac{\text{kip}}{\text{ft}}\right)(7 \text{ ft})\left(\frac{1}{2}\right) = 2.1 \text{ kips}$$

(b) $w_D + w_W$
(wind load distributed to four girts attached to the column)

NDS Supp Tbl 1C

The section properties of 8³/₄ in × 12 in glulam are

$$A = 105 \text{ in}^2$$
$$S_x = 210 \text{ in}^3$$
$$S_y = 153.1 \text{ in}^3$$
$$I_x = 1260 \text{ in}^4$$
$$I_y = 669.9 \text{ in}^4$$

For the combination dead load, snow load case (see Fig. 13.4(a)), the unbraced length with respect to the x-axis is different than the unbraced length with respect to the y-axis.

NDS Sec 3.7; NDS Eq 3.7-1

Determine the column stability factor, C_P, about the x-axis (see Fig. 13.4).

$$\left(\frac{\ell_e}{d}\right)_x = \left(\frac{K_e\ell}{d}\right)_x = \frac{(1.0)(22.5 \text{ ft})\left(12 \ \frac{\text{in}}{\text{ft}}\right)}{12 \text{ in}} = 22.5$$

$$E'_{x,\min} = E_{x,\min}C_MC_t = \left(930{,}000 \ \frac{\text{lbf}}{\text{in}^2}\right)(1.0)(1.0) = 930{,}000 \text{ lbf/in}^2$$

For glulam, $c = 0.9$.

$$F_{cE} = \frac{0.822E'_{x,\min}}{\left(\frac{\ell_e}{d}\right)_x^2} = \frac{(0.822)\left(930{,}000 \ \frac{\text{lbf}}{\text{in}^2}\right)}{(22.5)^2} = 1510 \text{ lbf/in}^2$$

For snow, $C_D = 1.15$.

$$F_c^* = F_cC_DC_MC_t = \left(1550 \ \frac{\text{lbf}}{\text{in}^2}\right)(1.15)(1.0)(1.0) = 1782.5 \text{ lbf/in}^2$$

$$\frac{F_{cEx}}{F_c^*} = \frac{1510 \ \frac{\text{lbf}}{\text{in}^2}}{1782.5 \ \frac{\text{lbf}}{\text{in}^2}} = 0.847$$

$$C_P = \frac{1 + \frac{F_{cEx}}{F_c^*}}{2c} - \sqrt{\left(\frac{1 + \frac{F_{cEx}}{F_c^*}}{2c}\right)^2 - \frac{\frac{F_{cEx}}{F_c^*}}{c}}$$

$$= \frac{1 + 0.847}{(2)(0.9)} - \sqrt{\left(\frac{1 + 0.847}{(2)(0.9)}\right)^2 - \frac{0.847}{0.9}} = 0.692$$

Determine the column stability factor, C_P, about the y-axis (see Fig. 13.4).

$$\left(\frac{\ell_e}{d}\right)_y = \left(\frac{K_e\ell}{d}\right)_y = \frac{(1.0)(8.5\text{ ft})\left(12\ \frac{\text{in}}{\text{ft}}\right)}{8.75\text{ in}} = 11.7$$

$$E'_{y,\min} = E_{y,\min}C_M C_t = \left(780{,}000\ \frac{\text{lbf}}{\text{in}^2}\right)(1.0)(1.0) = 780{,}000\text{ lbf/in}^2$$

$$F_{cEy} = \frac{0.822E'_{y,\min}}{\left(\frac{\ell_e}{d}\right)_y^2} = \frac{(0.822)\left(780{,}000\ \frac{\text{lbf}}{\text{in}^2}\right)}{(11.7)^2} = 4683.8\text{ lbf/in}^2$$

$$\frac{F_{cEy}}{F_c^*} = \frac{4683.8\ \frac{\text{lbf}}{\text{in}^2}}{1782.5\ \frac{\text{lbf}}{\text{in}^2}} = 2.63$$

$$C_P = \frac{1+\frac{F_{cEy}}{F_c^*}}{2c} - \sqrt{\left(\frac{1+\frac{F_{cEy}}{F_c^*}}{2c}\right)^2 - \frac{\frac{F_{cEy}}{F_c^*}}{c}} = \frac{1+2.63}{(2)(0.9)} - \sqrt{\left(\frac{1+2.63}{(2)(0.9)}\right)^2 - \frac{2.63}{0.9}} = 0.950$$

The x-axis produces the smaller value for the column stability factor, C_P. The allowable compressive stress is

$$F'_c = F_c C_D C_M C_t C_P = \left(1550\ \frac{\text{lbf}}{\text{in}^2}\right)(1.15)(1.0)(1.0)(0.692) = 1233.5\text{ lbf/in}^2$$

The allowable column load is

Fig 13.2

$$P_{\text{allow}} = F'_c A = \left(1233.5\ \frac{\text{lbf}}{\text{in}^2}\right)(105\text{ in}^2) = 129{,}521\text{ lbf} > w_D + w_S = 28{,}770\text{ lbf}\quad[\text{OK}]$$

Fig 13.4

For the combination dead load plus wind load case, the actual axial stress from dead load is

$$f_c = \frac{P}{A} = \frac{11{,}590\text{ lbf}}{105\text{ in}^2} = 110.4\text{ lbf/in}^2$$

As calculated previously from the combination dead load plus snow load case, the column stability factor, C_P, about the x-axis is 0.692, and the allowable compressive stress of the column is 1233.5 lbf/in^2. The axial stress ratio is

$$\frac{f_c}{F'_c} = \frac{110.4 \ \frac{\text{lbf}}{\text{in}^2}}{1233.5 \ \frac{\text{lbf}}{\text{in}^2}} = 0.09 \quad \text{[OK]}$$

Determine the bending from wind load.

Based on the distribution of wind load, on tributary areas, from the siding to four girts attached to the column, the moment at the girt is

Fig 13.4(b)

$$M_x = (6750 \text{ lbf} - 2100 \text{ lbf})(7 \text{ ft}) = 32{,}550 \text{ ft-lbf}$$

The actual bending stress is

$$f_{bx} = \frac{M_x}{S_x} = \frac{(32{,}550 \text{ ft-lbf})\left(12 \ \frac{\text{in}}{\text{ft}}\right)}{210 \text{ in}^3} = 1860 \text{ lbf/in}^2$$

NDS Tbl 2.3.1

The allowable bending stress is the smaller of $F'_{bx} = F_{bx}C_DC_MC_tC_L$ and $F'_{bx} = F_{bx}C_DC_MC_tC_V$.

NDS Sec 3.3.3; NDS Eq 3.3-6

Determine the beam stability factor, C_L. Lateral-torsional buckling can occur in the plane of the y-axis. The unbraced length, ℓ_u, is 8.5 ft (102 in).

$$\frac{\ell_u}{d} = \frac{102 \text{ in}}{12 \text{ in}} = 8.5 > 7 \quad \text{[OK]}$$

NDS Tbl 3.3.3

$$\ell_e = 1.63\ell_u + 3d = (1.63)(102 \text{ in}) + (3)(12 \text{ in}) = 202.26 \text{ in}$$

$$R_B = \sqrt{\frac{\ell_e d}{b^2}} = \sqrt{\frac{(202.26 \text{ in})(12 \text{ in})}{(8.75 \text{ in})^2}} = 5.63$$

$$E'_y = E_yC_MC_t = \left(780{,}000 \ \frac{\text{lbf}}{\text{in}^2}\right)(1.0)(1.0) = 780{,}000 \text{ lbf/in}^2$$

$$F_{bE} = \frac{1.20E'_{y,\text{min}}}{R_B^2} = \frac{(1.20)\left(780{,}000 \ \frac{\text{lbf}}{\text{in}^2}\right)}{(5.63)^2} = 29{,}530 \text{ lbf/in}^2$$

For wind/earthquake loads, $C_D = 1.6$.

$$F_b^* = F_{bx}C_D C_M C_t = \left(2400\ \frac{\text{lbf}}{\text{in}^2}\right)(1.6)(1.0)(1.0) = 3840\ \text{lbf/in}^2$$

$$\frac{F_{bE}}{F_b^*} = \frac{29{,}530\ \frac{\text{lbf}}{\text{in}^2}}{3840\ \frac{\text{lbf}}{\text{in}^2}} = 7.69$$

NDS Eq 3.3-6

$$C_L = \frac{1+\frac{F_{bE}}{F_b^*}}{1.9} - \sqrt{\left(\frac{1+\frac{F_{bE}}{F_b^*}}{1.9}\right)^2 - \frac{\frac{F_{bE}}{F_b^*}}{0.95}}$$
$$= \frac{1+7.69}{1.9} - \sqrt{\left(\frac{1+7.69}{1.9}\right)^2 - \frac{7.69}{0.95}}$$
$$= 0.99$$

NDS Sec 5.3.6; NDS Eq 5.3-1; NDS Supp Tbl 5A

Determine the volume factor, C_V.

$$C_V = \left(\frac{21}{L}\right)^{1/x}\left(\frac{12}{d}\right)^{1/x}\left(\frac{5.125}{b}\right)^{1/x} \le 1.0$$

For all wood species other than southern pine, $x = 10$.

$$L = 22.5\ \text{ft}$$

$$C_V = \left(\frac{21}{22.5\ \text{ft}}\right)^{1/10}\left(\frac{12}{12\ \text{in}}\right)^{1/10}\left(\frac{5.125}{8.75\ \text{in}}\right)^{1/10} = 0.9 < C_L = 0.99$$

Therefore, use C_V to determine the allowable bending stress.

$$F'_{bx} = F_{bx}C_D C_M C_t C_V = \left(2400\ \frac{\text{lbf}}{\text{in}^2}\right)(1.6)(1.0)(1.0)(0.94) = 3609\ \text{lbf/in}^2$$

From the combination dead load plus snow load analysis, $F_{cEx} = 1510\ \text{lbf/in}^2$.

NDS Eq. 3.9-3

NDS Eq. 3.9-3 requires that combined compression and bending stresses satisfy

$$\left(\frac{f_c}{F'_c}\right)^2 + \frac{f_{bx}}{F'_{bx}\left(1-\frac{f_c}{F_{cEx}}\right)} \le 1.0$$

Therefore,

$$\left(\frac{110.4 \ \frac{\text{lbf}}{\text{in}^2}}{1233.5 \ \frac{\text{lbf}}{\text{in}^2}} \right)^2 + \frac{1860 \ \frac{\text{lbf}}{\text{in}^2}}{\left(3609 \ \frac{\text{lbf}}{\text{in}^2}\right)\left(1 - \frac{110.4 \ \frac{\text{lbf}}{\text{in}^2}}{1510 \ \frac{\text{lbf}}{\text{in}^2}}\right)}$$

$$= 0.564 < 1.0 \quad \text{[OK]}$$

Use a 8³/4 in × 12 in 24F-V3 douglas fir column.

Practice Problem 4: Bolted Splice Connection

The splice connection shown uses a row of five ³/4 in bolts. The lumber is no. 2 grade eastern white pine. The load is caused by a dead load and a snow load. Assume that $C_M = 1.0$ and $C_t = 1.0$.

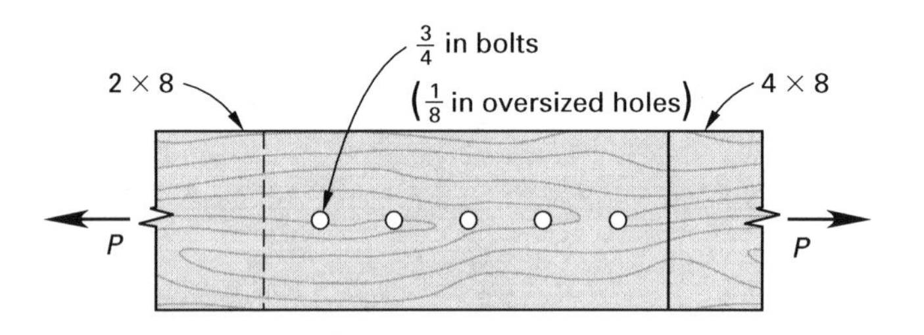

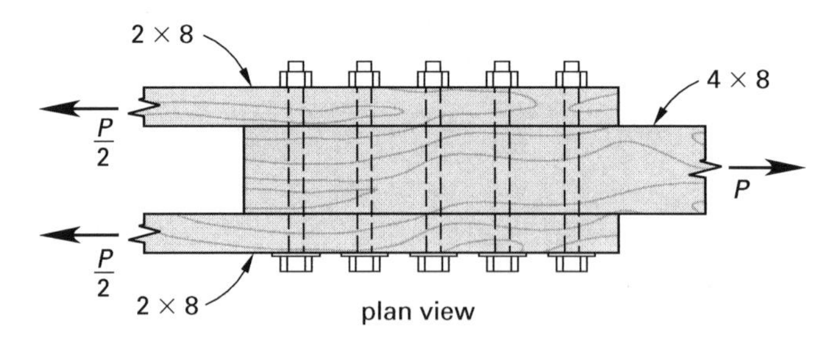

Determine the tension capacity of the splice and the number of ³/4 in bolts required. Assume bolt spacing and edge distance requirements are met for full design values.

REFERENCE

Solution (ASD Method)

1. Lumber capacity

Two 2 × 8s control.

NDS Supp Tbl 4A — For no. 2 eastern white pine, F_t is 275 lbf/in².

Adj Fac — C_M and C_t are both 1.0.

For 2 × 8 for F_t, C_F is 1.2.

NDS Tbl 2.3.2 — For snow, C_D is 1.15.

Therefore, the allowable stress is

NDS Tbl 4.3.1

$$F'_t = F_t C_D C_M C_t C_F = \left(275\ \frac{\text{lbf}}{\text{in}^2}\right)(1.15)(1.0)(1.0)(1.2) = 379.5\ \text{lbf/in}^2$$

The lumber capacity based on gross area is

$$P_{\text{allow}} = F'_t A_g = \left(379.5\ \frac{\text{lbf}}{\text{in}^2}\right)(1.5\ \text{in})(7.25\ \text{in})(2\ \text{members}) = 8254\ \text{lbf}$$

The lumber capacity based on net area is

$$P_{\text{allow}} = F'_t A_n = \left(379.5\ \frac{\text{lbf}}{\text{in}^2}\right)(1.5\ \text{in})\big(7.25\ \text{in} - (0.75\ \text{in} + 0.125\ \text{in})\big)(2\ \text{members}) = 7258\ \text{lbf} \quad \text{[controls]}$$

Note that while the gross area and the net area were checked, the net area will always be equal to or less than the gross area, and it will always control.

2. Bolt requirement

NDS Tbl 10.3.1

The allowable bolt design value is

$$Z' = Z C_D C_M C_t C_g C_\Delta$$

NDS Tbl 11F

For eastern softwoods, with all wood members loaded parallel to the grain, t_m being 3½ in, t_s being 1½ in, and with ¾ in bolts, $Z = Z_{\parallel} = 1990$ lbf.

NDS Sec 4.3

$$C_D = 1.0$$
$$C_M = 1.0$$
$$C_t = 1.0$$

NDS Tbl 10.3.6A

By interpolation (or use NDS Eq. 7.3-1 for more accuracy), the group action factor is

$C_g \approx 0.90$ for

$$A_s = (2)(1.5\ \text{in})(7.25\ \text{in}) = 21.75\ \text{in}^2$$
$$A_m = (3.5\ \text{in})(7.25\ \text{in}) = 25.375\ \text{in}^2$$
$$\frac{A_s}{A_m} = \frac{21.75\ \text{in}^2}{25.375\ \text{in}^2} = 0.857$$

There are five inline bolts.

NDS Sec 11.5.1

The geometry factor is $C_\Delta = 1.0$ if

$$\text{end distance} \geq 7D = (7)(\tfrac{3}{4}\ \text{in}) = 5.25\ \text{in}$$
$$\text{bolt spacing} \geq 4D = (4)(\tfrac{3}{4}\ \text{in}) = 3\ \text{in}$$
$$\text{edge distance} \geq 1.5D = (1.5)(\tfrac{3}{4}\ \text{in}) = 1.125\ \text{in}$$

NDS Tbl 10.3.1

$$\begin{aligned} Z' &= ZC_dC_MC_tC_gC_\Delta \\ &= (1990 \text{ lbf})(1.15)(1.0)(1.0)(0.90)(1.0) \\ &= 2060 \text{ lbf per bolt} \end{aligned}$$

The total number of 3/4 in bolts required is

$$N = \frac{\text{lumber capacity}}{\text{allowable bolt design value}} = \frac{7258 \text{ lbf}}{2060 \ \frac{\text{lbf}}{\text{bolt}}}$$

$$= 3.52 \text{ bolts}$$

Use four 3/4 in bolts for a tension capacity of 7258 lbf. Follow necessary requirements for bolt edge distance, end distance, and spacing. There is no need to recalculate for four bolts since C_g gets larger.

Practice Problem 5: Bolted Splice Connection with Metal Side Plates

The splice connection shown uses a row of 3/4 in bolts. The lumber is no. 2 grade eastern white pine. The size plates are 1/4 in × 2 in ASTM steel. The load is caused by a dead load and a snow load. Assume that C_M is 1.0 and C_t is 1.0.

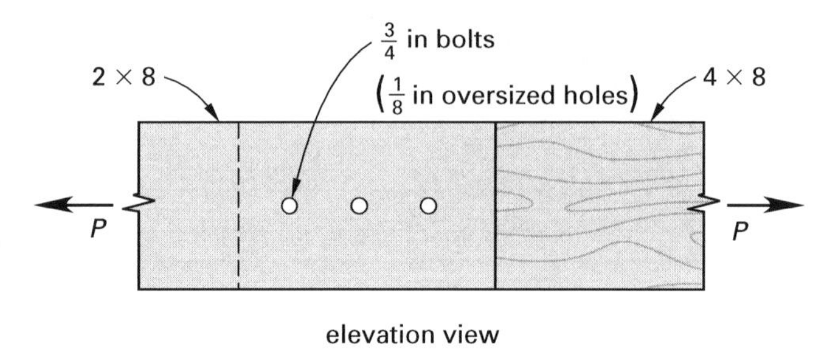

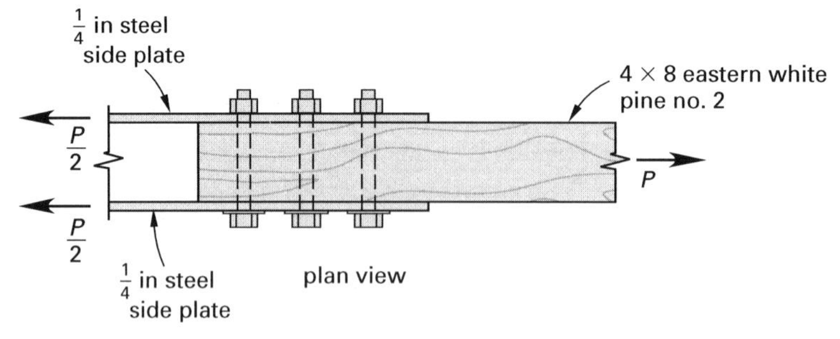

Determine the number of 3/4 in bolts required based on the tension capacity of the lumber.

REFERENCE

Solution (ASD Method)

NDS Tbl 11G

The allowable bolt design value is

$$Z' = ZC_DC_MC_tC_gC_\Delta$$

$$Z = Z_{\parallel} = 2660 \text{ lbf}$$

$$C_D = 1.15$$

$$C_M = 1.0$$

$$C_t = 1.0$$

The lumber capacity of 4×8 eastern white pine no. 2 is

NDS Supp Tbl 4A

Adj Fac

$$F_t = 275\ \text{lbf/in}^2$$
$$C_F = 1.2 \text{ for } 4 \times 8$$
$$\begin{aligned} F_t' &= F_t C_D C_M C_t C_F \\ &= \left(275\ \frac{\text{lbf}}{\text{in}^2}\right)(1.15)(1.0)(1.0)(1.2) \\ &= 380\ \text{lbf/in}^2 \end{aligned}$$
$$\begin{aligned} P_{\text{allow}} &= F_t' A_n \\ &= \left(380\ \frac{\text{lbf}}{\text{in}^2}\right)(3.5\ \text{in})\big(7.25\ \text{in} - (0.75\ \text{in} + 0.125\ \text{in})\big) \\ &= 8479\ \text{lbf} \end{aligned}$$

NDS Tbl 10.3.6C

The group action factor is $C_g \approx 0.97$ for

$$\begin{aligned} A_s &= (2)(0.25\ \text{in})(2\ \text{in}) \\ &= 1.0\ \text{in}^2 \end{aligned}$$
$$\begin{aligned} A_m &= (3.5\ \text{in})(7.25\ \text{in}) \\ &= 25.375\ \text{in}^2 \text{ for } 4 \times 8 \end{aligned}$$
$$\begin{aligned} \frac{A_m}{A_s} &= \frac{25.375\ \text{in}^2}{1.0\ \text{in}^2} \\ &= 25.375 \end{aligned}$$

Assume four bolts in a row. (Or, use NDS Eq. 7.3-1 for more accuracy.)

NDS Sec 11.5.1

The geometry factor, C_Δ, is 1.0 if the necessary bolt edge distance, end distance, and spacing requirements are met.

NDS Tbl 10.3.1

$$\begin{aligned} Z' &= Z C_P C_M C_t C_g C_\Delta \\ &= (2660\ \text{lbf})(1.15)(1.0)(1.0)(0.97)(1.0) \\ &= 2967\ \text{lbf per bolt} \end{aligned}$$

The number of $^3/_4$ in bolts required is

$$\begin{aligned} N &= \frac{8479\ \text{lbf}}{2967\ \frac{\text{lbf}}{\text{bolt}}} \\ &= 2.86\ \text{bolts} \end{aligned}$$

Therefore, use three $^3/_4$ in bolts. There is no need to recalculate with three bolts since C_g gets larger.

Practice Problem 6: Splice Connection with Nails

The splice connection shown uses a row of 50d common wire nails. The lumber is no. 2 grade eastern white pine. The load is caused by a dead load and a snow load. Assume that C_M is 1.0 and C_t is 1.0.

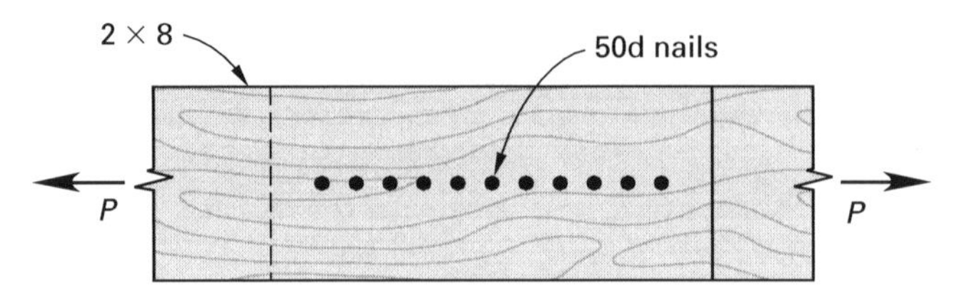

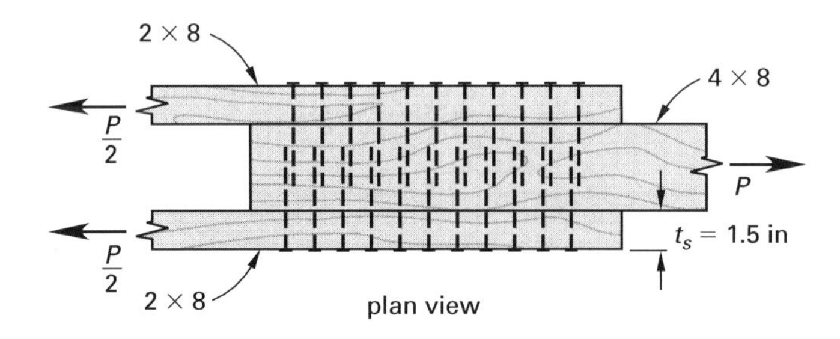

Determine the number of 50d common wire nails required. The total load, P, is 8254 lbf.

REFERENCE

Solution (ASD Method)

Determine the dimensions and design value for 50d wire nails.

NDS App Tbl L4

For eastern white pine, $t_s = 1.5$ in, and a 50d nail,

$L = 5.5$ in

$D = 0.244$ in

NDS Tbl 11N

$Z = 141$ lbf

The allowable design value for a 50d nail is

NDS Tbl 7.3.1

$Z' = ZC_DC_MC_tC_dC_{eg}C_{di}C_{tn}$

C_{eg}, C_{di}, and C_{tn} are not applicable.

For snow, $C_D = 1.15$.

For normal temperatures, $C_t = 1.0$.

For dry in-service conditions, $C_M = 1.0$.

NDS Sec 12.3.4

The penetration depth factor, C_d, is calculated as follows.

The actual penetration into the main 4 × 8 member is

$p = L - t_s = 5.5 \text{ in} - 1.5 \text{ in}$

$= 4.0 \text{ in}$

NDS Tbl 11N Ftn 3

$$6D = (6)(0.244 \text{ in}) = 1.46 \text{ in}$$

$$10D = (10)(0.244 \text{ in}) = 2.44 \text{ in}$$

$$\frac{p}{10D} \le 1.0$$

Since $6D \le p < 10D$ is not true, C_d is 1.0.

$$\begin{aligned} Z' &= ZC_DC_MC_tC_d \\ &= (141 \text{ lbf})(1.15)(1.0)(1.0)(1.0) \\ &= 162.2 \text{ lbf} \end{aligned}$$

The number of nails required for one-half of 8254 lbf is

$$\begin{aligned} N &= \frac{\frac{8254 \text{ lbf}}{2 \text{ side plates}}}{162.2 \ \frac{\text{lbf}}{\text{nail}}} \\ &= 25.4 \text{ nails} \end{aligned}$$

Use 26 50d nails on each side plate.

Practice Problem 7: Truss End Analysis

The end of a no. 1 eastern white pine truss is shown.

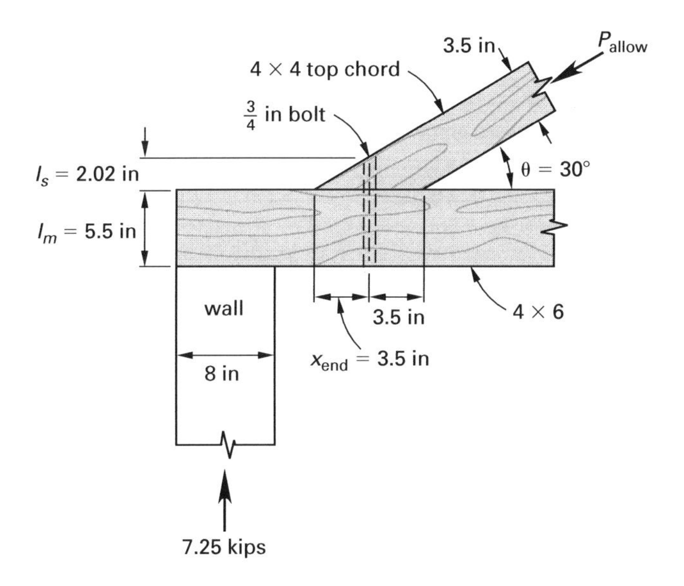

Determine the compressive load capacity of the 4×4 top chord as based on bolt capacity.

REFERENCE

Solution (ASD Method)

NDS Sec 11.5.1, NDS Tbls 11.5.1B, 11.3.9

Determine the geometry factor, C_Δ.

For the given end distance and loading at an angle, the minimum shear area for the full design value will be equivalent to $4D\ell_s$ for a parallel member in compression.

$$\begin{aligned}4D\ell_s &= (4)(0.75 \text{ in})(2.02 \text{ in}) \\ &= 6.06 \text{ in}^2\end{aligned}$$

For the reduced design value, the shear area is

$$\begin{aligned}2D\ell_s &= (2)(0.75 \text{ in})(2.02 \text{ in}) \\ &= 3.03 \text{ in}^2\end{aligned}$$

The actual equivalent shear area is (see illustration and NDS Fig. 11E).

$$\begin{aligned}\tfrac{1}{2}x_{\text{end}}\ell_s &= \left(\frac{1}{2}\right)(3.5 \text{ in})(2.02 \text{ in}) \\ &= 3.5 \text{ in}^2 > 3.03 \text{ in}^2 \quad \text{[OK]}\end{aligned}$$

$$\begin{aligned}C_\Delta &= \frac{\text{actual shear area}}{\text{shear area for full design}} = \frac{3.5 \text{ in}^2}{6.06 \text{ in}^2} \\ &= 0.578\end{aligned}$$

Determine the allowable bolt design value. For a 3/4 in bolt with $t_m = \ell_m = 5.5$ in, $t_s = \ell_s = 2.02$ in, and eastern white pine lumber,

$$Z' = Z_\parallel C_D C_M C_g C_\Delta$$

NDS Tbl 11A

$$\begin{aligned}Z_\parallel &= 1089 \text{ lbf} \quad \text{[by interpolation]} \\ C_D &= 1.15 \text{ for snow} \\ C_M &= 1.0 \text{ for dry in-service conditions} \\ C_g &= 1.0 \text{ for one bolt} \\ C_\Delta &= 0.578 \\ Z' &= Z_\parallel C_D C_M C_g C_\Delta \\ &= (1089 \text{ lbf})(1.15)(1.0)(1.0)(0.578) \\ &= 724 \text{ lbf} \quad \begin{bmatrix}\text{load capacity acting} \\ \text{perpendicular to bolt axis}\end{bmatrix}\end{aligned}$$

The maximum allowed compression in top chord based on bolt capacity is

$$\begin{aligned}P_{\text{allow}} &= \frac{Z'}{\cos\theta} = \frac{724 \text{ lbf}}{\cos 30^\circ} \\ &= 836 \text{ lbf}\end{aligned}$$

Appendix: Beam Formulas

1. Concentrated Load, P, in Simple Beams

A. At Any Point in Span

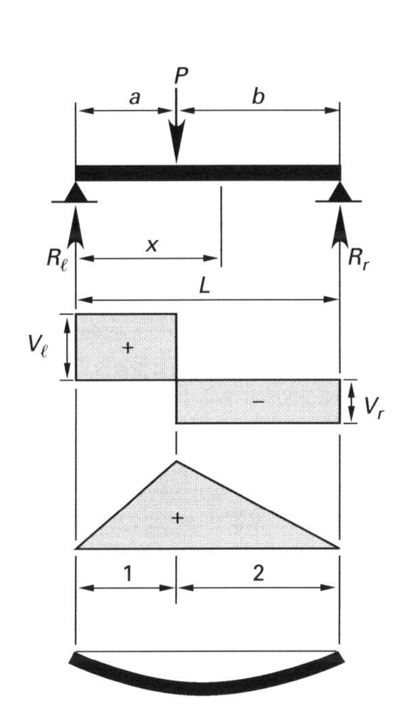

reactions: $R_\ell = \dfrac{Pb}{L}$

$$R_r = \frac{Pa}{L}$$

shears: $V_\ell = +\dfrac{Pb}{L}$

$$V_r = \frac{Pa}{L}$$

moments: $M_{x1} = \dfrac{Pbx}{L}$

$$M_{x2} = \frac{Pa(L-x)}{L}$$

$$M_{\max} = \frac{Pab}{L} \text{ at } x = a$$

deflections: $\Delta_{x1} = \left(\dfrac{Pb}{6EIL}\right)(L^2x - b^2x - x^3)$

$$\Delta_{x2} = \left(\frac{Pb}{6EIL}\right)\left(\left(\frac{L}{b}\right)(x-a)^3 + (L^2-b^2)x - x^3\right)$$

$$\Delta = \frac{Pa^2b^2}{3EIL} \text{ at } x = a$$

$$\Delta_{\max} = \left(\frac{0.06415Pb}{EIL}\right)(L^2-b^2)^{3/2} \text{ at } x = \sqrt{\frac{a(L+b)}{3}}$$

B. At Midspan

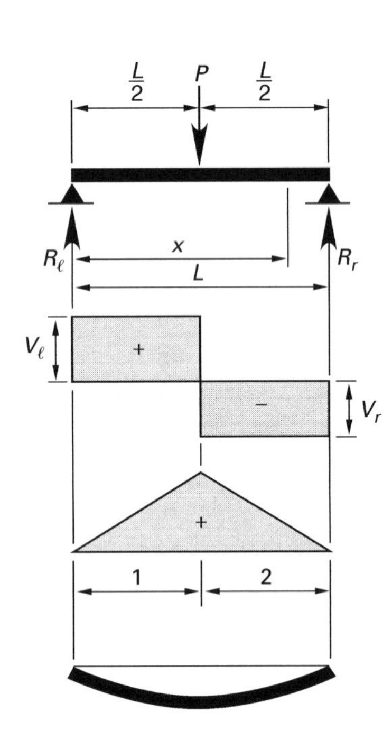

reactions: $R_\ell = R_r = \dfrac{P}{2}$

shears: $V_\ell = \dfrac{P}{2}$

$V_r = -\dfrac{P}{2}$

moments: $M_{x1} = \dfrac{Px}{2}$

$M_{x2} = \left(\dfrac{P}{2}\right)(L - x)$

$M_{\text{max}} = \dfrac{PL}{4}$

deflections: $\Delta_{x1} = \left(\dfrac{P}{48EI}\right)(3xL^2 - 4x^3)$

$\Delta_{\text{max}} = \dfrac{PL^3}{48EI}$ at $x = \dfrac{L}{2}$

C. Beam Fixed at Both Ends

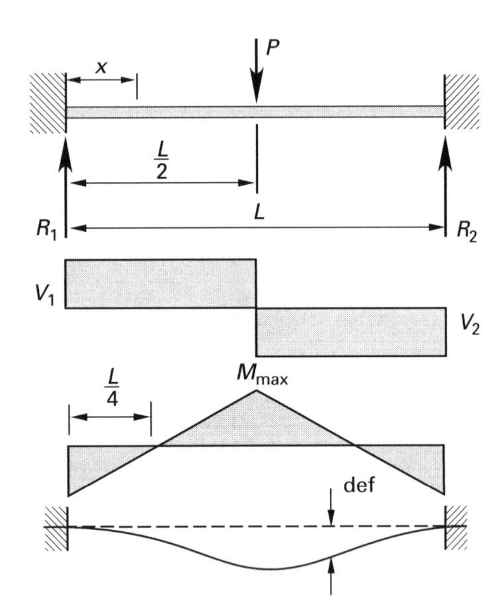

reactions: $R_1 = R_2 = \dfrac{P}{2}$

shears: $V_1 = +\dfrac{P}{2}$; $V_2 = -\dfrac{P}{2}$

moments:

$M_{\text{max}} = \dfrac{PL}{8}$, at center

$M_{\text{max}} = -\dfrac{PL}{8}$, at ends

deflections: $\Delta_{\text{max}} = \dfrac{PL^3}{192\,EI}$, at center

$\Delta = \left(\dfrac{Px^2}{48\,EI}\right)(3L - 4x),\ 0 \le x \le \dfrac{L}{2}$

D. Cantilever Beam

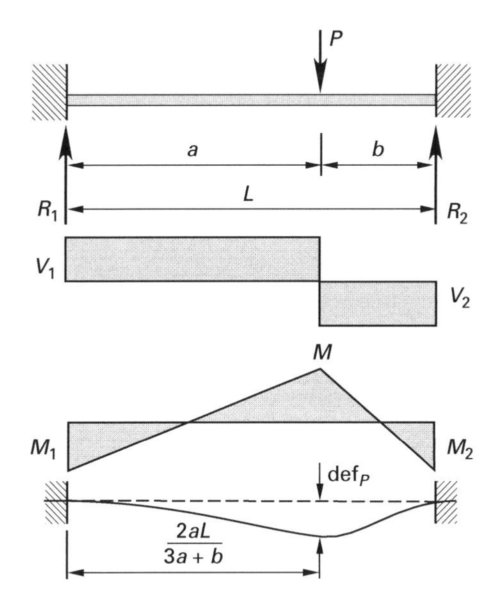

reactions: $R_1 = \left(\frac{Pb^2}{L^3}\right)(3a+b)$

$$R_2 = \left(\frac{Pa^2}{L^3}\right)(3b+a)$$

shears: $V_1 = R_1$; $V_2 = -R_2$

moments:

$$M_1 = -\frac{Pab^2}{L^2}, \text{ max. when } a < b$$

$$M_2 = -\frac{Pa^2b}{L^2}, \text{ max. when } a > b$$

$$M_P = +\frac{2Pa^2b^2}{L^3}, \text{ at point of load}$$

deflections: $\Delta_P = \frac{Pa^3b^3}{3\,EIL^3}$, at point of load

$$\Delta_{\max} = \frac{2Pa^3b^2}{3\,EI(3a+b)^2}, \text{ at } x = \frac{2aL}{3a+b}, \text{ for } a > b$$

2. Two Equal Spans: Concentrated Load

A. Concentrated Load at Center of One Span

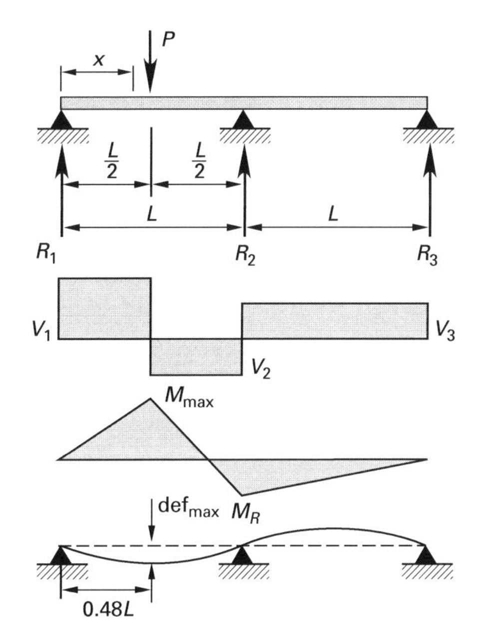

reactions: $R_1 = \frac{13}{32}P$, $R_2 = \frac{11}{16}P$, $R_3 = -\frac{3}{32}P$

shears: $V_1 = \frac{13}{32}P$, $V_2 = -\frac{19}{32}P$, $V_3 = \frac{3}{32}P$

moments:

$$M_{\max} = \frac{13}{64}PL, \text{ at point of load}$$

$$M_R = -\frac{3}{32}PL, \text{ at support}$$

deflections: $\Delta_{\max} = \frac{0.96\,PL^3}{64\,EI}$, at $x = 0.48L$

B. Concentrated Load at Any Point of One Span

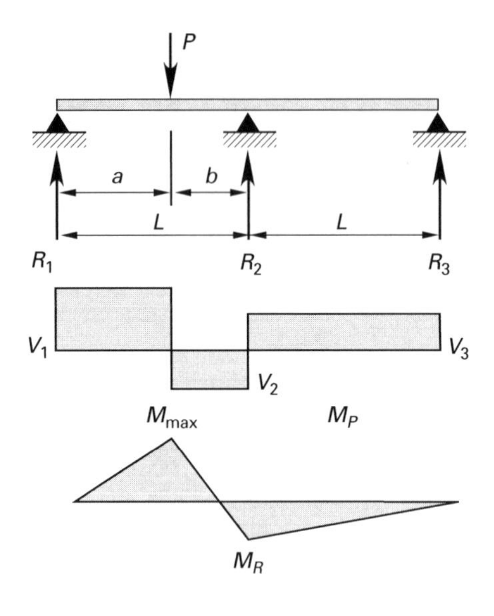

reactions: $R_1 = \left(\frac{Pb}{4L^3}\right)\left(4L^2 - a(L+a)\right)$

$$R_2 = \left(\frac{Pa}{2L^3}\right)\left(2L^2 + b(L+a)\right)$$

$$R_3 = \left(-\frac{Pab}{4L^3}\right)(L+a)$$

shears: $V_1 = \left(\frac{Pb}{4L^3}\right)\left(4L^2 - a(L+a)\right)$

$$V_2 = \left(-\frac{Pa}{4L^3}\right)\left(4L^2 + b(L+a)\right)$$

$$V_3 = \left(\frac{Pab}{4L^3}\right)(L+a)$$

moments: $M_{\text{max}} = \left(\frac{Pab}{4L^3}\right)\left(4L^2 - a(L+a)\right)$

$$M_R = \left(-\frac{Pab}{4L^2}\right)(L+a)$$

C. Continuous Beam of Two Equal Spans: Equal Concentrated Loads, P, at Center of Each Span

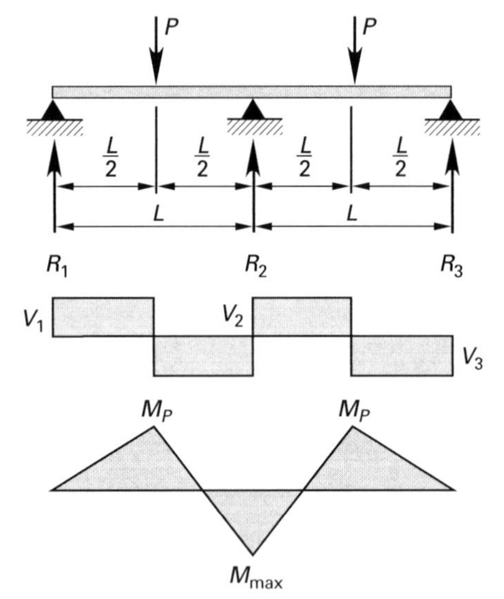

reactions: $R_1 = R_3 = \frac{5}{16}P$

$$R_2 = 1.375P$$

shears: $V_1 = -V_3 = \frac{5}{16}P$

$$V_2 = \pm\frac{11}{16}P$$

moments: $M_{\text{max}} = -\frac{6}{32}PL$, at R_2

$$M_P = \frac{5}{32}PL, \text{ at point of load}$$

D. Continuous Beam of Two Equal Spans: Concentrated Loads, P, at Third Points of Each Span

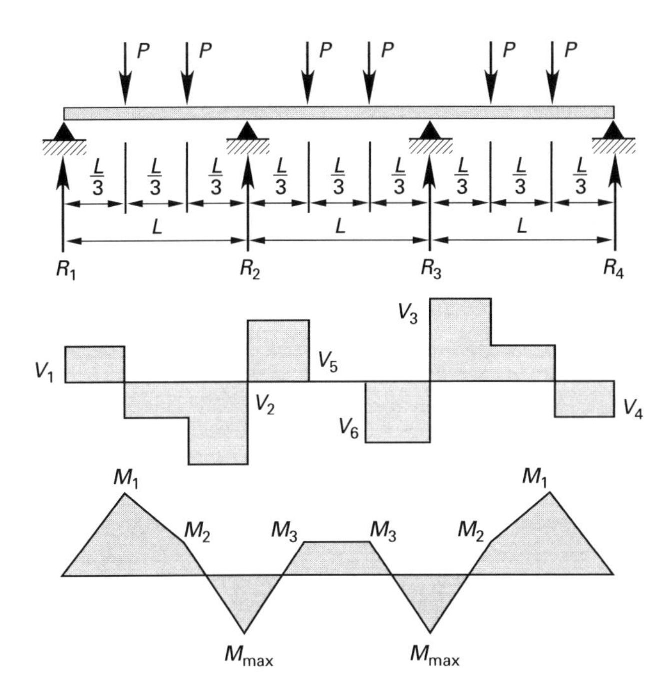

reactions: $R_1 = R_4 = \frac{11}{15}P$

$$R_2 = R_3 = \frac{34}{15}P$$

shears: $V_1 = -V_4 = \frac{11}{15}P$

$$V_3 = V_2 = \frac{9}{15}P$$

$$V_5 = -V_6 = P$$

moments: $M_{\text{max}} = -\frac{12}{45}PL$

$$M_1 = \frac{11}{45}PL$$

$$M_2 = \frac{7}{45}PL$$

$$M_3 = \frac{3}{45}PL$$

3. Uniformly Distributed Loads: One Span Beam

A. Simple Beam

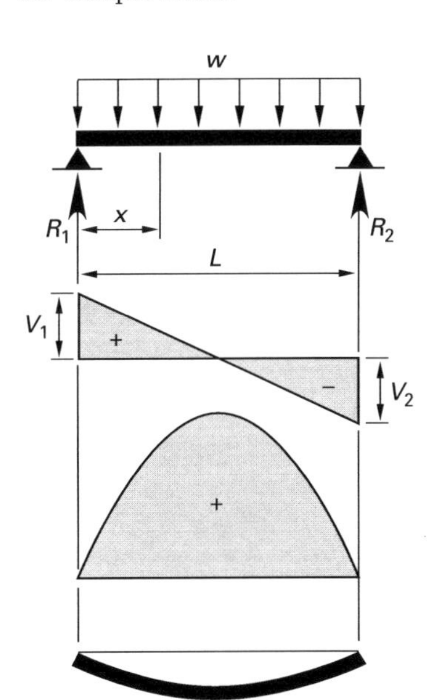

reactions: $R_1 = R_2 = \frac{wL}{2}$

shears: $V_1 = +\frac{wL}{2}$

$$V_2 = -\frac{wL}{2}$$

moments: $M = \left(\frac{w}{2}\right)(Lx - x^2)$

$$M_{\text{max}} = \frac{wL^2}{8}$$

deflections: $\Delta_x = \left(\frac{w}{24EI}\right)(L^3x - 2Lx^3 + x^4)$

$$\Delta_{\text{max}} = \frac{5wL^4}{384EI} \text{ at } x = \frac{L}{2}$$

B. Beam Fixed at Both Ends

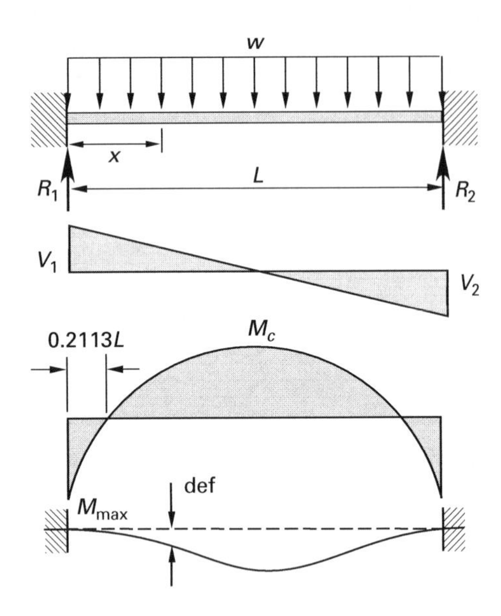

total load: $W = wL$

reactions: $R_1 = R_2 = \frac{W}{2}$

shears: $V_1 = +\frac{W}{2}$

$V_2 = -\frac{W}{2}$

maximum (negative) moment:

$$M_{\text{max}} = -\frac{wL^2}{12} = -\frac{WL}{12}, \text{ at end}$$

maximum (positive) moment:

$$M_c = \frac{wL^2}{24} = \frac{WL}{24}, \text{ at center}$$

deflections: $\Delta_{\text{max}} = \frac{wL^4}{384\,EI} = \frac{WL^3}{384\,EI}$, at center

$$\Delta = \left(\frac{wx^2}{24\,EI}\right)(L-x)^2, \ 0 \le x \le L$$

C. Cantilever Beam

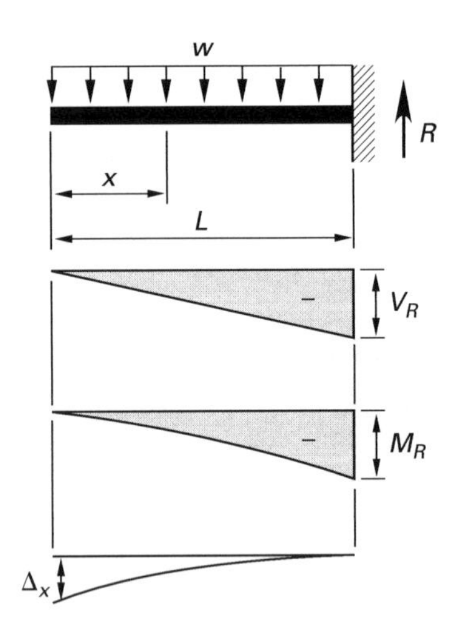

reactions: $R = wL$

shears: $V_x = -wx$

$V_{\text{max}} = -wL$

moments: $M_x = -\frac{wx^2}{2}$

$M_{\text{max}} = -\frac{wL^2}{2}$

deflections: $\Delta_x = \left(\frac{w}{24EI}\right)(3L^4 - 4L^3x + x^4)$

$\Delta_{\text{max}} = -\frac{wL^4}{8EI}$, at $x = 0$

4. Two Equal Spans: Uniform Load

A. Uniform Load on Both Spans, w

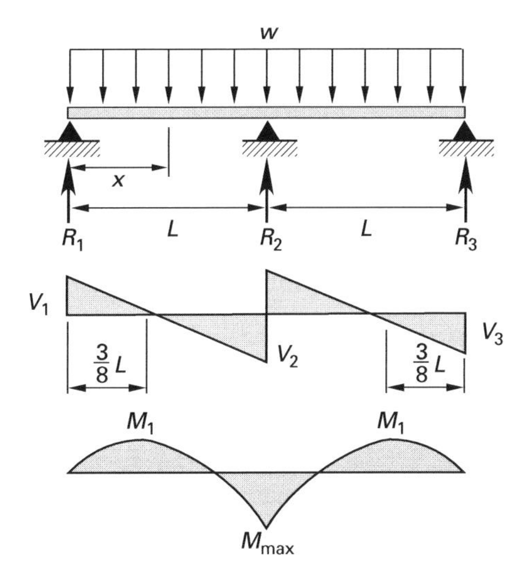

$$\text{reactions: } R_1 = R_3 = \frac{3}{8}wL$$

$$R_2 = \frac{10}{8}wL$$

$$\text{shears: } V_1 = -V_3 = \frac{3}{8}wL$$

$$V_2 = \pm\frac{5}{8}wL$$

$$\text{moments: } M_{\text{max}} = -\frac{1}{8}wL^2$$

$$M_1 = \frac{9}{128}wL^2$$

$$\text{deflections: } \Delta_{\text{max}} = 0.00541\left(\frac{wL^4}{EI}\right), \text{ at } x = 0.4215L$$

$$\Delta = \left(\frac{w}{48\,EI}\right)(L^3x - 3Lx^3 + 2x^4),\ 0 \le x \le L$$

B. Uniform Load on One Span

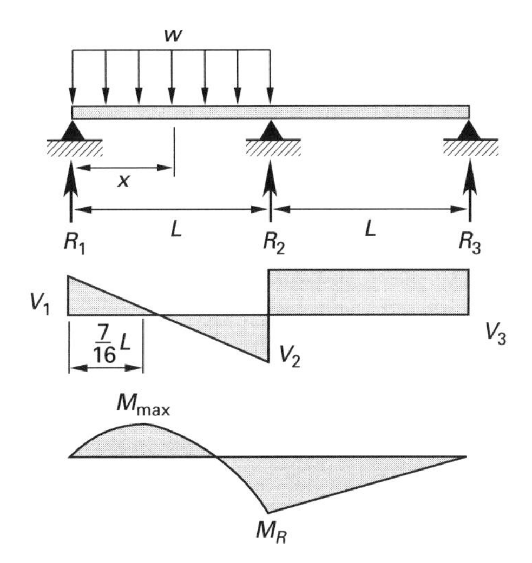

$$\text{reactions: } R_1 = \frac{7}{16}wL,\ R_2 = \frac{5}{8}wL,\ R_3 = -\frac{1}{16}wL$$

$$\text{shears: } V_1 = \frac{7}{16}wL,\ V_2 = -\frac{9}{16}wL,\ V_3 = \frac{1}{16}wL$$

$$\text{moments: } M_{\text{max}} = \frac{49}{512}wL^2, \text{ at } x = \frac{7}{16}L$$

$$M_R = -\frac{1}{16}wL^2, \text{ at } R_2$$

$$M = \left(\frac{wx}{16}\right)(7L - 8x),\ 0 \le x \le L$$

5. Beam Overhang Over One Support

A. Beam Overhang: Uniformly Distributed Load

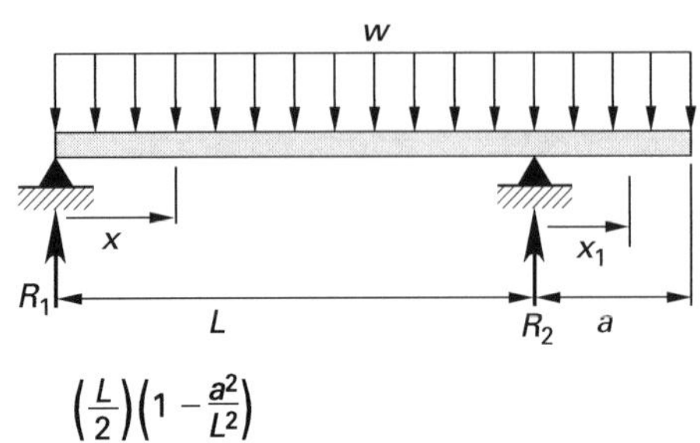

reactions: $R_1 = V_1 = \left(\frac{w}{2L}\right)(L^2 - a^2)$

$$R_2 = |V_2| + |V_3| = \left(\frac{w}{2L}\right)(L + a)^2$$

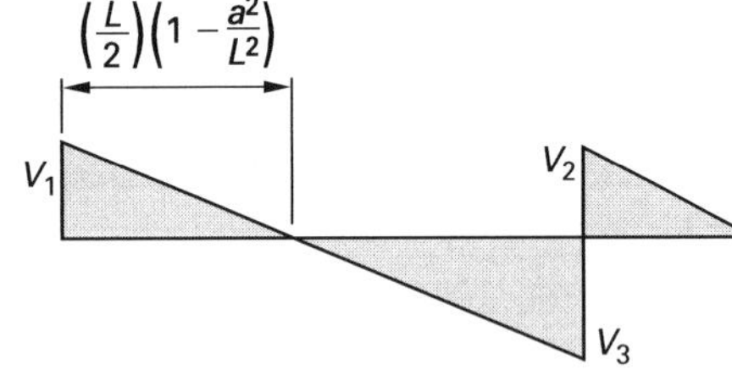

shears: $V_2 = wa$

$$V_3 = \left(\frac{-w}{2L}\right)(L^2 + a^2)$$

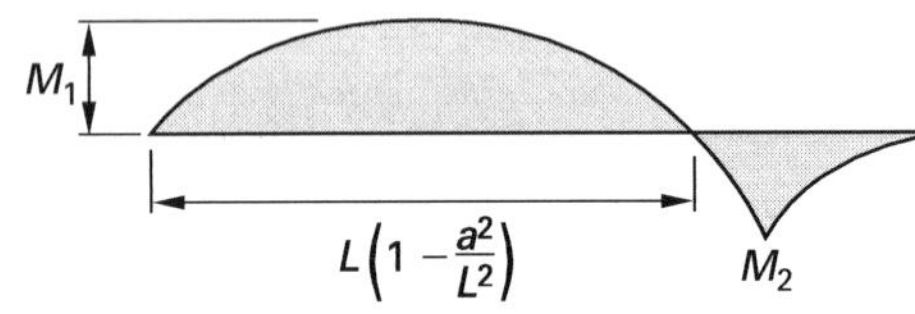

moments: $M_1 = \left(\frac{w}{8L^2}\right)(L + a)^2(L - a)^2$

$$M_2 = -\frac{wa^2}{2}$$

deflections: Δ_{x_1} (along overhang)

$$= \left(\frac{wx_1}{24EI}\right)(4a^2L - L^3 + 6a^2x_1 - 4ax_1^2 + x_1^3)$$

$$\Delta_{\text{overhanging tip}} = \left(\frac{wa}{24EI}\right)(4a^2L - L^3 + 3a^3)$$

Δ_x (between supports)

$$= \left(\frac{wx}{24EIL}\right)(L^4 - 2L^2x^2 + Lx^3 - 2a^2L^2 + 2a^2x^2)$$

B. Beam Overhang: Uniform Load on Overhang

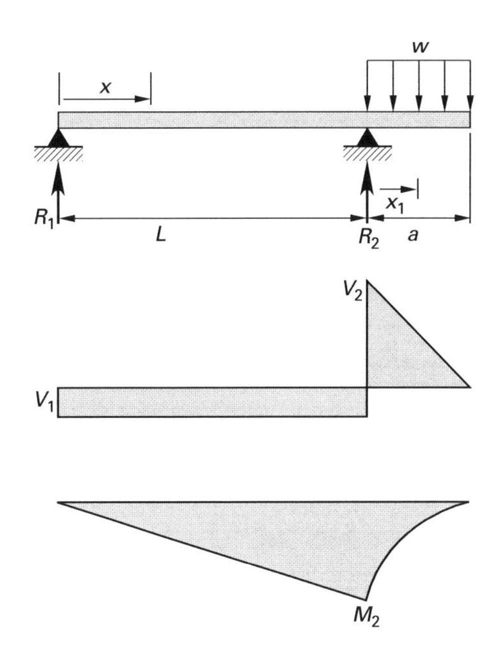

reactions: $R_1 = \frac{-wa^2}{2L}$

$$R_2 = |V_1| + |V_2| = \left(\frac{wa}{2L}\right)(2L + a)$$

shears: $V_1 = R_1$

$$V_2 = wa$$

moments: $M_2 = -\frac{wa^2}{2}$

deflections: Δ_x (between supports)

$$= \left(\frac{wa^2x}{12EIL}\right)(L^2 - x^2)$$

$$\Delta_{\text{max}} \left(\text{between supports at } x = \frac{L}{\sqrt{3}}\right)$$

$$= 0.03208\left(\frac{wa^2L^2}{EI}\right)$$

Δ_{x_1} (for overhang)

$$= \left(\frac{wx_1}{24EI}\right)(4a^2L + 6a^2x_1 - 4ax_1^2 + x_1^3)$$

$$\Delta_{\text{overhanging tip}} = \left(\frac{wa^3}{24EI}\right)(4L + 3a)$$

C. Beam Overhang: Concentrated Load Between Supports

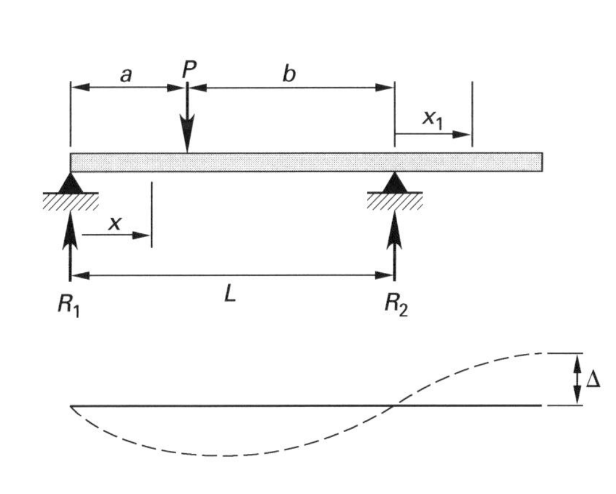

reactions: $R_1 = \frac{Pb}{L}$

$$R_2 = \frac{Pa}{L}$$

shears: Use case 1(A).

moments: Use case 1(A).

deflections: $\Delta_{\text{max}} \left(\text{at } x = \sqrt{\frac{a(a+2b)}{3}} \text{ when } a > b\right)$

$$= \frac{Pab(a+2b)\sqrt{3a(a+2b)}}{27EIL}$$

Δ_a (at point of load)

$$= \frac{Pa^2b^2}{3EIL}$$

$$\Delta_{x_1} = \left(\frac{Pabx_1}{6EIL}\right)(L + a)$$

D. Beam Overhang: Concentrated Load at Overhang

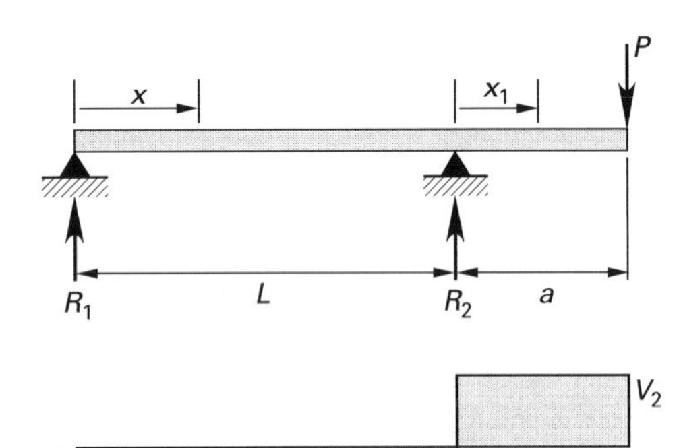

M

reactions: $R_1 = -\dfrac{Pa}{L}$

$$R_2 = \left(\frac{P}{L}\right)(L+a)$$

shears: $V_1 = R_1 = \dfrac{-Pa}{L}$

$$V_2 = P$$

moments: $M = -Pa$

deflections: $\Delta_x \text{ (between supports)} = \left(\dfrac{Pax}{6EIL}\right)(L^2 - x^2)$

$$\Delta_{\text{max}} \left(\text{between supports at } x = \frac{L}{\sqrt{3}}\right)$$

$$= 0.06415\left(\frac{PaL^2}{EI}\right)$$

$$\Delta_{x_1} \text{ (in overhang)}$$

$$= \left(\frac{Px_1}{6EI}\right)(2aL + 3ax_1 - x_1^2)$$

$$\Delta_{\text{overhanging tip}} = \left(\frac{Pa^2}{3EI}\right)(L+a)$$

E. Beam Overhang, One Support: Uniformly Distributed Load Between Supports

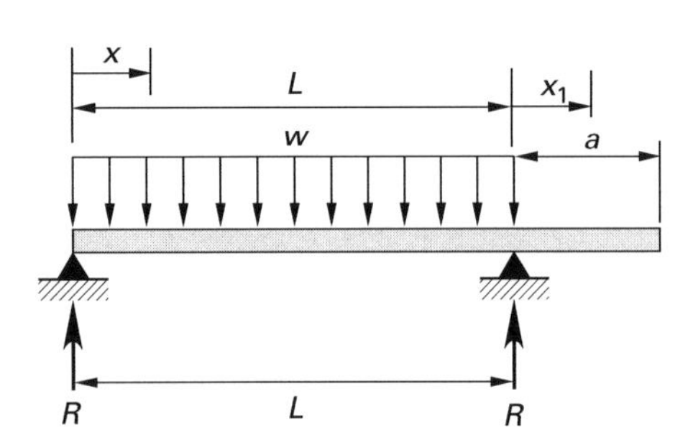

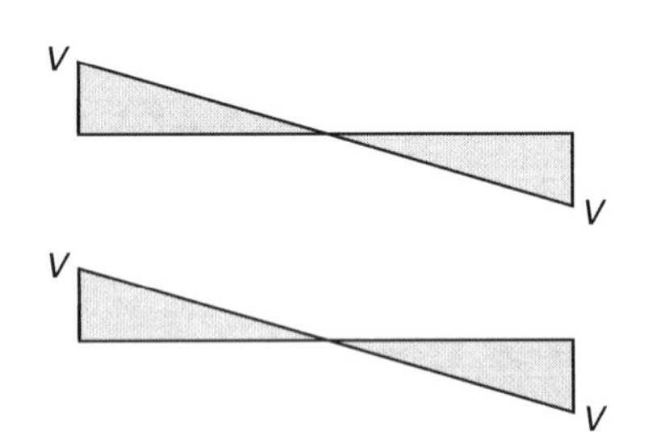

reactions: $R = \dfrac{wL}{2}$

shears: $V = \dfrac{wL}{2}$

moments: $M_{\text{max}} \text{ (at center)} = \dfrac{wL^2}{8}$

deflections: $\Delta_{\text{max}} \text{ (at center)} = \dfrac{5wL^4}{384EI}$

$$\Delta_x = \left(\frac{wx}{24EI}\right)(L^3 - 2Lx^2 + x^3)$$

$$\Delta_{x_1} = \frac{wL^3x_1}{24EI}$$

$$\Delta_{\text{overhanging tip}} = \frac{wL^3a}{24EI}$$

6. Cantilevered Beam: Two Equal Spans

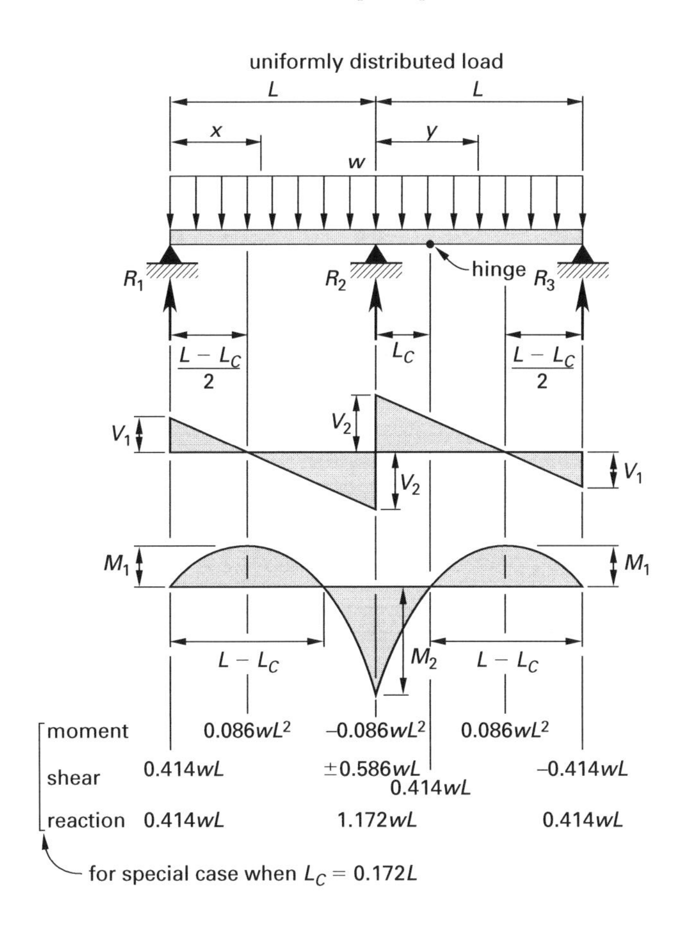

reactions: $R_1 = R_3 = \left(\frac{w}{2}\right)(L - L_C)$

$$R_2 = w(L + L_C)$$

shears: $V_1 = \left(\frac{w}{2}\right)(L - L_C)$

$$V_x = \left(\frac{w}{2}\right)(L - L_C - 2x)$$

$$V_2 = \left(\frac{w}{2}\right)(L + L_C)$$

$$V_y = \left(\frac{w}{2}\right)(L + L_C - 2y)$$

moments: $M_1 = \left(\frac{w}{8}\right)(L - L_C)^2$

$$M_2 = -\frac{wLL_C}{2}$$

$$M_x = \left(\frac{wx}{2}\right)(L - L_C - x)$$

$$M_y = \left(\frac{w}{2}\right)(y - L_C)(L - y)$$

For maximum positive moment equal to maximum negative moment, $L_C = 0.172L$. For this special case, the maximum deflection occurs in span between R_1 and R_2 and is

$$\Delta_{\max} = \left(\frac{0.370}{48}\right)\left(\frac{wL^4}{EI}\right)$$

(Note: The "x" subscript for M and V denotes moment and shear, respectively, for the span between R_1 and R_2. The "y" subscript for M and V is for the span between R_2 and R_3.)

7. Three Equal Spans: Uniform Load

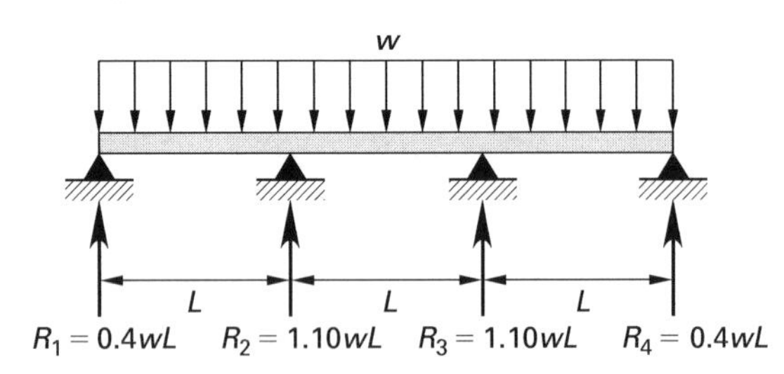

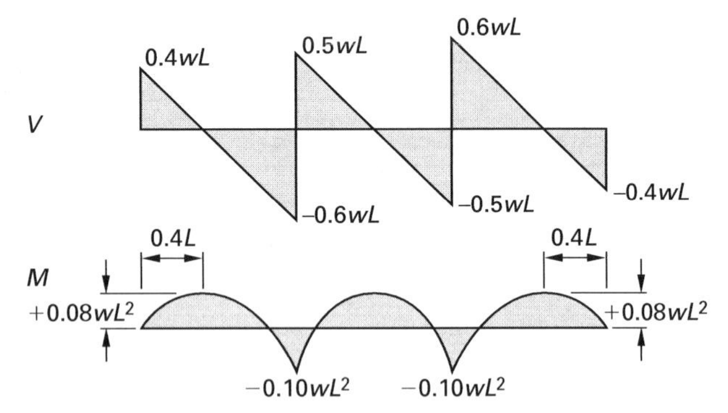

deflections: $\Delta_{\max} = \dfrac{wL^4}{145EI}$ at $0.446L$ from R_1 or R_4

8. Continuous Beam: Three Equal Spans (One End Span Unloaded)

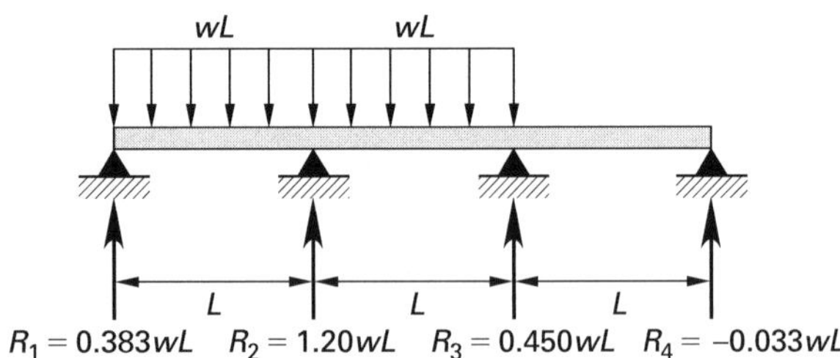

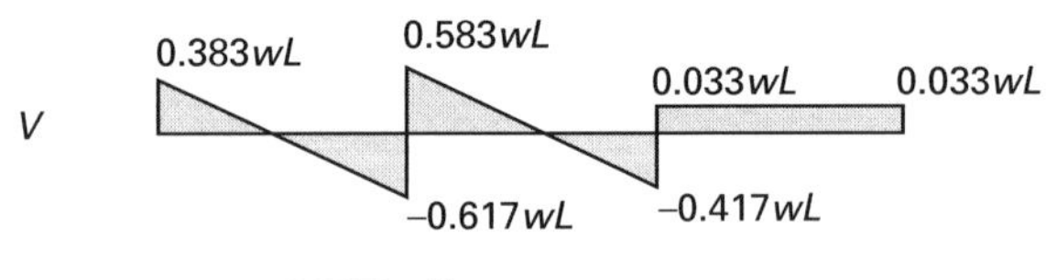

deflections: $\Delta_{\max}$ $(0.430L$ from $R_1) = \dfrac{0.0059wL^4}{EI}$

9. Continuous Beam: Three Equal Spans (End Spans Loaded)

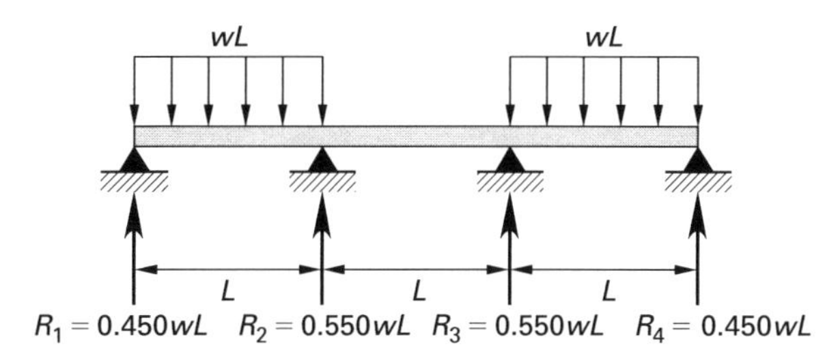

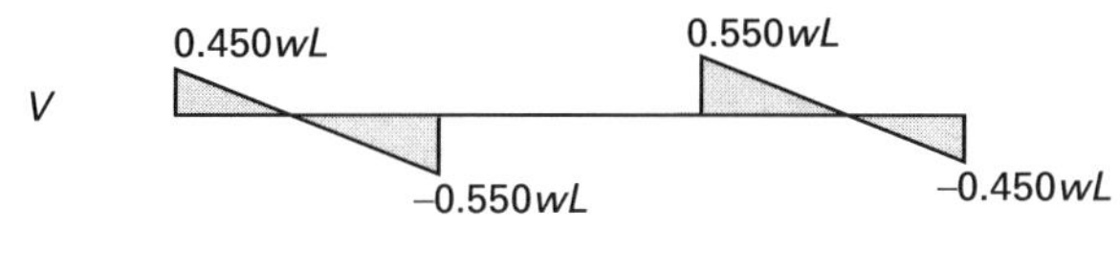

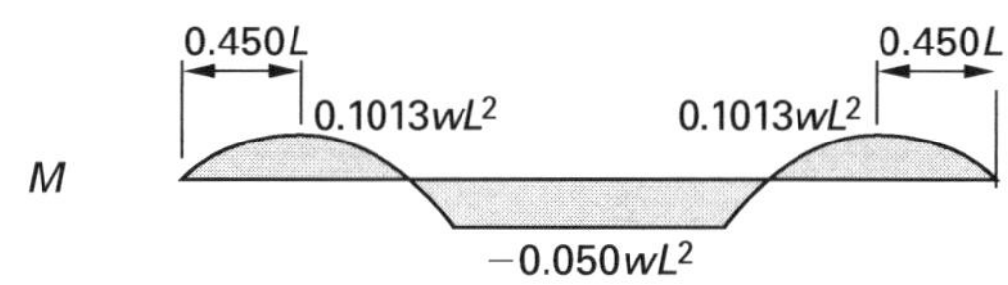

$$\text{deflections: } \Delta_{\text{max}}\ (0.479L \text{ from } R_1 \text{ or } R_4) = \frac{0.0099wL^4}{EI}$$

Index